Springer Finance

Editorial Board
M. Avellaneda
G. Barone-Adesi
M. Broadie
M.H.A. Davis
E. Derman
C. Klüppelberg
W. Schachermayer

Springer Finance

Springer Finance is a programme of books addressing students, academics and practitioners working on increasingly technical approaches to the analysis of financial markets. It aims to cover a variety of topics, not only mathematical finance but foreign exchanges, term structure, risk management, portfolio theory, equity derivatives, and financial economics.

For further volumes:
http://www.springer.com/series/3674

Archil Gulisashvili

Analytically Tractable Stochastic Stock Price Models

Archil Gulisashvili
Department of Mathematics
Ohio University
Athens, OH, USA

ISBN 978-3-642-43386-3 ISBN 978-3-642-31214-4 (eBook)
DOI 10.1007/978-3-642-31214-4
Springer Heidelberg New York Dordrecht London

Mathematics Subject Classification (2010): 91Gxx, 91G80, 91B25, 91G20

JEL Classification: GO2, G13

© Springer-Verlag Berlin Heidelberg 2012
Softcover reprint of the hardcover 1st edition 2012
This work is subject to copyright. All rights are reserved by the Publisher, whether the whole or part of the material is concerned, specifically the rights of translation, reprinting, reuse of illustrations, recitation, broadcasting, reproduction on microfilms or in any other physical way, and transmission or information storage and retrieval, electronic adaptation, computer software, or by similar or dissimilar methodology now known or hereafter developed. Exempted from this legal reservation are brief excerpts in connection with reviews or scholarly analysis or material supplied specifically for the purpose of being entered and executed on a computer system, for exclusive use by the purchaser of the work. Duplication of this publication or parts thereof is permitted only under the provisions of the Copyright Law of the Publisher's location, in its current version, and permission for use must always be obtained from Springer. Permissions for use may be obtained through RightsLink at the Copyright Clearance Center. Violations are liable to prosecution under the respective Copyright Law.
The use of general descriptive names, registered names, trademarks, service marks, etc. in this publication does not imply, even in the absence of a specific statement, that such names are exempt from the relevant protective laws and regulations and therefore free for general use.
While the advice and information in this book are believed to be true and accurate at the date of publication, neither the authors nor the editors nor the publisher can accept any legal responsibility for any errors or omissions that may be made. The publisher makes no warranty, express or implied, with respect to the material contained herein.

Printed on acid-free paper

Springer is part of Springer Science+Business Media (www.springer.com)

To Olga, Alex, and Misha

Preface

This book focuses primarily on applications of mathematical analysis to stock price models with stochastic volatility and more general stochastic asset price models. The central objective of the book is to characterize limiting behavior of several important functions associated with such models, e.g., stock price densities, call and put pricing functions, and implied volatilities.

Stock price models with stochastic volatility have been developed in the last decades to improve pricing and hedging performance of the classical Black–Scholes model and to account for certain imperfections in it. The main shortcoming of the Black–Scholes model is its constant volatility assumption. Statistical analysis of stock market data shows that the volatility of a stock is a time-dependent quantity. Moreover, it exhibits various random features. Stochastic volatility models address this randomness by assuming that both the stock price and the volatility are stochastic processes affected by different sources of risk. Unlike the Black–Scholes model, stock price models with stochastic volatility explain such stylized facts as the implied volatility smile and skew. They can also incorporate the leverage effect, that is, the tendency of the volatility of the stock to increase when the stock price decreases. Stochastic volatility models reflect the leverage effect by imposing the restriction that the stock price and the volatility are negatively correlated.

An important problem in mathematical finance is to describe the asymptotic behavior of the stock price density in a stochastic volatility model. Once we have a good understanding of how this density changes, we can estimate many other characteristics of the model, for example, left and right tails of stock return distributions, option pricing functions, and implied volatilities. Asymptotic formulas for distribution tails of stock returns in a stochastic volatility model can be used to analyze how well the model addresses the tail risk. In financial practice, the tail risk is defined as the probability that stock returns will move more than three standard deviations beyond the mean. The Black–Scholes model underestimates the tail risk, since the probability of extreme variations of stock returns in this model is negligible. This follows from the fact that distribution tails of stock returns in the Black–Scholes model decay like Gaussian density functions. In the present book, we obtain sharp asymptotic formulas with relative error estimates for stock price densities in three popular stock price models with stochastic volatility: the Hull–White model, the Stein–Stein model, and the Heston model. These formulas show that for the above-mentioned models, the stock price distributions have Pareto type tails, that is to say, the tails decay like regularly varying functions. As a consequence, the Hull–White, Stein–Stein, and Heston models estimate the probability of abrupt downward movements of stock prices (disastrous scenarios) better than the Black–Scholes model.

The implied volatility associated with the call pricing function in a stochastic asset price model may be poetically described as the reflection of this function in the Black–Scholes mirror. One can obtain the implied volatility by inverting the Black–Scholes call pricing function and composing the inverse function with the call pricing function of our interest. A substantial part of the present book discusses model free asymptotic formulas for the implied volatility at extreme strikes in general asset price models. Some of the reasons why such asymptotic formulas are important are the following. On the one hand, these formulas help to check whether the given stochastic asset price model produces a skewed volatility pattern often observed in real markets. On the other hand, since the implied volatility at extreme strikes is associated with out-of-the-money and in-the-money put and call options, the analysis of the implied volatility for large and small strikes quantifies the expectations and fears of investors of possible large upward or downward movements in asset prices. Note that buying out-of-the money put options has been a popular hedging strategy against negative tail risk.

The text is organized as follows. The main emphasis in Chaps. 1–7 is on special stochastic volatility models (the Hull–White, Stein–Stein, and Heston models). In Chap. 1, we consider stochastic processes, which play the role of volatility in these models, i.e., geometric Brownian motion, Ornstein–Uhlenbeck process, and Cox–Ingersoll–Ross process (Feller process). Chapter 2 introduces general correlated stock price models with stochastic volatility. It also discusses risk-neutral measures in such models. Chapter 3 is concerned with realized volatility and mixing distributions. For an uncorrelated stochastic volatility model, the mixing distribution is the law of the realized volatility, while for correlated models, mixing distributions are defined as joint distributions of various combinations of the variance of the stock price, the integrated volatility, and the integrated variance. Chapter 4 considers integral transforms of mixing distribution densities, and provides explicit formulas for the stock price density in terms of mixing distributions. In Chap. 5 we prove a Tauberian theorem for the two-sided Laplace transform, and also Abelian theorems for fractional integrals and for integral operators with log-normal kernels. The Tauberian theorem is used in Chap. 5 to characterize the asymptotics of mixing distributions by inverting their Laplace transforms approximately, while the Abelian theorem for fractional integrals is a helpful tool in the study of mixing distributions in the Hull–White model. In Chap. 6 we provide asymptotic formulas with error estimates for the stock price distribution densities in the Hull–White, Stein–Stein, and Heston models. For the correlated Heston model the proof of the asymptotic formula is based on affine principles, while in the absence of correlation an alternative proof of the asymptotic formula is given. In the latter proof the Abelian theorem for integral operators with log-normal kernels plays an important role. Finally, in

Chap. 7 we include a short exposition of the theory of regularly varying functions. This chapter also considers Pareto type distributions and their applications.

The second part of the book (Chaps. 8–11) is devoted to general call and put pricing functions in no-arbitrage setting and to the Black–Scholes implied volatility. In the beginning of Chap. 8 we prove a characterization theorem for call pricing functions, and at the end of this chapter we establish sharp asymptotic formulas with error estimates for the call pricing functions in the Hull–White, Stein–Stein, and Heston models. Chapter 8 also presents an analytical proof of the Black–Scholes call option pricing formula, which is arguably the most famous formula of mathematical finance. Chapter 9 introduces the notion of implied volatility (or "smile") and provides model free asymptotic formulas for the implied volatility at extreme strikes. One more topic discussed in Chap. 9 concerns certain symmetries hidden in option pricing models. The contents of Chap. 10 can be guessed from its title "More Formulas for Implied Volatility". It is shown in this chapter that R. Lee's moment formulas for the implied volatility and the tail-wing formulas due to S. Benaim and P. Friz can be derived from more general results established in Chap. 9. Chapter 10 also presents an important result obtained by E. Renault and N. Touzi, which can be shortly presented as follows: "The absence of correlation between the stock price and the volatility implies smile". The last section of Chap. 10 deals with J. Gatheral's SVI parameterization of implied variance. SVI parameterization provides a good approximation to implied variance observed in the markets and also to implied variance used in stochastic volatility models. Finally, in Chap. 11 we study implied volatility in models without moment explosions. Here we show that V.V. Piterbarg's conjecture concerning the limiting behavior of implied volatility in models without moment explosions is true in a modified form. Chapter 11 also studies smile asymptotics in various special models, e.g., the displaced diffusion model, the constant elasticity of variance model, SV1 and SV2 models introduced by L.C.G. Rogers and L.A.M. Veraart, and the finite moment log-stable models developed by P. Carr and L. Wu.

We will next make a brief comparison between the present book and the following related books: [Lew00, Gat06, H-L09], and [FPSS11]. It is easy to check that although the books on the previous list and the present book have the same main heroes (stochastic volatility models, stock price densities, option pricing functions, and implied volatilities), they differ substantially from each other with respect to the choice of special topics. For example, the book by A. Lewis [Lew00] deals with various methods of option pricing under stochastic volatility, and the topics covered in [Lew00] include the volatility of volatility series expansions, volatility explosions, and related corrections in option pricing formulas. The book by J. Gatheral [Gat06] is a rich source of information on implied and local volatilities in stochastic stock price models and on the asymptotic and dynamic behavior of volatility surfaces. In particular, the book [Gat06] discusses early results on smile asymptotics for large and small strikes. The book by P. Henry-Labordère [H-L09] uses powerful methods of differential geometry and mathematical physics to study the asymptotic behavior of implied volatility in local and stochastic volatility models. For instance, heat kernel expansions in Riemannian manifolds and Schrödinger semigroups with

Kato class potentials play an important role in [H-L09]. The book by J.-P. Fouque, G. Papanicolaou, R. Sircar, and K. Sølna [FPSS11] is devoted to pricing and hedging of financial derivatives in stochastic volatility models. In [FPSS11], regular and singular perturbation techniques are used to study small parameter asymptotics of option pricing functions and implied volatilities. The authors of [FPSS11] obtain first and second order approximations to implied volatility in single-factor and multifactor stochastic volatility models, and explain how to use these approximations to calibrate stochastic volatility models and price more complex derivative contracts. A more detailed comparison shows that a large part of the material appearing in the present book is not covered by the books on the list. Moreover, to the best of the author's knowledge, many of the results discussed in the present book have never been published before in book form. These results include sharp asymptotic formulas with error estimates for stock price densities, option pricing functions, and implied volatilities in special stochastic volatility models, and sharp model free asymptotic formulas for implied volatilities.

This book is aimed at a variety of people: researchers in the field of financial mathematics, professional mathematicians interested in applications of mathematical analysis to finance, and advanced graduate students thinking of a career in applied analysis or financial mathematics. It is assumed that the reader is familiar with basic definitions and facts from probability theory, stochastic differential equations, asymptotic analysis, and complex analysis. The book does not aspire to completeness, several important topics related to its contents have been omitted. For example, small and large maturity asymptotics of implied volatility, affine models, local martingale option pricing models, or applications of geometric methods to the study of implied volatility are not discussed in the book. The reader can find selected references to publications on the missing subjects in the sections "Notes and References" that conclude each chapter, or search the bibliography at the end of the book for additional reading.

Athens, Ohio
December 2011

Archil Gulisashvili

Acknowledgements

This book was first thought of as a lecture notes book. Its early versions were based on the contents of several summer courses on selected topics in financial mathematics that I gave at Bielefeld University. However, the book underwent significant transformations during my work on it. It has grown in size, and its final version looks more like a research monograph than a lecture notes book. I express my gratitude to Michael Röckner for extending to me the invitations to visit the International Graduate College "Stochastics and Real World Models" at Bielefeld University in the summers of 2007–2011 and for giving me the opportunity to spend several months in a very stimulating mathematical environment. I also thank all the members of the graduate college and the stochastic analysis group in the Department of Mathematics at Bielefeld University for their hospitality, and all the students who attended my courses at Bielefeld University.

The present book would not exist without Elias M. Stein. It has been a privilege and a pleasure working with him on a project concerning classical stochastic volatility models. A substantial part of this book has strongly benefited from his impact, and many results obtained in our joint papers are covered in the book. I express my deep gratitude to E.M. Stein for his friendship, collaboration, and encouragement.

I am indebted to many of my friends and colleagues who were of great help to me in my work on this book. I would like to single out Peter Laurence for his friendship, advice, valuable mathematical and bibliographical information, and stimulating discussions on some of the topics covered in the book. Peter Friz and Sean Violante for important remarks concerning the contents of the book. Peter Friz, Stefan Gerhold, and Stephan Sturm for their much enjoyable and fruitful collaboration on the joint project concerning stock price density asymptotics in the correlated Heston model. The results of this work are included in the book. René Carmona and Ronnie Sircar for interesting conversations and important advice during my early work on stochastic volatility models and the implied volatility. Roger Lee for being very generous in sharing with me his profound knowledge of stochastic volatility models and smile asymptotics. Josep Vives for his friendship and collaboration on the joint paper devoted to stochastic volatility models with jumps. Jean-Pierre Fouque, Josef Teichmann, and Martin Keller-Ressel for always interesting and informative conversations.

I am very grateful to my wife Olga for her constant support, encouragement, and infinite patience during the difficult years of my work on this book. I would also like to thank my home institution, Ohio University.

And of course I am very obliged to Catriona Byrne, Editorial Director at Springer-Verlag, for her support. She is definitely one of those people thanks to whom this book has become a reality. I am also grateful to Marina Reizakis and Donatas Akmanavičius at Springer-Verlag who helped me during the final stages.

Contents

Chapter 1
Volatility Processes

Volatility is an unobservable quantity that measures relative changes in the price of a stock. The direction of these changes is ignored, only their magnitude is taken into account. Volatility is often interpreted as the instantaneous standard deviation of stock returns. During the epochs of low volatility, the stock price does not change much, while large movements of the stock price may be expected during the times of high volatility. In the celebrated Black–Scholes model, the volatility is a positive constant. However, statistical analysis of financial data does not confirm the constant volatility assumption. To improve the performance of the Black–Scholes model, various stock price models with stochastic volatility have been developed in the last decades. The volatility of a stock in such a model is described by a non-negative stochastic process. For instance, in the Hull–White model, a geometric Brownian motion plays the role of stochastic volatility. In the Stein–Stein model, the volatility is represented by an Ornstein–Uhlenbeck process, or by the absolute value of this process, while in the Heston model, the volatility process is the square root of a Cox–Ingersoll–Ross process. In the present chapter, we discuss elementary properties of stochastic processes describing the volatility in classical stochastic stock price models.

1.1 Brownian Motion

This section introduces mathematical Brownian motion or the Wiener process, which is the main building block of stochastic modeling. In physics, Brownian motion is a chaotic movement of particles suspended in a fluid. It was named after the Scottish botanist Robert Brown who reported on such a movement in a pamphlet published in 1828. The modern mathematical theory of Brownian motion goes back to Norbert Wiener. Mathematical Brownian motion describes continuous random fluctuations. Before Wiener, many scientists made valuable contributions to the understanding of the physics and mathematics behind Brownian motion. If suffices to mention Thorvald N. Thiele, Louis Bachelier, Albert Einstein, and Marian Smolu-

A. Gulisashvili, *Analytically Tractable Stochastic Stock Price Models*,
Springer Finance, DOI 10.1007/978-3-642-31214-4_1,
© Springer-Verlag Berlin Heidelberg 2012

chowski. Regrettably, the work of Thiele and Bachelier was not sufficiently recognized by their contemporaries and became well known only much later.

We will next define the normal distribution and introduce mathematical Brownian motion. Note that in this book we mostly consider continuous real-valued stochastic processes indexed by the set $\{t : t \geq 0\}$.

Definition 1.1 Let $(\Omega, \mathcal{F}, \mathbb{P})$ be a probability space. A random variable X is called normally distributed with mean m and variance $v > 0$, if the distribution of X has a density ρ given by

$$\rho(x) = \frac{1}{\sqrt{2\pi v}} \exp\left\{-\frac{(y-m)^2}{2v}\right\}, \quad x \in \mathbb{R}.$$

If the conditions in Definition 1.1 hold, then we say that the random variable X is $N(m, v)$-distributed.

Definition 1.2 Let $(\Omega, \mathcal{F}, \mathbb{P})$ be a probability space. A standard Brownian motion is a continuous stochastic process W satisfying the following conditions:

- $W_0 = 0$ $\mathbb{P}$-a.s.
- The process W has independent increments. This means that for all systems of numbers $0 \leq \tau_1 \leq t_1 \leq \tau_2 \leq t_2 \leq \cdots \leq \tau_k \leq t_k$, the random variables $W_{t_1} - W_{\tau_1}, W_{t_2} - W_{\tau_2}, \ldots, W_{t_k} - W_{\tau_k}$ are independent.
- The process W has stationary increments, that is, for all $\tau < t$ and $h > 0$, the random variables $W_{t+h} - W_{\tau+h}$ and $W_t - W_\tau$ are identically distributed.
- For all $\tau < t$ the increment $W_t - W_\tau$ is $N(0, t - \tau)$-distributed.

It follows from Definition 1.2 that for a standard Brownian motion W, the mean and the variance of the random variable W_t with $t > 0$ are given by $m(W_t) = 0$ and $v(W_t) = t$, respectively. Moreover, for every $t > 0$ the distribution density of the random variable W_t is given by

$$\rho_t(y) = \frac{1}{\sqrt{2\pi t}} \exp\left\{-\frac{y^2}{2t}\right\}, \quad y \in \mathbb{R}.$$

Our next goal is to introduce Brownian motions with respect to a filtration. We will need several elementary definitions from probability theory.

Definition 1.3 Let Ω be a set, and let $\mathcal{A}$ be a family of subsets of Ω. The smallest σ-algebra which contains every set in $\mathcal{A}$ is denoted $\sigma(\mathcal{A})$ and called the σ-algebra generated by the family $\mathcal{A}$. For a collection $\{\mathcal{A}_\lambda\}_{\lambda \in \Lambda}$ of such families, the smallest σ-algebra which contains every set in every $\mathcal{A}_\lambda$ with $\lambda \in \Lambda$ is denoted $\sigma(\mathcal{A}_\lambda : \lambda \in \Lambda)$.

Definition 1.4 A filtration $\{\mathcal{F}_t\}_{t \geq 0}$ on a measurable space $(\Omega, \mathcal{F})$ is a family of sub-σ-algebras of $\mathcal{F}$ such that $\mathcal{F}_s \subset \mathcal{F}_t$ for all $0 \leq s < t < \infty$. The σ-algebra $\mathcal{F}_\infty$

associated with the filtration $\{\mathcal{F}_t\}_{t\geq 0}$ is defined by

$$\mathcal{F}_\infty = \sigma(F_t : t \geq 0).$$

A filtration $\{\mathcal{F}_t\}_{t\geq 0}$ is called right-continuous if

$$\mathcal{F}_t = \bigcap_{s:s>t} \mathcal{F}_s, \quad t \geq 0.$$

Definition 1.5 A filtration $\{\mathcal{F}_t\}_{t\geq 0}$ on a measure space $(\Omega, \mathcal{F}, \mathbb{P})$ is called complete if the σ-algebra $\mathcal{F}_0$ contains all the negligible sets of the completion of the σ-algebra $\mathcal{F}_\infty$ with respect to the measure $\mathbb{P}$. It is said that a filtration satisfies usual conditions if it is complete and right-continuous.

We will often use the notation $\{\mathcal{F}_t\}$ instead of $\{\mathcal{F}_t\}_{t\geq 0}$. Let $(\Omega, \mathcal{F}, \mathbb{P})$ be a probability space, and let $\{\mathcal{F}_t\}$ be a filtration. Then the system $(\Omega, \mathcal{F}, \{\mathcal{F}_t\}, \mathbb{P})$ is called a filtered probability space. It is said that a stochastic process X on Ω is adapted to the filtration $\{\mathcal{F}_t\}$ if for every $t \geq 0$ the random variable X_t is $\mathcal{F}_t$-measurable.

Definition 1.6 Let $(\Omega, \mathcal{F}, \{\mathcal{F}_t\}, \mathbb{P})$ be a filtered probability space. A stochastic process W on Ω is called a standard Brownian motion with respect to the filtration $\{\mathcal{F}_t\}$ if the following conditions hold:

- The process W is a standard Brownian motion.
- The process W is adapted to the filtration $\{\mathcal{F}_t\}$.
- For all t and s with $s < t$, the increment $W_t - W_s$ is independent from the σ-algebra $\mathcal{F}_s$.

Let X be a stochastic process on Ω. Then the filtration $\{\mathcal{F}_t\}$ such that

$$\mathcal{F}_t = \sigma(X_s : 0 \leq s \leq t)$$

for all $t \geq 0$ is called the filtration generated by X. It follows from Definitions 1.2 and 1.6 that any standard Brownian motion W is also a standard Brownian motion with respect to the filtration generated by W.

We will next give several definitions, which will be used throughout the book. It is assumed in these definitions that a filtered probability space $(\Omega, \mathcal{F}, \{\mathcal{F}_t\}, \mathbb{P})$ is given and all random variables are defined on Ω.

Definition 1.7 A random variable $T : \Omega \mapsto [0, \infty]$ is called a stopping time if $\{T \leq t\} \in \mathcal{F}_t$ for all $t \in [0, \infty]$.

Definition 1.8 For a stopping time T and a stochastic process X, the stopped process X^T is defined by $X_t^T = X_{t\wedge T}$, $t \geq 0$.

Definition 1.9 Let X be a continuous adapted process such that

$$\mathbb{E}\big[|X_t|\big] < \infty \quad \text{for all } t \geq 0.$$

The process X is called a continuous submartingale if

$$\mathbb{E}[X_t|\mathcal{F}_s] \geq X_s \quad \text{for all } 0 \leq s \leq t < \infty,$$

and a continuous supermartingale if

$$\mathbb{E}[X_t|\mathcal{F}_s] \leq X_s \quad \text{for all } 0 \leq s \leq t < \infty.$$

If the process X is simultaneously a continuous submartingale and supermartingale, then it is called a continuous martingale.

Definition 1.10 A continuous adapted process X is called a continuous local martingale if there exists an increasing sequence T_k, $k \geq 1$, of stopping times such that $\lim_{k\to\infty} T_k = \infty$ a.s. and the process

$$X_{t\wedge T_k}\mathbb{1}_{\{T_k>0\}}, \quad t > 0,$$

is a martingale for each k. A continuous adapted process X is called a continuous semimartingale if $X = X_1 + X_2$ where X_1 is a continuous local martingale and X_2 is a continuous adapted process of locally finite variation.

For more information the reader is referred to [RY04], Sect. 3 in Chap. I (stopping times and stopped processes), Sect. 1 in Chap. II (martingales), and Sect. 1 in Chap. IV (local martingales and semimartingales).

Let us denote by $\mathcal{M}_2$ the set of all continuous martingales M on the filtered probability space $(\Omega, \mathcal{F}, \{\mathcal{F}_t\}, \mathbb{P})$ such that $M_0 = 0$ $\mathbb{P}$ a.s. and

$$\mathbb{E}\big[M_t^2\big] < \infty \quad \text{for all } t \geq 0.$$

It is known that if $M \in \mathcal{M}_2$, then there exists a unique (up to indistinguishability) continuous increasing process $\langle M, M\rangle$ such that $\langle M, M\rangle_0 = 0$ $\mathbb{P}$ a.s. and the process $M^2 - \langle M, M\rangle$ is a martingale. The process $\langle M, M\rangle$ is called the quadratic variation of the process M. For any two processes $M \in \mathcal{M}_2$ and $N \in \mathcal{M}_2$, their quadratic covariation $\langle M, N\rangle$ is defined as follows:

$$\langle M, N\rangle_t = \frac{1}{4}\big[\langle M+N, M+N\rangle_t - \langle M-N, M-N\rangle_t\big], \quad t \geq 0.$$

It is not hard to see that the process $MN - \langle M, N\rangle$ is a martingale. We refer the reader to [KS91], Sect. 5.1, or [Pro04] for more information on the quadratic variation and covariation.

How to recognize a standard Brownian motion? An answer to this question was given by Paul Lévy.

Theorem 1.11 *Let $(\Omega, \mathcal{F}, \{\mathcal{F}_t\}, \mathbb{P})$ be a filtered probability space and let X be an $\{\mathcal{F}_t\}$-adapted stochastic process. Then X is a standard $\{\mathcal{F}_t\}$-Brownian motion if and only if X is a continuous local martingale starting at zero and such that its quadratic variation satisfies $\langle X, X\rangle_t = t$ for all $t \geq 0$.*

Theorem 1.11 is called Lévy's characterization theorem.

Definition 1.12 Let X be a stochastic process. Given $t \geq 0$, the marginal distribution ρ_t of the process X is the distribution of the random variable X_t.

Consider the stochastic process Y defined by $Y_t = y_0 + \sigma W_t$, where W is a standard Brownian motion and y_0 is a real number. The process Y is the unique solution to the following stochastic differential equation:

$$dY_t = \sigma\, dW_t, \quad Y_0 = y_0.$$

The mean of the random variable Y_t is given by $m(Y_t) = y_0$, and its variance satisfies $v(Y_t) = \sigma^2 t$. Moreover, for any $t \geq 0$ the marginal distribution density ρ_t of the process Y is given by

$$\rho_t(y) = \frac{1}{\sqrt{2\pi t}\sigma} \exp\left\{-\frac{|y - y_0|^2}{2\sigma^2 t}\right\}.$$

Therefore, Y_t is $N(y_0, \sigma^2 t)$-distributed.

Definition 1.13 An n-dimensional standard Brownian motion is a process $\vec{W}_t = (W_t^{(1)}, \ldots, W_t^{(1)})$, $t \geq 0$, on Ω with state space $\mathbb{R}^n$ such that its components are independent standard Brownian motions.

In this book we assume the reader is familiar with standard definitions and facts from the theory of stochastic differential equations. There are numerous excellent books presenting this theory, for example [IW81, KS91, Øks03, Pro04, RY04, CE05, All07, LL08].

Brownian motion starting at $y_0 \in \mathbb{R}$ and having drift μ is the unique solution to the following stochastic differential equation:

$$dY_t = \mu\, dt + \sigma\, dW_t, \quad Y_0 = y_0.$$

The solution is given by $Y_t = y_0 + \mu t + \sigma W_t$. The process Y represents a motion along the straight line $t \mapsto \mu t$ with random fluctuations. The mean of Y_t is given by $m(Y_t) = y_0 + \mu t$, while the variance is $v(Y_t) = \sigma^2 t$. The marginal distribution densities of Brownian motion with drift are as follows:

$$\rho_t(y) = \frac{1}{\sqrt{2\pi t}\sigma} \exp\left\{-\frac{|y - (y_0 + \mu t)|^2}{2\sigma^2 t}\right\}.$$

Hence, Y_t is $N(y_0 + \mu t, \sigma^2 t)$-distributed.

1.2 Geometric Brownian Motion

Geometric Brownian motion satisfies the following stochastic differential equation:

$$dY_t = \mu Y_t\, dt + \sigma Y_t\, dW_t, \quad Y_0 = y_0. \tag{1.1}$$

We will look for the solution to (1.1) that can be represented in the following form: $Y_t = y_0 \exp\{at + \sigma W_t\}$. Our goal is to choose the number a so that the process Y solves (1.1). Using the integration by parts formula and Itô's formula (see [RY04], Chap. IV, Sect. 3), we get

$$\begin{aligned} dY_t &= y_0 e^{at}\, de^{\sigma W_t} + y_0 a e^{at} e^{\sigma W_t}\, dt \\ &= \left(a + \frac{1}{2}\sigma^2\right) Y_t\, dt + \sigma Y_t\, dW_t. \end{aligned}$$

Now it is clear that if we take $a = \mu - \frac{\sigma^2}{2}$, then the resulting process

$$Y_t = y_0 \exp\left\{\left(\mu - \frac{\sigma^2}{2}\right)t + \sigma W_t\right\}, \quad t > 0, \tag{1.2}$$

solves equation (1.1).

Geometric Brownian motion plays the role of the volatility process in the Hull–White model (see Sect. 2.3 below). We will next calculate the mean and the variance of Y_t. It follows from (1.2) that

$$m(Y_t) = y_0 \exp\left\{\left(\mu - \frac{\sigma^2}{2}\right)t\right\} \mathbb{E}\left[\exp\{\sigma W_t\}\right]. \tag{1.3}$$

Moreover,

$$\begin{aligned} \mathbb{E}\left[\exp\{\sigma W_t\}\right] &= \frac{1}{\sqrt{2\pi t}} \int_{-\infty}^{\infty} e^{\sigma y} \exp\left\{-\frac{y^2}{2t}\right\} dy \\ &= \frac{1}{\sqrt{2\pi t}} \int_{-\infty}^{\infty} \exp\left\{-\frac{1}{2t}\left(y^2 - 2t\sigma y\right)\right\} dy \\ &= \frac{1}{\sqrt{2\pi t}} e^{\frac{t\sigma^2}{2}} \int_{-\infty}^{\infty} \exp\left\{-\frac{1}{2t}(y - t\sigma)^2\right\} dy = e^{\frac{t\sigma^2}{2}}. \end{aligned} \tag{1.4}$$

Therefore, (1.3) and (1.4) give $m(Y_t) = y_0 e^{\mu t}$. As a consequence of the previous formula, the long-run mean level of the process Y is given by

$$\lim_{t\to\infty} m(Y_t) = \begin{cases} 0, & \text{if } \mu < 0, \\ y_0, & \text{if } \mu = 0, \\ \infty, & \text{if } \mu > 0. \end{cases}$$

The variance of a geometric Brownian motion can be computed as follows. Using (1.2), we obtain

$$v(Y_t) = \mathbb{E}\left[Y_t^2\right] - \left(\mathbb{E}[Y_t]\right)^2 = y_0^2 \exp\left\{(2\mu - \sigma^2)t\right\}\mathbb{E}\left[\exp\{2\sigma W_t\}\right] - y_0^2 e^{2\mu t}.$$

Now (1.4) with 2σ instead of σ implies

$$v(Y_t) = y_0^2 \exp\left\{(2\mu - \sigma^2)t + 2t\sigma^2\right\} - y_0^2 e^{2\mu t} = y_0^2 e^{2\mu t}\left(e^{\sigma^2 t} - 1\right).$$

Finally,

$$v(Y_t) = y_0^2 e^{2\mu t}\left(e^{\sigma^2 t} - 1\right).$$

It follows that the long-run variance of Y_t is given by

$$\lim_{t\to\infty} v(Y_t) = \begin{cases} y_0^2, & \text{if } \mu < 0 \text{ and } \sigma^2 = 2|\mu|, \\ 0, & \text{if } \mu < 0 \text{ and } \sigma^2 < 2|\mu|, \\ \infty, & \text{otherwise.} \end{cases}$$

Our next goal is to compute the marginal distribution densities ρ_t, $t > 0$, of a geometric Brownian motion. An explicit formula for such a density is contained in the next lemma.

Lemma 1.14 *For every $y > 0$, the following formula holds*:

$$\rho_t(y) = \frac{1}{\sqrt{2\pi t}\sigma y} \exp\left\{-\frac{[\log y - (\log y_0 + (\mu - \frac{\sigma^2}{2})t)]^2}{2t\sigma^2}\right\}. \tag{1.5}$$

Proof Using formula (1.2) and the formula for the distribution density of Brownian motion with drift, we see that for every $\lambda > 0$,

$$\begin{aligned} \mathbb{P}[Y_t < \lambda] &= \mathbb{P}\left[\left(\mu - \frac{\sigma^2}{2}\right)t + \sigma W_t < \log\frac{\lambda}{y_0}\right] \\ &= \frac{1}{\sqrt{2\pi t}\sigma} \int_{-\infty}^{\log\frac{\lambda}{y_0}} \exp\left\{-\frac{[z - (\mu - \frac{\sigma^2}{2})t]^2}{2t\sigma^2}\right\} dz. \end{aligned} \tag{1.6}$$

Making the substitution $z = \log\frac{y}{y_0}$ in (1.6), we obtain

$$\mathbb{P}[Y_t < \lambda] = \frac{1}{\sqrt{2\pi t}\sigma} \int_0^{\lambda} \exp\left\{-\frac{[\log y - (\log y_0 + (\mu - \frac{\sigma^2}{2})t)]^2}{2t\sigma^2}\right\}\frac{dy}{y}. \tag{1.7}$$

Now it is clear that (1.7) implies (1.5).

This completes the proof of Lemma 1.14. □

Definition 1.15 A positive random variable U is called log-normally distributed if the random variable $\log U$ is normally distributed.

If the random variable $\log U$ is $N(m,\gamma^2)$-distributed, then the distribution density of U is

$$f(x;m,\gamma)=\frac{1}{\sqrt{2\pi}\gamma x}\exp\left\{-\frac{(\log x-m)^2}{2\gamma^2}\right\}.$$

It follows from formula (1.5) that the marginal distribution density ρ_t of a geometric Brownian motion is log-normal with

$$m=\log y_0+\left(\mu-\frac{\sigma^2}{2}\right)t$$

and $\gamma=\sqrt{t}\sigma$.

Formula (1.5) can be rewritten as follows:

$$\rho_t(y)=\frac{\sqrt{y_0e^{\mu t}}}{\sqrt{2\pi t}\sigma}\exp\left\{-\frac{\sigma^2 t}{8}\right\}y^{-\frac{3}{2}}\exp\left\{-\frac{(\log\frac{y}{y_0e^{\mu t}})^2}{2t\sigma^2}\right\}. \tag{1.8}$$

Using (1.8), we see that the density ρ_t possesses the following symmetry property:

$$\rho_t\left(\frac{(y_0e^{\mu t})^2}{y}\right)=\frac{y^3}{(y_0e^{\mu t})^3}\rho_t(y) \tag{1.9}$$

for all $y>0$. Formula (1.9) has important applications in the theory of uncorrelated stochastic volatility models. It will be shown in Sect. 9.8 that a similar formula holds for the distribution density of the stock price in a general uncorrelated stochastic volatility model.

1.3 Long-Time Behavior of Marginal Distributions

We can simplify formula (1.5) even more. Indeed, it not hard to see that

$$\rho_t(y)=\frac{1}{\sqrt{2\pi}\sigma}y_0^{-\frac{\mu}{\sigma^2}+\frac{1}{2}}t^{-\frac{1}{2}}\exp\left\{-\frac{(\sigma^2-2\mu)^2}{8\sigma^2}t\right\}y^{-\frac{3}{2}+\frac{\mu}{\sigma^2}}$$
$$\times\exp\left\{-\left[2t\sigma^2\right]^{-1}\left(\log\frac{y}{y_0}\right)^2\right\}. \tag{1.10}$$

It follows from (1.10) that $\lim_{t\to\infty}\rho_t(y)=0$ for all $y>0$. Hence ρ_t converges to zero pointwise on $(0,\infty)$ as $t\to\infty$. We will next describe the long-time behavior of marginal distributions of a geometric Brownian motion in the weak topology.

Denote by C_b the space of bounded continuous functions on $[0,\infty)$, by C^+ the space of bounded continuous functions ϕ on $[0,\infty)$, for which the limit $\varphi(\infty)=\lim_{y\to\infty}\phi(y)$ exists and is finite, and by C_0^+ the subspace of C^+ consisting of the functions φ with $\varphi(\infty)=0$.

Lemma 1.16 *The following statements hold*:

- *Let* $2\mu < \sigma^2$. *Then*

$$\lim_{t\to\infty}\int_0^\infty \varphi(y)\rho_t(y)\,dy = \varphi(0)$$

for every function $\phi \in C_b$.
- *Let* $2\mu = \sigma^2$. *Then*

$$\lim_{t\to\infty}\int_0^\infty \varphi(y)\rho_t(y)\,dy = \frac{1}{2}\big[\varphi(0)+\varphi(\infty)\big]$$

for every function $\varphi \in C^+$. *In particular*,

$$\lim_{t\to\infty}\int_0^\infty \varphi(y)\rho_t(y)\,dy = \frac{1}{2}\varphi(0)$$

for every function $\varphi \in C_0^+$.
- *Let* $2\mu > \sigma^2$. *Then*

$$\lim_{t\to\infty}\int_0^\infty \varphi(y)\rho_t(y)\,dy = \varphi(\infty)$$

for every function $\varphi \in C^+$. *In particular*,

$$\lim_{t\to\infty}\int_0^\infty \varphi(y)\rho_t(y)\,dy = 0$$

for every function $\varphi \in C_0^+$.

Proof Using (1.10), we see that

$$\int_0^\infty \varphi(y)\rho_t(y)\,dy = \frac{1}{\sqrt{2\pi}\sigma}y_0^{-\frac{\mu}{\sigma^2}+\frac{1}{2}}t^{-\frac{1}{2}}\exp\left\{-\frac{(\sigma^2-2\mu)^2}{8\sigma^2}t\right\}$$
$$\times\int_0^\infty \varphi(y)y^{-\frac{3}{2}+\frac{\mu}{\sigma^2}}\exp\left\{-[2t\sigma^2]^{-1}\left(\log\frac{y}{y_0}\right)^2\right\}dy. \tag{1.11}$$

Making the substitution $u = t^{-\frac{1}{2}}\sigma^{-1}\log\frac{y}{y_0}$ in the integral on the right-hand side of equality (1.11) and transforming the resulting expression, we obtain

$$\int_0^\infty \varphi(y)y^{-\frac{3}{2}+\frac{\mu}{\sigma^2}}\exp\left\{-[2t\sigma^2]^{-1}\left(\log\frac{y}{y_0}\right)^2\right\}dy$$
$$= \sqrt{t}\sigma y_0^{-\frac{1}{2}+\frac{\mu}{\sigma^2}}\exp\left\{-\frac{1}{2}t\sigma^2\left(-\frac{1}{2}+\frac{\mu}{\sigma^2}\right)^2\right\}$$

$$\times \int_{-\infty}^{\infty} \varphi(y_0 \exp\{\sqrt{t}\sigma u\}) \exp\left\{-\frac{1}{2}\left(u - \sqrt{t}\sigma\left(-\frac{1}{2} + \frac{\mu}{\sigma^2}\right)\right)^2\right\} du$$

$$= \sqrt{t}\sigma y_0^{-\frac{1}{2}+\frac{\mu}{\sigma^2}} \exp\left\{\frac{1}{2} t\sigma^2 \left(-\frac{1}{2} + \frac{\mu}{\sigma^2}\right)^2\right\}$$

$$\times \int_{-\infty}^{\infty} \varphi\left(y_0 \exp\left\{\sqrt{t}\sigma\left[w + \sqrt{t}\sigma\left(-\frac{1}{2} + \frac{\mu}{\sigma^2}\right)\right]\right\}\right) e^{-\frac{w^2}{2}} dw.$$

Next, using (1.11) we see that

$$\int_0^{\infty} \varphi(y)\rho_t(y)\, dy$$

$$= \frac{1}{\sqrt{2\pi}} \int_{-\infty}^{\infty} \varphi\left(y_0 \exp\left\{-\sqrt{t}\sigma w + t\sigma^2\left(-\frac{1}{2} + \frac{\mu}{\sigma^2}\right)\right\}\right) e^{-\frac{w^2}{2}} dw. \tag{1.12}$$

Now let $t \to \infty$ in (1.12). It is not hard to see, using the Lebesgue dominated convergence theorem, that Lemma 1.16 holds. □

Remark 1.17 Denote by M the space of finite Borel measures on $[0, \infty)$, and by $\widetilde{M}$ the space of finite Borel measures on $[0, \infty]$. Then Lemma 1.16 can be interpreted as follows:

- If $2\mu < \sigma^2$, then $\rho_t \to \delta_0$ as $t \to \infty$ in the weak topology $\sigma(M, C_b)$ associated with the duality (M, C_b).
- If $2\mu = \sigma^2$, then $\rho_t \to \frac{1}{2}[\delta_0 + \delta_\infty]$ as $t \to \infty$ in the topology $\sigma(\widetilde{M}, C^+)$. In particular, $\rho_t \to \frac{1}{2}\delta_0$ as $t \to \infty$ in the topology $\sigma(M, C_0^+)$.
- If $2\mu > \sigma^2$, then $\rho_t \to \delta_\infty$ as $t \to \infty$ in the topology $\sigma(\widetilde{M}, C^+)$. In particular, $\rho_t \to 0$ as $t \to \infty$ in the topology $\sigma(M, C_0^+)$.

1.4 Ornstein–Uhlenbeck Processes

The absolute value of an Ornstein–Uhlenbeck process plays the role of stochastic volatility in the uncorrelated Stein–Stein model (see Sect. 2.3 below). Let $q > 0$, $m \geq 0$, and $\sigma > 0$, and consider the following stochastic differential equation:

$$dY_t = q(m - Y_t)\, dt + \sigma\, dZ_t, \quad Y_0 = y_0. \tag{1.13}$$

The coefficients in (1.13) satisfy the linear growth and the Lipschitz conditions. Hence, there exists the unique strong solution Y of (1.13) (see, e.g. [RY04, LL08]). The process Y is called the Ornstein–Uhlenbeck process. We will also use the symbol $Y(q, m, \sigma, y_0)$ to denote the Ornstein–Uhlenbeck process when it is important to specify the values of the model parameters. The next statement provides an explicit expression for the Ornstein–Uhlenbeck process.

Lemma 1.18 *The following formula holds*:

$$Y_t(q,m,\sigma,y_0)=e^{-qt}y_0+\left(1-e^{-qt}\right)m+\sigma e^{-qt}\int_0^t e^{qu}\,dZ_u. \tag{1.14}$$

Proof Using the integration by parts formula, we see that the process Y defined by (1.14) satisfies

$$dY_t=-qy_0e^{-qt}dt+qme^{-qt}dt-\sigma qe^{-qt}\int_0^t e^{qu}\,dZ_u+\sigma\,dZ_t. \tag{1.15}$$

On the other hand, we have

$$\begin{aligned}&q(m-Y_t)\,dt+\sigma\,dZ_t\\&\quad=q\left(m-e^{-qt}y_0-\left(1-e^{-qt}\right)m-\sigma e^{-qt}\int_0^t e^{qu}\,dZ_u\right)dt+\sigma\,dZ_t.\end{aligned} \tag{1.16}$$

Since the right-hand sides of (1.15) and (1.16) are equal, the process Y satisfies the stochastic differential equation (1.13).

This completes the proof of Lemma 1.18. □

Remark 1.19 It follows from Lemma 1.18 that

$$Y_t(q,m,\sigma,y_0)=Y_t\left(q,0,\sigma,y_0+\left(e^{qt}-1\right)m\right). \tag{1.17}$$

It is not hard to see using (1.14) that the mean and the variance of Y_t satisfy

$$m(Y_t)=e^{-qt}y_0+\left(1-e^{-qt}\right)m,\qquad v(Y_t)=\frac{\sigma^2}{2q}\left(1-e^{-2qt}\right). \tag{1.18}$$

Hence, the random variable Y_t is normally distributed with mean and variance given in (1.18).

Let ρ_t denote the distribution density of Y_t. We have

$$\rho_t(y)=\frac{\sqrt{q}}{\sigma\sqrt{\pi(1-e^{-2qt})}}\exp\left\{-\frac{q[y-(e^{-qt}y_0+(1-e^{-qt})m)]^2}{\sigma^2(1-e^{-2qt})}\right\}. \tag{1.19}$$

It is also true that

$$\lim_{t\to\infty}m(Y_t)=m\quad\text{and}\quad\lim_{t\to\infty}v(Y_t)=\frac{\sigma^2}{2q}.$$

Moreover, for every $y\in\mathbb{R}$,

$$\lim_{t\to\infty}\rho_t(y)=\frac{\sqrt{q}}{\sigma\sqrt{\pi}}\exp\left\{-\frac{q(y-m)^2}{\sigma^2}\right\}.$$

Therefore, the long-time marginal distributions of Y are normal with mean m and variance $\frac{\sigma^2}{2q}$. It is also easy to prove that

$$\lim_{t\to\infty}\int_{\mathbb{R}}\varphi(y)\rho_t(y)\,dy = \frac{\sqrt{q}}{\sigma\sqrt{\pi}}\int_{\mathbb{R}}\varphi(y)\exp\left\{-\frac{q(y-m)^2}{\sigma^2}\right\}dy$$

for all $\varphi \in C_b$.

1.5 Ornstein–Uhlenbeck Processes and Time-Changed Brownian Motions

The next lemma shows that there is a relation between an Ornstein–Uhlenbeck process and a standard Brownian motion.

Lemma 1.20 *Let Y be the Ornstein–Uhlenbeck process satisfying*

$$dY_t = q(m - Y_t)\,dt + \sigma\,dW_t, \quad Y_0 = y_0,$$

and let

$$\phi(t) = \sigma^2(2q)^{-1}\left(e^{2qt} - 1\right).$$

Then there exists an $\mathcal{F}_{\phi^{-1}(t)}$-standard Brownian motion B such that the following formula holds:

$$Y_t(q, m, \sigma, y_0) = e^{-qt}\left(y_0 + \left(e^{qt} - 1\right)m + B_{\phi(t)}\right), \quad t \ge 0. \tag{1.20}$$

Proof It follows from Lemma 1.18 that

$$Y_t(q, m, \sigma, y_0) = e^{-qt}\left[y_0 + \left(e^{qt} - 1\right)m + M_t\right], \quad t \ge 0, \tag{1.21}$$

where

$$M_t = \int_0^t \sigma e^{qu}\,dZ_u, \quad t \ge 0.$$

The process M is a continuous $\mathcal{F}_t$-martingale and its quadratic variation is given by

$$\langle M, M\rangle_t = \int_0^t \sigma^2 e^{2qs}\,ds = \phi(t), \quad t \ge 0.$$

It is clear that the function ϕ is smooth, strictly increasing, and such that $\phi(0) = 0$. Consider the following process:

$$B_t = M\left(\phi^{-1}(t)\right), \quad t \ge 0. \tag{1.22}$$

This process is a continuous $\mathcal{F}_{\phi^{-1}(t)}$-martingale such that $B_0 = 0$ a.s. Its quadratic variation is given by

$$\langle B, B\rangle_t = \int_0^{\phi^{-1}(t)} \sigma^2 e^{qs}\, ds = t, \quad t > 0.$$

By Lévy's characterization theorem, the process B is an $\mathcal{F}_{\phi^{-1}(t)}$-standard Brownian motion. It follows from (1.22) that $M_t = B_{\phi(t)}$. Now (1.21) shows that (1.20) holds.
This completes the proof of Lemma 1.20. □

Remark 1.21 Lemma 1.20 illustrates in a very simple case how to represent a martingale as a time-changed Brownian motion (see [KS91], p. 174, for a general representation theorem).

If $a = 0$, then we have $\phi(t) = \sigma^2 t$, $t > 0$, and formula (1.20) becomes

$$Y_t(0, m, \sigma, y_0) = y_0 + B_{t\sigma^2}, \quad t \ge 0,$$

where B_t is a $\mathcal{F}_{\phi^{-1}(t)}$-standard Brownian motion.

1.6 Absolute Value of an Ornstein–Uhlenbeck Process

Let Y be the Ornstein–Uhlenbeck process discussed in Sect. 1.4. Put $U = |Y|$. We will next compute the mean, the variance, and the marginal distributions of the process U.

Let us denote by η_t the marginal distribution density of U_t (it will be shown below that the density η_t exists). Then, for every $\lambda \ge 0$ we have

$$\mathbb{P}[U_t \le \lambda] = \mathbb{P}[-\lambda \le Y_t \le \lambda] = \int_{-\lambda}^{\lambda} \rho_t(y)\, dy = \int_0^{\lambda} \big[\rho_t(y) + \rho_t(-y)\big]\, dy,$$

where ρ_t is the distribution density of Y_t given by (1.19). It is not hard to see that

$$\eta_t(y) = \rho_t(y) + \rho_t(-y), \quad y > 0. \tag{1.23}$$

Our first goal is to compute the mean of U_t, using (1.23) and the formula

$$\frac{1}{\sqrt{2\pi}\gamma} \int_0^{\infty} y \exp\left\{-\frac{(y-a)^2}{2\gamma^2}\right\} dy = \frac{\gamma}{\sqrt{2\pi}} \exp\left\{-\frac{a^2}{2\gamma^2}\right\} + \frac{a}{\sqrt{2\pi}} \int_{-\frac{a}{\gamma}}^{\infty} \exp\left\{-\frac{u^2}{2}\right\} du.$$

We have

$$m\big(|Y_t|\big) = \frac{2\gamma}{\sqrt{2\pi}} \exp\left\{-\frac{a^2}{2\gamma^2}\right\} + \frac{a}{\sqrt{2\pi}} \int_{-\frac{a}{\gamma}}^{\frac{a}{\gamma}} \exp\left\{-\frac{u^2}{2}\right\} du,$$

where

$$a = e^{-qt} y_0 + (1 - e^{-qt}) m \quad \text{and} \quad \gamma^2 = \frac{\sigma^2}{2q}(1 - e^{-2qt}).$$

Moreover, the formulas

$$v(|Y_t|) = \mathbb{E}[Y_t^2] - (\mathbb{E}[|Y_t|])^2$$

and

$$\frac{1}{\sqrt{2\pi}\gamma} \int_{-\infty}^{\infty} y^2 \exp\left\{-\frac{(y-a)^2}{2\gamma^2}\right\} dy = \gamma^2 + a^2$$

imply

$$v(|Y_t|) = a^2 + \gamma^2 - \left[\frac{\sqrt{2}\gamma}{\sqrt{\pi}} \exp\left\{-\frac{a^2}{2\gamma^2}\right\} + \frac{a}{\sqrt{2\pi}} \int_{-\frac{a}{\gamma}}^{\frac{a}{\gamma}} \exp\left\{-\frac{u^2}{2}\right\} du\right]^2,$$

where $a = e^{-qt} y_0 + (1 - e^{-qt}) m$ and $\gamma^2 = \frac{\sigma^2}{2q}(1 - e^{-2qt})$.

The long-time behavior of the quantities computed above can be determined from the following formulas:

$$\lim_{t\to\infty} m(|Y_t|) = \frac{\sigma}{\sqrt{\pi q}} \exp\left\{-\frac{qm^2}{\sigma^2}\right\} + \frac{m}{\sqrt{2\pi}} \int_{-\frac{\sqrt{2q}m}{\sigma}}^{\frac{\sqrt{2q}m}{\sigma}} \exp\left\{-\frac{u^2}{2}\right\} du,$$

$$\lim_{t\to\infty} v(|Y_t|) = m^2 + \frac{\sigma^2}{2q} - \left[\frac{\sigma}{\sqrt{\pi q}} \exp\left\{-\frac{qm^2}{\sigma^2}\right\} + \frac{m}{\sqrt{2\pi}} \int_{-\frac{\sqrt{2q}m}{\sigma}}^{\frac{\sqrt{2q}m}{\sigma}} \exp\left\{-\frac{u^2}{2}\right\} du\right]^2,$$

and

$$\lim_{t\to\infty} \eta_t(y) = \frac{\sqrt{q}}{\sigma\sqrt{\pi}} \left[\exp\left\{-\frac{q(y-m)^2}{\sigma^2}\right\} + \exp\left\{-\frac{q(y+m)^2}{\sigma^2}\right\}\right]$$

for all $y > 0$.

1.7 Squared Bessel Processes and CIR Processes

In the Heston model (see Sect. 2.3), the volatility follows the process $\sqrt{Y}$ where Y is a Cox–Ingersoll–Ross process (a CIR process). The process Y is the unique solution to the following stochastic differential equation:

$$dY_t = (a - bY_t)\,dt + c\sqrt{Y_t}\,dW_t, \quad Y_0 = y_0. \tag{1.24}$$

The existence and uniqueness of solutions to (1.24) follow from the Yamada–Watanabe theorem (see, e.g., Theorem 3.2 in [IW81], see also [RY04]). The CIR process is a non-negative stochastic process. We assume that $a \geq 0$, $b \geq 0$, and $c > 0$. If $b > 0$, then (1.24) can be rewritten in the mean-reverting form as follows:

$$dY_t = q(m - Y_t)\,dt + c\sqrt{Y_t}\,dW_t, \quad Y_0 = y_0, \tag{1.25}$$

where $q = b$ and $m = \frac{a}{b}$.

An important special case of a CIR process is the squared Bessel process.

Definition 1.22 For every $\delta \geq 0$ and $y_0 \geq 0$ the unique strong solution T of the equation

$$dT_t = \delta\,dt + 2\int_0^t \sqrt{T_s}\,dW_s, \quad T_0 = y_0, \tag{1.26}$$

is called the squared Bessel process of dimension δ starting at y_0.

It is clear that the process T in (1.26) is a CIR process. Indeed, we can take $a = \delta$, $b = 0$, and $c = 2$ in (1.24). The number $\nu = \frac{\delta}{2} - 1$ is called the index of the squared Bessel process T.

Throughout the book, we will denote the process T introduced in Definition 1.22 by $BESQ^{\delta}_{y_0}$. The integrated form of (1.26) is the following:

$$T_t = y_0 + \delta t + 2\int_0^t \sqrt{T_s}\,dW_s. \tag{1.27}$$

Taking the expectations on both sides of (1.27), we see that the mean of the random variable T_t is given by

$$m(T_t) = y_0 + \delta t. \tag{1.28}$$

For the variance of T_t, we have

$$v(T_t) = 4\int_0^t \mathbb{E}[T_s]\,ds = 4y_0 t + 2\delta t^2. \tag{1.29}$$

The marginal distributions of the process $BESQ^{\delta}_{y_0}$ will be computed in the next sections.

1.8 Squared Bessel Processes and Sums of the Squares of Independent Brownian Motions

Let $\delta = n$, where n is a positive integer. Suppose $x \geq 0$ and consider Brownian motion $\vec{W}_t = (x_1 + W_t^1, \dots, x_n + W_t^n)$ in $\mathbb{R}^n$ starting at the point $\vec{x} = (x_1, \dots, x_n)$ and such that

$$x_1^2 + \cdots + x_n^2 = x. \tag{1.30}$$

Put

$$M_t^{(n)} = |\vec{W}_t|^2 = \sum_{i=1}^{n} \left(x_i + W_t^i\right)^2, \quad t \geq 0. \tag{1.31}$$

Theorem 1.23 *For every integer $n > 0$ and every $x \geq 0$ the following equality holds: $BESQ_x^n = M^{(n)}$, where the process $M^{(n)}$ is defined by* (1.31), *and the initial condition $\vec{x}$ for the process $\vec{W}$ is given by* (1.30).

Proof Applying the multi-dimensional Itô formula to the process $M_t^{(n)}$ defined by (1.31), we get

$$dM_t^{(n)} = n\,dt + 2\sum_{i=1}^{n} W_t^i\,dW_t^i.$$

Therefore,

$$dM_t^{(n)} = n\,dt + 2\sqrt{M_t^{(n)}} \sum_{i=1}^{n} \frac{W_t^i}{\sqrt{M_t^{(n)}}}\,dW_t^i. \tag{1.32}$$

The possibility of dividing by $M_t^{(n)}$ in (1.32) can be justified as follows. We have

$$\left\{t : M_t^{(n)} = 0\right\} \subset \left\{t : W_t^1 = -x_1\right\}$$

$\mathbb{P}$-a.s. on Ω. Note that it is known that the latter set is a perfect set in $\mathbb{R}$ of Lebesgue measure zero (see, e.g., [IM65], Sect. 1.7).

Consider the process $\beta^{(n)}$ defined by

$$\beta_t^{(n)} = \sum_{i=1}^{n} \int_0^t \frac{W_s^i}{\sqrt{M_s^{(n)}}}\,dW_s^i, \tag{1.33}$$

and denote by $\{\mathcal{F}_t^{(n)}\}$ the filtration generated by the n-dimensional Brownian motion $\vec{W}$.

Lemma 1.24 *For every $n \geq 1$ the process $\beta^{(n)}$ defined by* (1.33) *is a one-dimensional standard $\{\mathcal{F}_t^{(n)}\}$-Brownian motion.*

Proof The process $\beta^{(n)}$ is continuous and adapted to the filtration $\{\mathcal{F}_t^{(n)}\}$. Moreover, the following formula holds for the quadratic variation of the process $\beta^{(n)}$:

$$\left\langle \beta^{(n)}, \beta^{(n)} \right\rangle_t = \sum_{i=1}^{n} \int_0^t \frac{(W_s^i)^2}{M_s^{(n)}}\,ds = t. \tag{1.34}$$

It follows from (1.33) and (1.34) that the process $\beta^{(n)}$ is a $(\{\mathcal{F}_t^{(n)}\}, \mathbb{P})$-martingale. By Lévy's characterization theorem, $\beta^{(n)}$ is a standard $\{\mathcal{F}_t^{(n)}\}$-Brownian motion.

This completes the proof of Lemma 1.24. □

Let us return to the proof of Theorem 1.23. Note that (1.32), (1.33), and Lemma 1.24 imply that

$$dM_t^{(n)} = n\,dt + 2\sqrt{M_t^{(n)}}\,d\beta_t^{(n)}, \quad M_0^{(n)} = x.$$

Therefore, Theorem 1.23 holds by the uniqueness result for solutions to the stochastic differential equation defining a squared Bessel process.

Our next goal is to compute the marginal distributions of the process $BESQ_{y_0}^{\delta}$. This will be accomplished in the next sections. It will be shown that the random variable $BESQ_{y_0}^{\delta}(t)$ is distributed according to a noncentral chi-square law. □

1.9 Chi-Square Distributions

Definition 1.25 A random variable U is distributed as chi-square with n degrees of freedom provided that

$$Q = \sum_{i=1}^{n} X_i^2, \tag{1.35}$$

where X_i, $1 \le i \le n$, are independent $N(0,1)$-distributed random variables.

It will be shown below that the chi-square distribution with n degrees of freedom has a density which depends on n and does not depend on the representation of U in (1.35). This density will be denoted χ_n^2.

Definition 1.26 A Gamma distribution with scale parameter $\theta > 0$ and shape parameter $k > 0$ is the distribution whose density is given by

$$\Gamma(y;k,\theta) = \frac{1}{\theta^k \Gamma(k)} y^{k-1} e^{-\frac{y}{\theta}} \mathbb{1}_{\{y \ge 0\}}.$$

In Definition 1.26, the symbol $\Gamma(k)$ stands for the Gamma function defined by

$$\Gamma(k) = \int_0^{\infty} y^{k-1} e^{-y}\,dy, \quad k > 0.$$

Lemma 1.27 *The chi-square distribution with n degrees of freedom admits a density χ_n^2 satisfying*

$$\chi_n^2 = \Gamma\left(\cdot;\frac{n}{2},2\right). \tag{1.36}$$

Proof We will prove Lemma 1.27 for $n = 1$ and $n = 2$, and leave the rest of the prove as an exercise for the reader.

Let $n = 1$. Then for every $\lambda \geq 0$ we have

$$\mathbb{P}(X_1^2 \leq \lambda) = \frac{1}{\sqrt{2\pi}} \int_{-\sqrt{\lambda}}^{\sqrt{\lambda}} e^{-\frac{u^2}{2}} du = \frac{1}{\sqrt{2\pi}} \int_0^{\lambda} z^{-\frac{1}{2}} e^{-\frac{z}{2}} dz.$$

Therefore,

$$\chi_1^2(y) = \frac{1}{\sqrt{2\pi}} y^{-\frac{1}{2}} e^{-\frac{y}{2}} \mathbb{1}_{\{y \geq 0\}} = \frac{1}{\sqrt{2}\Gamma(\frac{1}{2})} y^{-\frac{1}{2}} e^{-\frac{y}{2}} \mathbb{1}_{\{y \geq 0\}}, \tag{1.37}$$

which establishes (1.36) with $n = 1$.

In order to find χ_n^2, we have to convolve n copies of χ_1^2. Let $n = 2$. Then using (1.37), we see that

$$\begin{aligned}
\chi_2^2(y) &= \frac{1}{2[\Gamma(\frac{1}{2})]^2} \int_{-\infty}^{\infty} z^{-\frac{1}{2}} e^{-\frac{z}{2}} \mathbb{1}_{\{z \geq 0\}} (y - z)^{-\frac{1}{2}} e^{-\frac{y-z}{2}} \mathbb{1}_{\{y - z \geq 0\}} dz \\
&= \frac{1}{2[\Gamma(\frac{1}{2})]^2} e^{-\frac{y}{2}} \int_0^1 u^{-\frac{1}{2}} (1 - u)^{-\frac{1}{2}} du.
\end{aligned} \tag{1.38}$$

The last integral in (1.38) is a special case of the Beta function.

Definition 1.28 Let $\mathrm{Re}(x) > 0$ and $\mathrm{Re}(y) > 0$. Then the Beta function B is defined by

$$B(x, y) = \int_0^1 t^{x-1} (1 - t)^{y-1} dt.$$

The Beta function can be represented in terms of the Gamma function (see [AAR99], p. 5, or [SSh03]). The following formula holds for all $x \in \mathbb{C}$ and $y \in \mathbb{C}$ with $\mathrm{Re}(x) > 0$ and $\mathrm{Re}(y) > 0$:

$$B(x, y) = \frac{\Gamma(x)\Gamma(y)}{\Gamma(x + y)}. \tag{1.39}$$

Now using (1.38) and (1.39) with $r = s = \frac{1}{2}$, we see that (1.36) holds for $n = 2$.

The proof of Lemma 1.27 is thus completed. □

1.10 Noncentral Chi-Square Distributions

The distribution of a squared Bessel process is related to the noncentral chi-square distribution.

Definition 1.29 A random variable U is distributed as the noncentral chi-square with the number of degrees of freedom n and the parameter of noncentrality λ, if U can be represented as follows:

$$U = \sum_{i=1}^{n} \left(\frac{X_i}{\sigma_i} \right)^2, \tag{1.40}$$

where X_i, $1 \le i \le n$, is a finite sequence of independent normally distributed random variables with means m_i and variances σ_i^2. The parameter λ is defined by

$$\lambda = \sum_{i=1}^{n} \left(\frac{m_i}{\sigma_i} \right)^2.$$

It is clear that we can restrict ourselves to the case where $\sigma_i = 1$, $1 \le i \le n$. Indeed, put $Y_i = \frac{X_i}{\sigma_i}$, $1 \le i \le n$. Then (1.40) becomes

$$U = \sum_{i=1}^{n} Y_i^2. \tag{1.41}$$

Here Y_i is a normally distributed variable with mean $\lambda_i = \frac{m_i}{\sigma_i}$ and variance $\sigma_i^2 = 1$. It is clear that the value of λ does not change after such a transformation.

The parameter λ is a measure of noncentrality. Indeed, λ is equal to the difference of the mean of U which is $n + \lambda$ and the mean of the corresponding central chi-square distribution which is n.

Suppose the random variable U admits a representation such as in (1.41) with Y_i independent and $N(\lambda_i, 1)$-distributed. We will next show that the distribution of U depends only on n and $\lambda = \sum_{i=1}^{n} \lambda_i^2$. Indeed, for every Borel set $A \subset [0, \infty)$ we have

$$\begin{aligned} \mathbb{P}(U \in A) &= \frac{1}{(2\pi)^{\frac{n}{2}}} \int_{\mathbb{R}^n} \mathbb{1}_{\{\sum_{i=1}^{n} x_i^2 \in A\}} \exp\left\{ -\frac{1}{2} \sum_{i=1}^{n} (x_i - \lambda_i)^2 \right\} dx_1 \cdots dx_n \\ &= \frac{1}{(2\pi)^{\frac{n}{2}}} e^{-\frac{\lambda}{2}} \int_{\mathbb{R}^n} f(x_1, \dots, x_n) \exp\left\{ -\sum_{i=1}^{n} \lambda_i x_i \right\} dx_1 \cdots dx_n, \end{aligned} \tag{1.42}$$

where the function f is defined by

$$f(x_1, \dots, x_n) = \mathbb{1}_{\{\sum_{i=1}^{n} x_i^2 \in A\}} \exp\left\{ -\frac{1}{2} \sum_{i=1}^{n} x_i^2 \right\}.$$

Since the function f is invariant with respect to all orthogonal transformations of $\mathbb{R}^n$, the Laplace transform in (1.42) is also invariant with respect to all such transformations. Therefore, (1.42) shows that the noncentral chi-square distribution depends only on n and λ.

Our next goal is to find the distribution density $\chi^2(\cdot; n, \lambda)$ of the chi-square distribution with the number of degrees of freedom n and the noncentrality parameter λ. We will need the following definition.

Definition 1.30 The modified Bessel function of the first kind I_ν where $\nu \in \mathbb{R}$ is defined as follows:

- For a real number ν that is different from a negative integer,

$$I_\nu(y) = \sum_{n=0}^{\infty} \frac{1}{n!\Gamma(n+\nu+1)} \left(\frac{y}{2}\right)^{\nu+2n}, \quad y \in \mathbb{R}.$$

- For a negative integer ν, $I_\nu(y) = I_{-\nu}(y)$, $y \in \mathbb{R}$.

The I-Bessel function I_ν is a solution to the following ordinary differential equation:

$$y^2 \frac{d^2u}{dy^2} + y\frac{du}{dy} - (y^2 + \nu^2)u = 0. \tag{1.43}$$

If ν is not an integer, then $(I_\nu, I_{-\nu})$ is a fundamental system of solutions of (1.43). If ν is an integer, then (I_ν, K_ν) is a fundamental system, where K_ν is the modified Bessel function of the second kind (see [AAR99]), or [Wat95] for more information on Bessel functions).

Theorem 1.31 *The following formula holds for all integers $n \geq 1$ and all $\lambda > 0$:*

$$\chi^2(y; n, \lambda) = \frac{1}{2}\left(\frac{y}{\lambda}\right)^{\frac{n}{4}-\frac{1}{2}} e^{-\frac{y+\lambda}{2}} I_{\frac{n}{2}-1}(\sqrt{\lambda y}) \mathbb{1}_{\{y \geq 0\}}. \tag{1.44}$$

Proof We will first prove formula (1.44) for $n = 1$. Let $c \geq 0$. Then we have

$$\begin{aligned}
\mathbb{P}(X_1^2 \leq c) &= \frac{1}{\sqrt{2\pi}} \int_{-\sqrt{c}}^{\sqrt{c}} \exp\left\{-\frac{(y-\lambda_1)^2}{2}\right\} dy \\
&= \frac{1}{2\sqrt{2\pi}} \int_0^c \left[\exp\left\{-\frac{(\sqrt{u}-\lambda_1)^2}{2}\right\} + \exp\left\{-\frac{(\sqrt{u}+\lambda_1)^2}{2}\right\}\right] \frac{du}{\sqrt{u}}.
\end{aligned}$$

It follows that

$$\chi^2(y; 1, \lambda) = \frac{1}{\sqrt{2\pi}} y^{-\frac{1}{2}} e^{-\frac{y+\lambda}{2}} \cosh\sqrt{\lambda y}, \quad y > 0. \tag{1.45}$$

Now it is clear that in order to prove formula (1.44) with $n = 1$, we have to check that

$$\frac{1}{2}\lambda^{\frac{1}{4}} y^{-\frac{1}{4}} e^{-\frac{y+\lambda}{2}} I_{-\frac{1}{2}}(\sqrt{\lambda y}) = \frac{1}{\sqrt{2\pi}} y^{-\frac{1}{2}} e^{-\frac{y+\lambda}{2}} \cosh\sqrt{\lambda y},$$

or

$$I_{-\frac{1}{2}}(\sqrt{\lambda y}) = \frac{1}{\sqrt{2\pi}}(\lambda y)^{-\frac{1}{4}}\left[e^{\sqrt{\lambda y}} + e^{-\sqrt{\lambda y}}\right].$$

The previous formula follows from the equality

$$I_{-\frac{1}{2}}(u) = \frac{1}{\sqrt{2\pi}}u^{-\frac{1}{2}}\left[e^{u} + e^{-u}\right], \tag{1.46}$$

which can be established by comparing the power series representation for the function $I_{-\frac{1}{2}}$ (see Definition 1.30) with that of the function on the right-hand side of (1.46).

Our next goal is to prove formula (1.44) for $n \geq 2$. Let $U = \sum_{i=1}^{n} Y_i^2$, where for $1 \leq i \leq n$ the random variable Y_i is $N(\lambda_i, 1)$-distributed. Put $\lambda = \sum_{i=1}^{n} \lambda_i^2$. Since the chi-square distribution depends only on n and λ, we can assume with no loss of generality that $\lambda_1 = \sqrt{\lambda}$ and $\lambda_2 = \cdots = \lambda_n = 0$.

Set $V = \sum_{i=2}^{n} Y_i^2$. Then V is a central chi-square variable with the number of degrees of freedom $(n-1)$. It follows that the density $\chi^2(\cdot; n, \lambda)$ is the convolution of the densities $\chi^2(\cdot; 1, \lambda)$ and χ^2_{n-1}. Since formula (1.45) is equivalent to the formula

$$\chi^2(y; 1, \lambda) = 2^{-\frac{1}{2}} e^{-\frac{y+\lambda}{2}} \sum_{i=0}^{\infty} \frac{2^{-2i}}{i!\Gamma(i+\frac{1}{2})}\lambda^i y^{-\frac{1}{2}+i}\mathbb{1}_{\{y\geq 0\}}$$

and moreover

$$\chi^2_{n-1}(y) = \frac{1}{2^{\frac{n-1}{2}}\Gamma(\frac{n-1}{2})} y^{\frac{n-3}{2}} e^{-\frac{y}{2}}\mathbb{1}_{\{y\geq 0\}}$$

(see formula (1.36)), we have

$$\begin{aligned}
&\chi^2(y; n, \lambda) \\
&= \frac{1}{2^{\frac{n}{2}}\Gamma(\frac{n-1}{2})}\int_0^y e^{-\frac{z+\lambda}{2}}\left(\sum_{i=0}^{\infty}\frac{2^{-2i}}{i!\Gamma(i+\frac{1}{2})}\lambda^i z^{-\frac{1}{2}+i}\right)(y-z)^{\frac{n-3}{2}} e^{-\frac{y-z}{2}}\,dz \\
&= \frac{1}{2^{\frac{n}{2}}\Gamma(\frac{n-1}{2})}\sum_{i=0}^{\infty}\frac{2^{-2i}}{i!\Gamma(i+\frac{1}{2})}\lambda^i y^{\frac{n}{2}-1+i} B\left(\frac{n-1}{2}, i+\frac{1}{2}\right).
\end{aligned}$$

Next using Definition 1.30 and (1.39), we obtain formula (1.44) for $n \geq 2$.

This completes the proof of Theorem 1.31. □

In the next definition, we replace the integer n in formula (1.44) by a real number $\delta \geq 0$.

Definition 1.32 Let $\delta \geq 0$ and $\lambda > 0$. The function

$$\chi^2(y;\delta,\lambda) = \frac{1}{2}\left(\frac{y}{\lambda}\right)^{\frac{\delta}{4}-\frac{1}{2}} e^{-\frac{y+\lambda}{2}} I_{\frac{\delta}{2}-1}(\sqrt{\lambda y})\mathbb{1}_{\{y\geq 0\}}$$

is called the noncentral chi-square distribution density with δ degrees of freedom and the index of noncentrality λ.

1.11 Marginal Distributions of Squared Bessel Processes. Formulations

In order to find an explicit formula for the marginal distribution density ρ_t of a squared Bessel process of integral dimension, we note that Theorem 1.23 implies the following equality:

$$BESQ^n_{y_0}(t) = t\sum_{i=1}^{n} Y_i^2, \tag{1.47}$$

where $Y_i = t^{-\frac{1}{2}}(x_i + W_t^i)$, $1 \leq i \leq n$, and $\sum_{i=1}^n x_i^2 = y_0$. It is not hard to see using (1.47) and Theorem 1.31 that the marginal distribution density ρ_t of the process $BESQ^n_{y_0}$ is given by

$$\rho_t(y) = \frac{1}{2t}\left(\frac{y}{y_0}\right)^{\frac{n}{4}-\frac{1}{2}} e^{-\frac{y+y_0}{2t}} I_{\frac{n}{2}-1}\left(\frac{\sqrt{y_0 y}}{t}\right)\mathbb{1}_{\{y\geq 0\}},$$

$$\rho_t(y) = \frac{1}{t}\chi^2\left(\frac{y}{t}; n, \frac{y_0}{t}\right).$$

It will be established next that similar equalities hold for nonintegral dimensions.

Theorem 1.33

(a) *Let $\delta > 0$. Then the marginal distribution μ_t of the process $BESQ^\delta_0$ admits a density ρ_t given by*

$$\rho_t(y) = \frac{1}{(2t)^{\frac{\delta}{2}}\Gamma(\frac{\delta}{2})} y^{\frac{\delta}{2}-1} e^{-\frac{y}{2t}}\mathbb{1}_{\{y\geq 0\}}. \tag{1.48}$$

(b) *Let $\delta > 0$ and $y_0 > 0$. Then the marginal distribution μ_t of the process $BESQ^\delta_{y_0}$ admits a density ρ_t given by*

$$\rho_t(y) = \frac{1}{2t}\left(\frac{y}{y_0}\right)^{\frac{\delta}{4}-\frac{1}{2}} e^{-\frac{y+y_0}{2t}} I_{\frac{\delta}{2}-1}\left(\frac{\sqrt{y_0 y}}{t}\right)\mathbb{1}_{\{y\geq 0\}}. \tag{1.49}$$

(c) *Let $y_0 > 0$. Then the marginal distribution μ_t of the process $BESQ^0_{y_0}$ is a probability measure defined by*

$$\mu_t(A) = e^{-\frac{y_0}{2t}} \delta_0(A) + \frac{1}{2t} \int_{A \cap [0,\infty)} \left(\frac{y}{y_0} \right)^{-\frac{1}{2}} e^{-\frac{y+y_0}{2t}} I_1 \left(\frac{\sqrt{y_0 y}}{t} \right) dy \qquad (1.50)$$

for all Borel sets A in $\mathbb{R}$.

(d) *The marginal distribution of the process $BESQ^0_0$ is given by $\mu_t = \delta_0$ for all $t \geq 0$.*

Remark 1.34 Part (d) of Theorem 1.33 is immediate since $BESQ^0_0(t) = 0$, $t \geq 0$. Moreover, equality (1.49) can be rewritten in the following form:

$$\rho_t(y) = \frac{1}{t} \chi^2 \left(\frac{y}{t}; \delta, \frac{y_0}{t} \right).$$

Theorem 1.33 will be proven in the next two sections. We follow [RY04] in the presentation of the proof of this theorem.

1.12 Laplace Transforms of Marginal Distributions

In order to find an explicit formula for the marginal distribution μ_t of the squared Bessel process $BESQ^\delta_{y_0}$, we will first compute the Laplace transform

$$L\mu_t(\lambda) = \mathbb{E}\left[\exp\left\{-\lambda BESQ^\delta_{y_0}(t)\right\}\right], \quad \lambda > 0,$$

of the measure μ_t, and then invert this transform.

Lemma 1.35 *The Laplace transform of the marginal distribution μ_t of the process $BESQ^\delta_{y_0}$ with $y_0 \geq 0$ and $\delta \geq 0$ is given by the following formula:*

$$L\mu_t(\lambda) = (1 + 2t\lambda)^{-\frac{\delta}{2}} \exp\left\{ -\frac{y_0 \lambda}{1 + 2t\lambda} \right\} \qquad (1.51)$$

for all $\lambda > 0$.

Proof Lemma 1.35 will be derived from the following general result.

Theorem 1.36 *Let ν be a finite Borel measure on $[0,\infty)$ with compact support. Then for all $x \geq 0$ and $\delta \geq 0$, there exist numbers $A^x_\nu > 0$ and $B^\delta_\nu > 0$ such that A^x_ν depends only on x and ν, B^δ_ν depends only on δ and ν, and*

$$\mathbb{E}\left[\exp\left\{ -\int_0^\infty BESQ^\delta_x(s)\, d\nu(s) \right\}\right] = A^x_\nu B^\delta_\nu. \qquad (1.52)$$

Proof Let W and W' be independent standard Brownian motions on a probability space $(\Omega, \mathcal{F}, \mathbb{P})$, and let

$$Z = BESQ_x^\delta \quad \text{and} \quad Z' = BESQ_{x'}^{\delta'}$$

be independent squared Bessel processes driven by W and W', respectively. Put $X = Z + Z'$. The next lemma states that squared Bessel processes satisfy an additivity condition with respect to the parameters x and δ.

Lemma 1.37 *The following equality holds*: $X = BESQ_{x+x'}^{\delta+\delta'}$.

Proof The additivity property of squared Bessel processes formulated in Theorem 1.36 states that the process X has the same finite-dimensional distributions as the process $BESQ_{x+x'}^{\delta+\delta'}$. First note that the initial conditions for the processes X and $BESQ_{x+x'}^{\delta+\delta'}$ coincide. Moreover,

$$dX_t = (\delta + \delta')\, dt + 2\left(\sqrt{Z_t}\, dW_t + \sqrt{Z_t'}\, dW_t'\right). \tag{1.53}$$

Let W_t'' be a third standard Brownian motion, independent of W_t and W_t'. Define a stochastic process γ by

$$d\gamma_t = \mathbb{1}_{\{X_t>0\}} \frac{\sqrt{Z_t}\, dW_t + \sqrt{Z_t'}\, dW_t'}{\sqrt{X_t}} + \mathbb{1}_{\{X_t=0\}}\, dW_t'', \quad \gamma_0 = 0.$$

Then we have

$$\sqrt{X_t}\, d\gamma_t = \sqrt{Z_t}\, dW_t + \sqrt{Z_t'}\, dW_t'. \tag{1.54}$$

Next, using (1.53) and (1.54), we see that

$$dX_t = (\delta + \delta')\, dt + 2\sqrt{X_t}\, d\gamma_t. \tag{1.55}$$

It remains to prove that γ is a standard Brownian motion. It is clear that the process γ is a continuous martingale starting at 0 with

$$\langle \gamma, \gamma \rangle_t = \mathbb{E} \int_0^t \left[\mathbb{1}_{\{X_s>0\}} \frac{Z_s + Z_s'}{X_s} + \mathbb{1}_{\{X_s=0\}} \right] ds = t.$$

By Lévy's characterization theorem, the process γ is a standard Brownian motion. Now, using (1.55) and the uniqueness theorem for stochastic differential equations defining Bessel processes, we see that the process $X = BESQ_x^\delta + BESQ_{x'}^{\delta'}$ has the same law as the process $BESQ_{x+x'}^{\delta+\delta'}$.

This completes the proof of Lemma 1.37. □

We will next return to the proof of Theorem 1.36. Denote the left-hand side of formula (1.52) by $\phi(x, \delta)$. Then using Lemma 1.37, we see that

$$\begin{aligned}
&\phi(x+x',\delta+\delta') \\
&\quad = \mathbb{E}\left[\exp\left\{-\int_0^\infty BESQ_{x+x'}^{\delta+\delta'}(s)\,d\nu(s)\right\}\right] \\
&\quad = \mathbb{E}\left[\exp\left\{-\int_0^\infty \left(BESQ_x^{\delta}(s)+BESQ_{x'}^{\delta'}(s)\right)d\nu(s)\right\}\right] \\
&\quad = \mathbb{E}\left[\exp\left\{-\int_0^\infty BESQ_x^{\delta}(s)\,d\nu(s)\right\}\right]\mathbb{E}\left[\exp\left\{-\int_0^\infty BESQ_{x'}^{\delta'}(s)\,d\nu(s)\right\}\right].
\end{aligned}$$

Therefore,

$$\phi(x+x',\delta+\delta') = \phi(x,\delta)\phi(x',\delta'). \tag{1.56}$$

It follows from equality (1.56) that $\phi(x,\delta)=\phi(x,0)\phi(0,\delta)$. Now it is clear that Theorem 1.36 holds if we take $A_\nu^x=\phi(x,0)$ and $B_\nu^\delta=\phi(0,\delta)$. □

The following list contains several simple properties of the function ϕ:

- The functions $x\mapsto\phi(x,0)$ and $\delta\mapsto\phi(0,\delta)$ are positive.
- $\phi(x,\delta)\le 1$ (follows from the definition of ϕ).
- The functions $x\mapsto\phi(x,0)$ and $\delta\mapsto\phi(0,\delta)$ are multiplicative, that is,

$$\phi(x+x',0)=\phi(x,0)\phi(x',0) \quad \text{and} \quad \phi(0,\delta+\delta')=\phi(0,\delta)\phi(0,\delta')$$

  (these equalities follow from (1.56)).
- $\phi(0,0)=1$.

Lemma 1.38 *For every $\delta\ge 0$ ($x\ge 0$) the function*

$$x\mapsto\phi(x,\delta) \quad \left(\delta\mapsto\phi(x,\delta)\right)$$

is decreasing. In addition,

$$\phi(x,\delta)=\phi\left(\frac{x}{\delta},1\right)^{\delta} \tag{1.57}$$

for all $x\ge 0$ and $\delta>0$, and

$$\phi(x,\delta)=\phi\left(1,\frac{\delta}{x}\right)^{x} \tag{1.58}$$

for all $x>0$ and $\delta\ge 0$.

Proof Let $\epsilon>0$. Then (1.56) and the fact that the function ϕ never exceeds 1 imply that $\phi(x+\varepsilon,\delta)=\phi(x,\delta)\phi(\varepsilon,0)<\phi(x,\delta)$. This establishes the fact that the function $x\mapsto\phi(x,\delta)$ is decreasing. The proof of this property for the function $\delta\mapsto\phi(x,\delta)$ is similar.

It is not hard to show using (1.56) that

$$\phi(x,r)=\phi\left(\frac{x}{r},1\right)^r \tag{1.59}$$

where $r>0$ is a rational number. Next, let δ be an irrational number, and put

$$A=\{r>0: r \text{ rational and } r>\delta\}, \qquad B=\{r>0: r \text{ rational and } r<\delta\}.$$

Then, using equality (1.59) and the fact that the function $x\mapsto\phi(x,\delta)$ is decreasing, we see that

$$\begin{aligned}\phi(x,\delta)&\ge\sup_{r\in A}\{\phi(x,r)\}=\sup_{r\in A}\left\{\phi\left(\frac{x}{r},1\right)^r\right\}\\&\ge\lim_{r\downarrow\delta}\left\{\phi\left(\frac{x}{\delta},1\right)^r\right\}=\phi\left(\frac{x}{\delta},1\right)^\delta\end{aligned}\tag{1.60}$$

and

$$\begin{aligned}\phi(x,\delta)&\le\inf_{r\in B}\{\phi(x,r)\}=\inf_{r\in B}\left\{\phi\left(\frac{x}{r},1\right)^r\right\}\\&\le\lim_{r\uparrow\delta}\left\{\phi\left(\frac{x}{\delta},1\right)^r\right\}=\phi\left(\frac{x}{\delta},1\right)^\delta.\end{aligned}\tag{1.61}$$

It follows from (1.60) and (1.61) that equality (1.57) holds. The proof of equality (1.58) is similar.

This completes the proof of Lemma 1.38. □

Now we are well equipped to finish the proof of Lemma 1.35. Let $t>0$, and choose $\nu=\lambda\delta_t$ in Theorem 1.36, where δ_t is the Dirac measure concentrated at t. Then we have

$$\begin{aligned}\phi(x,1)&=\mathbb{E}_{\sqrt{x}}\left[\exp\{-\lambda W_t^2\}\right]\\&=\frac{1}{\sqrt{2\pi t}}\int_{-\infty}^{\infty}e^{-\lambda y^2}\exp\left\{-\frac{(\sqrt{x}-y)^2}{2t}\right\}dy\\&=\frac{1}{\sqrt{2\pi t}}\exp\left\{-\frac{x}{2t}\right\}\int_{-\infty}^{\infty}e^{\frac{\sqrt{x}}{t}y}\exp\left\{-\frac{1+2t\lambda}{2t}y^2\right\}dy\\&=\frac{1}{\sqrt{2\pi}\sqrt{1+2t\lambda}}\exp\left\{-\frac{x}{2t}\right\}\int_{-\infty}^{\infty}\exp\left\{\frac{\sqrt{x}}{\sqrt{t}\sqrt{1+2t\lambda}}u\right\}\exp\left\{-\frac{u^2}{2}\right\}du\\&=\frac{1}{\sqrt{1+2t\lambda}}\exp\left\{-\frac{x\lambda}{1+2t\lambda}\right\}.\end{aligned}\tag{1.62}$$

Now it follows from (1.57) and (1.62) that formula (1.51) holds.

This completes the proof of Lemma 1.35. □

1.13 Marginal Distributions of Squared Bessel Processes. Proofs

Proof of Theorem 1.33. The proof splits into three parts.

Case $y_0 = 0$, $\delta > 0$. It follows from (1.51) that

$$\mathbb{E}\big[\exp\{-\lambda BESQ_0^\delta(t)\}\big] = (2t\lambda + 1)^{-\frac{\delta}{2}} \tag{1.63}$$

for all $\lambda > 0$. Inverting the Laplace transform, we see that $\rho_t(y) = \Gamma(y; \frac{\delta}{2}, 2t)$, where $\Gamma(\cdot; \frac{\delta}{2}, 2t)$ is the Gamma density (see Definition 1.26). This establishes formula (1.48).

Case $y_0 > 0$, $\delta > 0$. Put $A(\lambda) = (1 + 2t\lambda)^{-1}$. Then formula (1.51) implies

$$\begin{aligned}\mathbb{E}\big[\exp\{-\lambda BESQ_{y_0}^\delta(t)\}\big] &= A(\lambda)^{\frac{\delta}{2}} \exp\Big\{-\frac{y_0}{2t}\Big\} \exp\Big\{\frac{y_0 A(\lambda)}{2t}\Big\} \\ &= \exp\Big\{-\frac{y_0}{2t}\Big\} \sum_{n=0}^{\infty} \frac{y_0^n}{(2t)^n n!} A(\lambda)^{n+\frac{\delta}{2}}.\end{aligned}$$

The function $A(\lambda)^{n+\frac{\delta}{2}}$ is the Laplace transform of the function

$$\frac{y^{n+\frac{\delta}{2}-1}}{(2t)^{n+\frac{\delta}{2}} \Gamma(n+\frac{\delta}{2})} e^{-\frac{y}{2t}}.$$

Therefore,

$$\rho_t(y) = e^{-\frac{y_0+y}{2t}} \sum_{n=0}^{\infty} \frac{y_0^n y^{n+\frac{\delta}{2}-1}}{n!\Gamma(n+\frac{\delta}{2})(2t)^{2n+\frac{\delta}{2}}}. \tag{1.64}$$

Recall that the I-Bessel function is defined by

$$I_\nu(z) = \sum_{n=0}^{\infty} \frac{(\frac{z}{2})^{\nu+2n}}{n!\Gamma(n+\nu+1)} \tag{1.65}$$

if ν differs form a negative integer, and by $I_\nu = I_{-\nu}$ otherwise (see Definition 1.30). It follows from (1.65) that

$$\Big(\frac{y}{y_0}\Big)^{\frac{\nu}{2}} \frac{1}{2t} I_\nu\Big(\frac{\sqrt{y_0 y}}{t}\Big) = \sum_{n=0}^{\infty} \frac{y_0^n y^{\nu+n}}{n!\Gamma(n+\nu+1)(2t)^{\nu+2n+1}}. \tag{1.66}$$

Next, we see that (1.64) and (1.66) give

$$\rho_t(y) = \frac{1}{2t}\Big(\frac{y}{y_0}\Big)^{\frac{\nu}{2}} e^{-\frac{y_0+y}{2t}} I_\nu\Big(\frac{\sqrt{y_0 y}}{t}\Big)$$

where $\nu = \frac{\delta}{2} - 1$. This establishes formula (1.49).

Case $y_0 > 0$, $\delta = 0$. In this case, formula (1.51) gives

$$\mathbb{E}\big[\exp\{-\lambda BESQ^0_{y_0}(t)\}\big] = \exp\left\{-\frac{y_0}{2t}\right\}\sum_{n=0}^{\infty}\frac{y_0^n}{(2t)^n n!}A(\lambda)^n. \tag{1.67}$$

The expression on the right-hand side of (1.67) is the Laplace transform of the measure μ_t, whose singular component is $\exp\{-\frac{y_0}{2t}\}\delta_0$. The Radon–Nikodym derivative of the measure μ_t with respect to the Lebesgue measure on $[0, \infty)$ equals

$$\begin{aligned}
&\exp\left\{-\frac{y_0+y}{2t}\right\}\sum_{n=1}^{\infty}\frac{y_0^n y^{n-1}}{(2t)^{2n} n!\Gamma(n)} \\
&= \frac{1}{2t}\left(\frac{y}{y_0}\right)^{-\frac{1}{2}}\exp\left\{-\frac{y_0+y}{2t}\right\}\sum_{n=1}^{\infty}\frac{y_0^{n-\frac{1}{2}} y^{n-\frac{1}{2}}}{(2t)^{2n-1} n!\Gamma(n)} \\
&= \frac{1}{2t}\left(\frac{y}{y_0}\right)^{-\frac{1}{2}}\exp\left\{-\frac{y_0+y}{2t}\right\}\sum_{n=0}^{\infty}\frac{y_0^{n+\frac{1}{2}} y^{n+\frac{1}{2}}}{(2t)^{2n+1} (n+1)!\Gamma(n+1)} \\
&= \frac{1}{2t}\left(\frac{y}{y_0}\right)^{-\frac{1}{2}}\exp\left\{-\frac{y_0+y}{2t}\right\}I_1\left(\frac{\sqrt{y_0 y}}{t}\right).
\end{aligned}$$

It follows that the distribution of the random variable $BESQ^0_{y_0}(t)$ is a measure μ_t given by

$$\begin{aligned}
\mu_t(A) = \exp\left\{-\frac{y_0}{2t}\right\}\delta_0(A) &+ \frac{1}{2t}y_0^{\frac{1}{2}}\exp\left\{-\frac{y_0}{2t}\right\} \\
&\times \int_{A\cap[0,\infty)} y^{-\frac{1}{2}}\exp\left\{-\frac{y}{2t}\right\}I_1\left(\frac{\sqrt{y_0 y}}{t}\right)dy
\end{aligned}$$

for all Borel sets in $[0, \infty)$. This establishes formula (1.50) and completes the proof of Theorem 1.33. □

Remark 1.39 Theorem 1.33 can be reformulated as follows:

- For $\delta > 0$ the distribution density of $BESQ^\delta_0$ is given by

$$\rho_t(y) = \Gamma\left(y; \frac{\delta}{2}, 2t\right).$$

- For $y_0 > 0$ and $\delta > 0$ the distribution density of $BESQ^\delta_{y_0}$ is given by

$$\rho_t(y) = \frac{1}{t}\chi^2\left(\frac{y}{y_0}; \delta, \frac{y_0}{t}\right).$$

- For $y_0 > 0$ the distribution of $BESQ^0_{y_0}$ is a Borel measure μ_t on $[0, \infty)$, whose singular component is equal to $\exp\{-\frac{y_0}{2t}\}\delta_0$ and the Radon–Nikodym derivative ρ_t with respect to the Lebesgue measure is given by

$$\rho_t(y) = \frac{1}{t}\chi^2\left(\frac{y}{t}; 0, \frac{y_0}{t}\right).$$

1.14 Time-Changed Squared Bessel Processes and CIR Processes

In this section, we explain how squared Bessel processes are related to CIR processes.

Theorem 1.40 *Let Y be the CIR process satisfying*

$$dY_t = (a - bY_t)\,dt + c\sqrt{Y_t}\,dZ_t, \quad Y_0 = y_0 \quad a.s. \tag{1.68}$$

where $y_0 \geq 0$, $a \geq 0$, $b \neq 0$, and $c > 0$, and let T be the squared Bessel process solving the equation

$$dT_t = \frac{4a}{c^2}\,dt + 2\sqrt{T_t}\,d\widetilde{Z}_t, \quad T_0 = y_0 \quad a.s.$$

Then the following equality is valid:

$$Y_t = e^{-bt}\,T\left(\frac{c^2}{4b}(e^{bt} - 1)\right), \quad t > 0. \tag{1.69}$$

Remark 1.41 Equality (1.69) can be rewritten in the following form:

$$Y_t = e^{-bt}\,BESQ^{\frac{4a}{c^2}}_{y_0}\left(\frac{c^2}{4b}(e^{bt} - 1)\right), \quad t > 0.$$

If $b = 0$, then (1.69) becomes

$$Y_t = BESQ^{\frac{4a}{c^2}}_{y_0}\left(\frac{c^2}{4}t\right), \quad t > 0.$$

Proof of Theorem 1.40. Put $\phi(t) = \frac{c^2}{4b}(e^{bt} - 1)$. Then we have

$$\phi'(t) = \frac{c^2}{4}e^{bt} \quad \text{and} \quad \sqrt{\phi'(t)} = \frac{c}{2}e^{\frac{bt}{2}}.$$

Let $\widetilde{B}$ be an $\mathcal{F}_{\phi(t)}$-standard Brownian motion, e.g.,

$$\widetilde{B}_t = \int_0^{\phi(t)} \sqrt{(\phi^{-1})'(u)}\,dR_u, \quad t > 0,$$

where R_u is a standard Brownian motion (use Lévy's characterization theorem to establish this statement). Put

$$M_t = \int_0^t \sqrt{\phi'(u)}\, d\widetilde{B}_u. \tag{1.70}$$

Then M_t is a continuous $\mathcal{F}_{\phi(t)}$-martingale starting at 0. It follows from Lévy's characterization theorem that the process $\widetilde{Z}$ defined by $\widetilde{Z}_t = M_{\phi^{-1}(t)}$, $t > 0$, is a standard $\mathcal{F}_t$-Brownian motion. We also have the following formula for the quadratic variation of M:

$$\langle M, M\rangle_t = \phi(t), \quad t > 0.$$

Let T be a squared Bessel process satisfying

$$T_t = y_0 + \frac{4a}{c^2}t + 2\int_0^t \sqrt{T_s}\, d\widetilde{Z}_s, \tag{1.71}$$

and put

$$\widetilde{Y}_t = e^{-bt} T_{\phi(t)} \tag{1.72}$$

Then the integration by parts formula applied to (1.72) gives

$$d\widetilde{Y}_t = -b\widetilde{Y}_t\, dt + e^{-bt} d[T_{\phi(t)}]. \tag{1.73}$$

It follows from (1.71) that

$$T_{\phi(t)} = y_0 + \frac{4a}{c^2}\phi(t) + 2\int_0^{\phi(t)} \sqrt{T_s}\, d\widetilde{Z}_s.$$

Hence,

$$d[T_{\phi(t)}] = \frac{4a}{c^2}\phi'(t)\, dt + 2d\left[\int_0^{\phi(t)} \sqrt{T_s}\, d\widetilde{Z}_s\right]$$

and

$$e^{-bt} d[T_{\phi(t)}] = a\, dt + 2e^{-bt} d\left[\int_0^{\phi(t)} \sqrt{T_s}\, d\widetilde{Z}_s\right]. \tag{1.74}$$

The next result (Theorem 1.44 below) explains how to change time in a continuous martingale. We need this theorem to continue the proof of Theorem 1.40. We will first discuss measurability properties of stochastic processes, and then formulate Theorem 1.44. More details can be found in [KS91], Sect. 3.4.

Let $(\Omega, \mathcal{F})$ be a measurable space and let X_t, $t \geq 0$, be an $\mathbb{R}^d$-valued stochastic process on Ω. For a topological space H denote by $\mathcal{B}(H)$ the Borel σ-algebra of subsets of H.

Definition 1.42 The stochastic process X is called measurable if the mapping $(t, \omega) \mapsto X_t(\omega)$ from $[0, \infty) \times \Omega$ into $\mathbb{R}^d$ is $\mathcal{B} \otimes \mathcal{F}/\mathcal{B}(R^d)$-measurable.

Definition 1.43 Let $(\Omega, \mathcal{F}, \{\mathcal{F}_t\})$ be a measure space equipped with the filtration $\{\mathcal{F}_t\}$, and let X_t, $t \ge 0$, be an $\mathbb{R}^d$-valued stochastic process on Ω. The process X is called progressively measurable with respect to the filtration $\{\mathcal{F}_t\}$ if for every $t \ge 0$, the mapping $(t, \omega) \mapsto X_t(\omega)$ from $[0, t] \times \Omega$ into $\mathbb{R}^d$ is $\mathcal{B}([0, t]) \otimes \mathcal{F}_t / \mathcal{B}(\mathbb{R}^d)$-measurable.

The notion of progressive measurability of stochastic processes was first introduced and studied in [CD65]. We refer the reader to ([CD65, KS91, RY04, GvC06]) for more information about measurability and progressive measurability of stochastic processes.

It is clear that if the process X is progressively measurable, then it is measurable and adapted. Moreover, if the process X is adapted to the filtration $\{\mathcal{F}_t\}$ and has right-continuous (left-continuous) sample paths $t \mapsto X_t(\omega)$, then X is progressively measurable (see [KS91], Proposition 1.13). It is also known that if the process X is progressively measurable with respect to the filtration $\{\mathcal{F}_t\}$ and T is a stopping time, then the stopped process X^T (see Definition 1.8) is progressively measurable with respect to the same filtration (see [KS91], Proposition 2.18).

Theorem 1.44 *Let $\{\mathcal{G}_t\}$ be a filtration and let M be a continuous $\{\mathcal{G}_t\}$-martingale such that its quadratic variation satisfies $\lim_{t\to\infty} \langle M, M\rangle_t = \infty$ a.s. Define, for each $0 \le s < \infty$, the stopping time $U(s)$ by the following formula:*

$$U(s) = \inf\bigl\{t : 0 \le t \le T, \langle M, M\rangle_t > s\bigr\},$$

and consider the time-changed process $B_s = M_{U(s)}$, $s \ge 0$. Then B_s is an $\mathcal{G}_{U(s)}$-standard Brownian motion for which $M_t = B_{\langle M,M\rangle_t}$, $t \ge 0$. In addition, let X be a progressively measurable process with respect to the filtration $\{\mathcal{G}_t\}$ such that

$$\int_0^\infty X_t^2 \, d\langle M, M\rangle_t < \infty$$

a.s., and put $Y_t = X_{U(t)}$, $t > 0$. Then the process Y is $\mathcal{G}_{U(t)}$-adapted and satisfies the conditions

$$\int_0^\infty Y_s^2 ds < \infty \quad \text{and} \quad \int_0^t X_v \, dM_v = \int_0^{\langle M,M\rangle_t} Y_u \, dB_u \quad \text{a.s.}$$

Let us return to the proof of Theorem 1.40. Suppose M_t is defined by (1.70), and take $\mathcal{G}_t = \mathcal{F}_{\phi(t)}$, $Y_t = \sqrt{T_t}$, and $X_t = Y_{\langle M,M\rangle_t} = \sqrt{T_{\phi(t)}}$, $t > 0$. Then, applying Theorem 1.44, we obtain

$$\int_0^{\phi(t)} \sqrt{T_s} \, d\widetilde{Z}_s = \int_0^t \sqrt{T_{\phi(s)}} \sqrt{\phi'(s)} \, d\widetilde{B}_t. \tag{1.75}$$

It follows from (1.73), (1.74), and (1.75) that the process $\widetilde{Y}$ defined by (1.72) satisfies

$$d\widetilde{Y}_t = (a - b\widetilde{Y}_t)\,dt + 2e^{-bt}\sqrt{T_{\phi(t)}}\sqrt{\phi'(t)}\,d\widetilde{B}_t$$
$$= (a - b\widetilde{Y}_t)\,dt + c\sqrt{\widetilde{Y}_t}\,d\widetilde{B}_t.$$

Finally, using the uniqueness theorem, we see that the process $\widetilde{Y}$ in (1.72) is a copy of the CIR process Y satisfying (1.68).

This completes the proof of Theorem 1.40. □

1.15 Marginal Distributions of CIR Processes

Let Y be a CIR process satisfying

$$dY_t = (a - bY_t)\,dt + c\sqrt{Y_t}\,dZ_t, \quad Y_0 = y_0 \quad \text{a.s.}$$

where $a \geq 0$, $b \geq 0$, $c > 0$, and $y_0 > 0$. Then (1.28) and Theorem 1.40 show that the mean of Y_t is given by

$$m(Y_t) = e^{-bt}\left(y_0 - \frac{a}{b}\right) + \frac{a}{b}.$$

Moreover, it follows from (1.29) and Theorem 1.40 that the following formula holds for the variance of Y_t:

$$v(Y_t) = e^{-2bt}\left[\frac{ac^2}{2b^2}(e^{bt} - 1)^2 + \frac{y_0c^2}{b}(e^{bt} - 1)\right].$$

In addition, it is not hard to see that

$$\lim_{t\to\infty} m(Y_t) = \frac{a}{b} \quad \text{and} \quad \lim_{t\to\infty} v(Y_t) = \frac{ac^2}{2b^2}.$$

Theorems 1.33 and 1.40 imply the following assertion that describes the marginal distributions of CIR processes.

Theorem 1.45

(a) *Let $a > 0$ and $b > 0$. Then the marginal distribution μ_t of the CIR process Y admits a density ρ_t given by*

$$\rho_t(y) = \exp\left\{b\left(1 + \frac{\nu}{2}\right)t\right\}\frac{2b}{c^2(e^{bt} - 1)}\left(\frac{y}{y_0}\right)^{\frac{\nu}{2}}$$
$$\times \exp\left\{-\frac{2b(y_0 + e^{bt}y)}{c^2(e^{bt} - 1)}\right\}I_\nu\left(\frac{4b\sqrt{e^{bt}y_0y}}{c^2(e^{bt} - 1)}\right), \quad y > 0, \qquad (1.76)$$

where $\nu = \frac{2a}{c^2} - 1$.

(b) *Let $a > 0$ and $b = 0$. Then the marginal distribution μ_t of the CIR process Y admits a density ρ_t given by*

$$\rho_t(y) = \frac{2}{c^2 t}\left(\frac{y}{y_0}\right)^{\frac{\nu}{2}} \exp\left\{-\frac{2(y_0+y)}{c^2 t}\right\} I_\nu\left(\frac{4\sqrt{y_0 y}}{c^2 t}\right), \quad y > 0,$$

where $\nu = \frac{2a}{c^2} - 1$.

(c) *Let $a = 0$ and $b > 0$. Then the marginal distribution μ_t of the CIR process Y is a probability measure given by*

$$\mu_t(A) = \exp\left\{-\frac{2by_0}{c^2(e^{bt}-1)}\right\}\delta_0(A) + \frac{2b\sqrt{y_0}}{c^2(e^{bt}-1)} \exp\left\{\frac{bt}{2}\right\} \times \int_{A\cap[0,\infty)} y^{-\frac{1}{2}} \exp\left\{-\frac{2b(y_0+e^{bt}y)}{c^2(e^{bt}-1)}\right\} I_1\left(\frac{4b\sqrt{e^{bt}y_0 y}}{c^2(e^{bt}-1)}\right) dy$$

for all Borel sets A in $\mathbb{R}$.

(d) *Let $a = 0$ and $b = 0$. Then the marginal distribution μ_t of the CIR process Y is a probability measure given by*

$$\mu_t(A) = \exp\left\{-\frac{2y_0}{c^2 t}\right\}\delta_0(A) + \frac{2\sqrt{y_0}}{c^2 t} \int_{A\cap[0,\infty)} y^{-\frac{1}{2}} \exp\left\{-\frac{2(y_0+y)}{c^2 t}\right\} I_1\left(\frac{4\sqrt{y_0 y}}{c^2 t}\right) dy$$

for all Borel sets A in $\mathbb{R}$.

It follows from Theorem 1.45 that the marginal distribution density ρ_t of a CIR process with $a > 0$ and $b > 0$ is expressed in terms of the χ^2-density as follows:

$$\rho_t(y) = \frac{4be^{bt}}{c^2(e^{bt}-1)} \chi^2\left(\frac{4be^{bt}y}{c^2(e^{bt}-1)}; \frac{4a}{c^2}, \frac{4by_0}{c^2(e^{bt}-1)}\right), \quad y > 0. \tag{1.77}$$

We will next find an explicit formula for the long-time distribution density of a CIR process with $a > 0$ and $b > 0$ by passing to the limit as $t \to \infty$ in formula (1.76). We will take into account that the following asymptotic formula holds for the I-Bessel function I_ν with $\nu \neq -1, -2, \ldots$:

$$I_\nu(z) \sim \frac{1}{\Gamma(\nu+1)}\left(\frac{1}{2}z\right)^\nu \quad \text{as } z \downarrow 0 \tag{1.78}$$

(see [Wat95]). It follows from (1.76) and (1.78) that

$$\lim_{t\to\infty} \rho_t(y) = \Gamma\left(y; \frac{2a}{c^2}, \frac{c^2}{2b}\right) \tag{1.79}$$

for all $y > 0$. The same conclusion is valid in the weak sense. Therefore, the long-time distribution density of the CIR process with $a \neq 0$ and $b \neq 0$ is the Γ-density with the parameters described in (1.79).

Remark 1.46 It is clear that the results obtained in Sect. 1.15 hold for a CIR process solving the stochastic differential equation in the mean-reverting form (see (1.25)). It suffices to replace a by qm and b by q. For instance, formula (1.77) becomes

$$\rho_t(y) = \frac{4qe^{qt}}{c^2(e^{qt}-1)}\chi^2\left(\frac{4qe^{qt}y}{c^2(e^{qt}-1)}; \frac{4qm}{c^2}, \frac{4qy_0}{c^2(e^{qt}-1)}\right), \quad y > 0. \tag{1.80}$$

Remark 1.47 In the classical paper of Cox, Ingersoll, and Ross, the process Y determined from (1.25) is used to model a random behavior of the interest rate (see [CIR85]). In mathematics, the CIR process is called the Feller process after W. Feller who studied the diffusion equation associated with this process. Formula (1.77) was formulated in [CIR85] without proof. The following is an excerpt from [CIR85], describing the conditional distribution of the process Y: "The probability density of the interest rate at time s, conditional on its value at the current time, t, is given by

$$f\big(r(s), s; r(t), t\big) = ce^{-u-v}\left(\frac{v}{u}\right)^{\frac{q}{2}} I_q\big(2(uv)^{\frac{1}{2}}\big), \tag{1.81}$$

where

$$c = \frac{2\kappa}{\sigma^2(1-e^{-\kappa(s-t)})}, \qquad u = cr(t)e^{-\kappa(s-t)}, \qquad v = cr(s), \qquad q = \frac{2\kappa\theta}{\sigma^2} - 1,$$

and $I_q(\cdot)$ is the modified Bessel function of the first kind of order q. The distribution function is the noncentral chi-square, $\chi^2[2cr(s); 2q+2, 2u]$, with $2q+2$ degrees of freedom and parameter of noncentrality $2u$ proportional to the current spot rate". To reconcile the notation used in [CIR85] with our notation, we take $t = 0$, $s = t$, $\kappa = q$, $\theta = m$, $q = \nu$, $\sigma = c$, $r(t) = y_0$, and $r(s) = y$. The symbols on the right-hand side of the previous equalities are from [CIR85], while those on the right-hand side are used in the present book. It is not hard to see, using the equality

$$\frac{4qe^{qt}}{c^2(e^{qt}-1)}\chi^2\left(\frac{4qe^{qt}y}{c^2(e^{qt}-1)}; \frac{4qm}{c^2}, \frac{4qy_0}{c^2(e^{qt}-1)}\right)$$
$$= \frac{4q}{c^2(1-e^{-qt})}\chi^2\left(\frac{4qy}{c^2(1-e^{-qt})}; \frac{4qm}{c^2}, \frac{4qe^{-qt}y_0}{c^2(1-e^{-qt})}\right),$$

that formulas (1.80) and (1.81) are identical.

1.16 Ornstein–Uhlenbeck Processes and CIR Processes

Let Z be a standard Brownian motion on $(\Omega, \mathcal{F}, \mathbb{P})$. Recall that an Ornstein–Uhlenbeck process with long-run mean zero is the unique strong solution to the stochastic differential equation

$$dY_t = -qY_t\,dt + \sigma\,dZ_t, \qquad Y_0 = y_0.$$

Put $T_t = Y_t(q, 0, \sigma, y_0)^2$, $t \geq 0$. Then Itô's formula gives

$$dT_t = \left(\sigma^2 - 2qT_t\right)dt + 2\sigma Y_t\,dZ_t.$$

It can be shown, using Lévy's characterization theorem, that the process $\widetilde{Z}$ defined by $d\widetilde{Z}_t = \operatorname{sign}(Y_t)\,dZ_t$ is a standard Brownian motion. Therefore

$$dT_t = \left(\sigma^2 - 2qT_t\right)dt + 2\sigma\sqrt{T_t}\,d\widetilde{Z}_t,$$

and it follows that the process $Y_t(q, 0, \sigma, y_0)^2$ has the same law as the CIR process $\widetilde{Y}_t(\sigma^2, 2q, 2\sigma, y_0^2)$. In addition, (1.17) implies that for every m and $t > 0$ the random variables

$$Y_t(q, m, \sigma, y_0)^2 \quad \text{and} \quad \widetilde{Y}_t\left(\sigma^2, 2q, 2\sigma, \left(y_0 + \left(e^{qt} - 1\right)m\right)^2\right)$$

are equally distributed.

1.17 Notes and References

- The reader can consult [Kah97, Kah98, Kah06, Dup06], Chaps. 2–4 of [Nel67], and Chap. 2 of [JL00] for interesting accounts of the history of Brownian motion.
- Louis Bachelier is deservedly recognized as the founding father of financial mathematics. In his dissertation [B1900], entitled "Théorie de la spéculation", Bachelier derived the transition law of Brownian motion process, and used this process to study the stock and option markets. More about Bachelier's life and his visionary scientific achievements can be found in [CKBC00, Taq02, CK02]. For information on predecessors of Bachelier, see [Gir02].
- There are numerous books about Albert Einstein. We single out only the scientific biography of Einstein [Pai05] written by A. Pais. Einstein's celebrated publications on quantitative theory of Brownian motion can be found in [Ein56]. We would also like to mention [Ein08], which is a collection of papers, discussing Einstein's legacy in science, art, and culture.
- In 1906, Polish physicist Marian Smoluchowski developed a theory of Brownian motion independently of Einstein. For a biography of M. Smoluchowski and an account of his work, we refer the reader to [CKSI00]. The original work of Smoluchowski on Brownian motion can be found in [S1906].
- Thorvald N. Thiele was a Danish scientist with a wide range of interests. Thiele's research spanned astronomy, mathematics, actuarial science, and statistics. In his work on the method of least squares in statistics, published in 1880, Thiele derived mathematical Brownian motion and established its properties. The book [Lau02] is a good source for biographical and scientific information about T.N. Thiele.

- A rigorous mathematical theory of Brownian motion was constructed by N. Wiener in [Wie23] and [Wie24]. Wiener's fundamental work laid the foundation of stochastic analysis and stochastic modeling.
- Brownian motion with drift is not a reasonable stochastic model for the stock price because the price process in this model takes negative values. An idea to use the exponential of Brownian motion with drift (a geometric Brownian motion) as the stock price process goes back to M.F.M. Osborne [Osb59] and P.A. Samuelson [Sam65], see also [Sam02]. Note that early empirical observations of the behavior of stock returns in [Ken53] and [Rob59] led the authors to the conclusion that stock returns are normally distributed. F. Black and M. Scholes used the Osborne–Samuelson model of the stock price in their famous work (see [BS73]) on option pricing models.
- An account of the history of the Ornstein–Uhlenbeck process can be found in [Jac96]. For the original work of Ornstein and Uhlenbeck, see [OU30].
- For information on Bessel processes, see the books [RY04] and [JYC09] the survey [G-JY03], and the papers [CS04, Duf05, GY93]. Theorem 1.36 is due to J. Pitman and M. Yor (see [PY82]). Lemma 1.37 can be found in [SW73].

Chapter 2
Stock Price Models with Stochastic Volatility

Modern models of the stock price use stochastic processes to describe the volatility of the stock. Allowing the volatility to be random makes the models more flexible and better adapted to the realities of the financial world. Stochastic volatility models succeed in explaining various features of the market, which the classical Black–Scholes model fails to predict, e.g., the implied volatility smile and skew, and heavy tails of stock price distributions. Important examples of stochastic volatility models are the Hull–White, the Stein–Stein, and the Heston models. It was mentioned earlier that the volatility in these models is distributed according to the log-normal, the Gaussian, and the noncentral chi-square law, respectively. Standard Brownian motions driving the stock price and the volatility in a stochastic volatility model may be dependent or independent. If the former condition holds, then the model is called correlated, while in the latter case, the term "uncorrelated model" is used. Correlated stochastic volatility models are more important in practice, since observations of financial markets suggest that there exists a negative correlation between the stock price and the volatility (the so-called leverage effect). On the other hand, uncorrelated models are simpler than the correlated ones because, in a certain sense, they are mixtures of Black–Scholes models. Our main goal in the present chapter is to introduce stochastic volatility models and discuss risk-neutral measures associated with them.

2.1 Stochastic Volatility

In their famous work on option pricing theory, Black and Scholes used the following stochastic differential equation to model the random behavior of the stock price:

$$\begin{cases} dX_t = \mu X_t \, dt + \sigma X_t \, dW_t, \\ X_0 = x_0 \end{cases} \tag{2.1}$$

(see [BS73]). Equation (2.1) was suggested by M.F.M. Osborne [Osb59] and P.A. Samuelson (see [Sam02]).

A. Gulisashvili, *Analytically Tractable Stochastic Stock Price Models*,
Springer Finance, DOI 10.1007/978-3-642-31214-4_2,

© Springer-Verlag Berlin Heidelberg 2012

The constants $\mu \in \mathbb{R}$ and $\sigma > 0$ in (2.1) are called the drift and the volatility of the stock, respectively. The drift can be interpreted as the instantaneous mean of the stock price, while the volatility is its instantaneous standard deviation. The stock price process X in the Osborne–Samuelson model is a geometric Brownian motion defined by

$$X_t = x_0 \exp\left\{\left(\mu - \frac{1}{2}\sigma^2\right)t + \sigma W_t\right\},$$

and the stock price distribution density $\rho_t^{(\sigma)}$ is given by

$$\rho_t^{(\sigma)}(x) = \frac{1}{\sqrt{2\pi t}x\sigma} \exp\left\{-\frac{(\log \frac{x}{x_0 e^{\mu t}} + \frac{t\sigma^2}{2})^2}{2t\sigma^2}\right\}, \quad x > 0, \tag{2.2}$$

for all $t > 0$ and $\sigma > 0$ (see Sect. 1.2).

A general class of stochastic volatility models was introduced and studied in [FPS00]. A stock price model belongs to this class if it can be described by the following system of stochastic differential equations:

$$\begin{cases} dX_t = \mu X_t \, dt + f(Y_t) X_t \, dW_t, \\ dY_t = b(Y_t) \, dt + \sigma(Y_t) \, dZ_t. \end{cases} \tag{2.3}$$

In (2.3), the symbol f stands for a positive continuous function on $\mathbb{R}$, while W and Z are standard one-dimensional Brownian motions. The stock price is modeled by the process X, while the volatility of the stock is described by the process $f(Y)$. The initial conditions for the processes X and Y will be denoted by x_0 and y_0, respectively. The first equation in (2.3) is called the stock price equation, while the second equation is the volatility equation. Note that we consider only time-homogeneous volatility equations. Exactly as in the Black–Scholes model, the drift coefficient μ in (2.3) is the instantaneous mean of the stock price, while the volatility process $f(Y)$ is interpreted as the instantaneous standard deviation (this time it is random). The model in (2.3) is called uncorrelated if Brownian motions W and Z driving the stock price and the volatility equations, respectively, are independent. Correlated models will be discussed in the next section.

2.2 Correlated Stochastic Volatility Models

Consider the model described by (2.3). It will be assumed in the sequel that Brownian motions W and Z are such that

$$d\langle W, Z\rangle_t = \rho \, dt \tag{2.4}$$

for some number ρ with $\rho \in [-1, 1]$. The constant ρ in (2.4) is called the correlation coefficient of W and Z. If the model in (2.3) satisfies condition (2.4) with $\rho \neq 0$, then it is called a correlated stochastic volatility model.

Suppose for some $\rho \in [-1, 1]$,

$$W = \sqrt{1-\rho^2}\widetilde{W} + \rho Z, \tag{2.5}$$

where $\widetilde{W}$ is a standard Brownian motion independent of Z. Then it is easy to see that condition (2.4) holds for W and Z. Similarly, if for some $\rho \in [-1, 1]$,

$$Z = \sqrt{1-\rho^2}\widetilde{Z} + \rho W, \tag{2.6}$$

where $\widetilde{Z}$ is a standard Brownian motion independent of W, then condition (2.4) holds for W and Z. We will next prove that the converse statements are also true.

Lemma 2.1 *Let W and Z be standard Brownian motions satisfying condition* (2.4). *Then there exist a standard Brownian motion $\widetilde{W}$ independent of Z and a standard Brownian motion $\widetilde{Z}$ independent of W such that* (2.5) *and* (2.6) *hold.*

Proof We will prove only the existence of $\widetilde{W}$ for which (2.5) holds. The proof of (2.6) is similar. Let us start with the case where $\rho \in (-1, 1)$. Put

$$\widetilde{W} = \frac{1}{\sqrt{1-\rho^2}} W - \frac{\rho}{\sqrt{1-\rho^2}} Z.$$

It is not hard to see that $\langle \widetilde{W}, Z \rangle_t = 0$ and the equality in (2.5) holds. It remains to prove that the process $\widetilde{W}$ is a standard Brownian motion. First we notice that $\langle \widetilde{W}, \widetilde{W} \rangle_t = t$. Moreover, the process $\widetilde{W}$ is a continuous martingale and $\widetilde{W}_0 = 0$ $\mathbb{P}$a.s. Applying the Lévy characterization theorem, we see that the process $\widetilde{W}$ is a standard Brownian motion.

Next, let $\rho = 1$ (the case where $\rho = -1$ is similar). Then (2.4) with $\rho = 1$ implies $\langle W - Z, W - Z \rangle_t = 0$ for all $t \geq 0$. It follows that the process $(W - Z)^2$ is a continuous martingale with initial condition 0, and hence the process W is indistinguishable from the process Z.

This completes the proof of Lemma 2.1. □

It is clear that under condition (2.5), the two-dimensional stochastic differential equation in (2.3) can be rewritten as follows:

$$\begin{cases} dX_t = \mu X_t \, dt + f(Y_t) X_t \big(\sqrt{1-\rho^2} \, d\widetilde{W}_t + \rho \, dZ_t \big), \\ dY_t = b(Y_t) \, dt + \sigma(Y_t) \, dZ_t \end{cases} \tag{2.7}$$

(use representation (2.5)). Similarly, (2.6) gives

$$\begin{cases} dX_t = \mu X_t \, dt + f(Y_t) X_t \, dW_t, \\ dY_t = b(Y_t) \, dt + \sigma(Y_t) \big(\sqrt{1-\rho^2} \, d\widetilde{Z}_t + \rho \, dW_t \big). \end{cases} \tag{2.8}$$

Observations show that the price of a stock has a tendency to go up when the volatility goes down and to go down when the volatility goes up. This feature of stock prices is called the leverage effect. To account for the leverage effect, the stochastic volatility models used in practice assume negative correlation between the stock price and the volatility.

The next definition introduces several classes of functions.

Definition 2.2

1. Let $g : \mathbb{R}^d \mapsto \mathbb{R}^1$. It is said that g satisfies the linear growth condition if
$$|g(x)| \le (c_1 + c_2\|x\|_d)$$
for some $c_1 > 0$, $c_2 > 0$, and all $x \in \mathbb{R}^d$.
2. A function g such as in part 1 satisfies the Lipschitz condition if
$$|g(x) - g(y)| \le c\|x - y\|_d$$
for some $c > 0$ and all $x,\ y \in \mathbb{R}^d$.
3. Let $g : \mathbb{R}^1 \mapsto \mathbb{R}^1$. It is said that g satisfies the Yamada–Watanabe condition if there exists a strictly increasing function $\rho : [0, \infty) \mapsto [0, \infty)$ with $\int_{0+} \rho(x)^{-2} dx = \infty$ and such that
$$|g(x) - g(y)| \le \rho(\|x - y\|_d)$$
for all $x,\ y \in \mathbb{R}^1$.

Our next goal is to study the solvability of the two-dimensional equation in (2.7). For this equation the drift coefficients are given by
$$b_1(x_1, x_2) = \mu x_1, \qquad b_2(x_1, x_2) = b(x_2)$$
and the diffusion coefficients are defined by
$$\sigma_{11}(x_1, x_2) = \sqrt{1 - \rho^2} x_1 f(x_2), \qquad \sigma_{12}(x_1, x_2) = \rho x_1 f(x_2),$$
$$\sigma_{21}(x_1, x_2) = 0, \quad \text{and} \quad \sigma_{22}(x_1, x_2) = \sigma(x_2).$$
Similarly, for (2.8), we have
$$b_1(x_1, x_2) = \mu x_1, \qquad b_2(x_1, x_2) = b(x_2),$$
$$\sigma_{11}(x_1, x_2) = x_1 f(x_2), \qquad \sigma_{12}(x_1, x_2) = 0,$$
$$\sigma_{21}(x_1, x_2) = \rho\sigma(x_2) \quad \text{and} \quad \sigma_{22}(x_1, x_2) = \sqrt{1 - \rho^2}\sigma(x_2).$$

It follows that even in an important special case where $f(u) = |u|$, the coefficient σ_{11} in (2.7) does not satisfy the linear growth condition. Since this condition is assumed in standard solvability and uniqueness theorems for stochastic differential equations (see, e.g., [IW81]), we cannot apply these theorems to establish the unique strong solvability of Eq. (2.7). Fortunately, a special structure of this equation allows us to solve it utilizing an indirect approach (see Lemmas 2.3 and 2.4 below). For the information on the concepts and facts used in the formulations and proofs of these lemmas, we refer the reader to the books on stochastic differential equations, mentioned at the end of Sect. 1.1.

Lemma 2.3 *Suppose the following conditions are satisfied*:

1. *Strong existence and pathwise uniqueness for the second equation in* (2.7).
2. *The linear growth condition for the function f in* (2.7).

Then strong existence and pathwise uniqueness hold for the two-dimensional stochastic differential equation in (2.7).

The next lemma provides sufficient conditions for the strong unique solvability of the volatility equation in (2.7).

Lemma 2.4 *Suppose the following assumptions hold*:

1. *The functions b and σ satisfy the linear growth condition.*
2. *The function b satisfies the Lipschitz condition.*
3. *The function σ either satisfies the Lipschitz condition, or it is continuous and satisfies the Yamada–Watanabe condition.*

Then strong existence and pathwise uniqueness are valid for the volatility equation in (2.7).

Proof Strong solvability and pathwise uniqueness for a stochastic differential equation under the linear growth and Lipschitz conditions for the coefficients was established by K. Itô (see Proposition 1.9 in [CE05]). To finish the proof of Lemma 2.4 suppose assumptions 1 and 2 and the second part of assumption 3 hold for the functions b and σ. Then weak solvability of the volatility equation in (2.7) can be obtained from assumption 1 and the continuity assumption for b and σ, using Skorokhod's theorem (see Proposition 1.13 in [CE05]) and localization principles, while pathwise uniqueness for the volatility equation follows from the Yamada–Watanabe theorem (see Proposition 1.12 in [CE05]). Finally, we use the fact that weak solvability and pathwise uniqueness imply strong solvability (see, e.g., Theorem 1.7 in Chap. IX in [RY04]) to complete the proof of Lemma 2.4. □

Remark 2.5 Under the conditions in Lemma 2.4, the unique strong solution Y to the volatility equation in (2.7) satisfies the square integrability condition, that is,

$$\mathbb{E}\left[\int_0^T Y_s^2 \, ds\right] < \infty \quad \text{for all } T > 0$$

(see Theorem 2.4 in Chap. IV of [IW81]).

Proof of Lemma 2.3 The proof proceeds in three steps. First, we find the solution Y of the volatility equation in (2.7), then use the process $f(Y)$ as the volatility process in the stock price equation, and finally solve the stock price equation. More details will be provided below.

By the second assumption in Lemma 2.3, there exists a unique process Y satisfying the volatility equation in the strong sense. Moreover, the process

$$M_t = \mu t + \int_0^t f(Y_s)\, dW_s, \quad t \geq 0,$$

is a continuous $\{\mathcal{F}_t\}$-semimartingale (the second term on the right-hand side is a local martingale), where $\{\mathcal{F}_t\}$ is the filtration generated by W and Z (the filtration of a two-dimensional standard Brownian motion). It follows that the first equation in (2.7) is a linear stochastic differential equation of the following form:

$$dX_t = X_t\, dM_t. \tag{2.9}$$

It is known that such equations are uniquely solvable up to indistinguishability, and the unique solution can be represented as an exponential functional. In our case, the following formula holds:

$$X_t = x_0 \exp\left\{\mu t - \frac{1}{2}\int_0^t f(Y_s)^2\, ds + \int_0^t f(Y_s)\, dW_s\right\}. \tag{2.10}$$

Formula (2.10) is a special case of the Doléans–Dade formula (see [RY04], Sect. IX-2). It is easy to see that the processes X and Y are $\{\mathcal{F}_t\}$-adapted, and the pair (X, Y) is a strong solution to the system in (2.7), satisfying $X_0 = x_0$, $Y_0 = y_0$ $\mathbb{P}$-a.s.

We will next show that pathwise uniqueness holds for the system in (2.7). Indeed, let $(\Omega, \mathcal{F}, \{\mathcal{F}_t\}, \mathbb{P})$ be a filtered probability space and let (W, Z) be a pair of $\{\mathcal{F}_t\}$-Brownian motions on Ω. Consider two strong solutions $(X^{(1)}, Y^{(1)})$ and $(X^{(2)}, Y^{(2)})$ to (2.7) with the same initial condition. Then $Y^{(1)}$ is indistinguishable from $Y^{(2)}$ by pathwise uniqueness for the second equation. Moreover, the linear equation (2.9) is the same for the processes $Y^{(1)}$ and $Y^{(2)}$. Since the processes $X^{(1)}$ and $X^{(2)}$ are solutions to this equation having the same initial conditions, they are indistinguishable. This completes the proof of pathwise uniqueness for the system in (2.7). □

2.3 Hull–White, Stein–Stein, and Heston Models

In the present section we discuss three stochastic volatility models, which will play an important role in this book.

The Hull–White model was introduced in [HW87]. The stock price process X and the volatility process Y in this model satisfy the following system of stochastic differential equations:

$$\begin{cases} dX_t = \mu X_t\,dt + Y_t X_t\,dW_t, \\ dY_t = \nu Y_t\,dt + \xi Y_t\,dZ_t, \end{cases} \tag{2.11}$$

where $\mu,\ \nu \in \mathbb{R}$ and $\xi > 0$. It is assumed that standard Brownian motions W and Z in (2.11) are such that

$$dW_t = \sqrt{1-\rho^2}\,d\widetilde{W}_t + \rho\,dZ_t, \tag{2.12}$$

where $\widetilde{W}$ is a standard Brownian motion independent of Z. The correlation coefficient ρ satisfies $-1 \le \rho \le 1$. The parameter μ is the drift coefficient of the stock price, ν is the drift coefficient of the volatility, and ξ is the volatility-of-volatility parameter. For the Hull–White model, we have

$$b(x) = \nu x, \quad \sigma(x) = \xi x, \quad \text{and} \quad f(u) = u.$$

The Stein–Stein model is defined as follows:

$$\begin{cases} dX_t = \mu X_t\,dt + Y_t X_t\,dW_t, \\ dY_t = q(m - Y_t)\,dt + \sigma\,dZ_t, \end{cases} \tag{2.13}$$

where $\mu \in \mathbb{R}$, $q \ge 0$, $m \ge 0$, $\sigma > 0$. This model was introduced in [SS91] under the assumption that the correlation coefficient ρ equals zero. The correlated Stein–Stein model is obtained when standard Brownian motions W and Z in (2.13) satisfy (2.12). The parameter μ in (2.13) is the drift coefficient of the stock price, q is the speed of mean-reversion for the volatility process, m is the long-run mean of the volatility process, and σ is the volatility-of-volatility parameter. For the Stein–Stein model, we have

$$b(x) = q(m - x), \quad \sigma(x) = \sigma, \quad \text{and} \quad f(u) = u.$$

Note that in the Stein–Stein model given by (2.13), the volatility is described by an arithmetic Ornstein–Uhlenbeck process Y, which is not positive. However, negative volatility does not cause any conceptual or computational problems (see [LS08]). It is important to mention that in the absence of correlation, a similar model with the volatility process $|Y|$, that is, the model given by

$$\begin{cases} dX_t = \mu X_t\,dt + |Y_t| X_t\,dW_t, \\ dY_t = q(m - Y_t)\,dt + \sigma\,dZ_t, \end{cases} \tag{2.14}$$

does not differ much from the model described by (2.13). More precisely, it can be shown that for every $t \ge 0$ the marginal distribution densities of the stock price processes in (2.13) and (2.14) coincide. The previous statement follows from the definition of the mixing distribution in Sect. 3.2 and formula (3.5).

The volatility of the stock was interpreted in the original paper [SS91] of E.M. Stein and J. Stein as an Ornstein–Uhlenbeck process reflected at zero. However, in [BR94] C.A. Ball and A. Roma came to the conclusion that the correct interpretation of the volatility in the uncorrelated Stein–Stein model is not the reflected Ornstein–Uhlenbeck process $\widetilde{Y}$, but rather the absolute value $|Y|$ of the Ornstein–Uhlenbeck process Y. Note that the transition density of the process $\widetilde{Y}$ differs significantly from the transition density of the process $|Y|$. The latter density is given by (1.23), while the former one satisfies a special Volterra type integral equation (see [RS87]).

Next we turn our attention to the Heston model (see [Hes93]). This model is defined as follows:

$$\begin{cases} dX_t = \mu X_t \, dt + \sqrt{Y_t} X_t \, dW_t, \\ dY_t = (a - bY_t) \, dt + c\sqrt{Y_t} \, dZ_t, \end{cases} \tag{2.15}$$

where $\mu \in \mathbb{R}$, $a \geq 0$, $b \geq 0$, and $c > 0$. If $b \neq 0$, then the second equation in (2.15) can be rewritten in mean-reverting form:

$$\begin{cases} dX_t = \mu X_t \, dt + \sqrt{Y_t} X_t \, dW_t, \\ dY_t = q(m - Y_t) \, dt + c\sqrt{Y_t} \, dZ_t. \end{cases} \tag{2.16}$$

It is assumed in (2.15) and (2.16) that condition (2.12) holds. The interpretation of the model parameters, appearing in (2.16), is the same as in the case of the Stein–Stein model. For the Heston model,

$$b(x) = q(m - x), \quad \sigma(x) = c\sqrt{x}, \quad \text{and} \quad f(u) = \sqrt{u}.$$

Finally, note that the functions b and σ in the volatility equations in (2.11) and (2.14) satisfy the linear growth and the Lipschitz condition. On the other hand, for the volatility equation in (2.16) the functions b and σ are continuous and satisfy the linear growth condition. Moreover, the function b satisfies the Lipschitz condition, and the function σ satisfies the Yamada–Watanabe condition. It follows from Lemmas 2.3 and 2.4 that the equations describing the Hull–White, Stein–Stein, and Heston models are strongly solvable and pathwise uniqueness holds for them.

2.4 Relations Between Stock Price Densities in Stein–Stein and Heston Models

Let us consider the correlated Stein–Stein model with the long-run mean m of the volatility process equal to zero. More precisely, we assume that

$$\begin{cases} dX_t = \mu X_t \, dt + Y_t X_t \left(\sqrt{1 - \rho^2} \, d\widetilde{W}_t + \rho \, dZ_t\right), \\ dY_t = -qY_t \, dt + \sigma \, dZ_t, \end{cases} \tag{2.17}$$

where $\mu \in \mathbb{R}$, $q \geq 0$, $\sigma > 0$, $-1 \leq \rho \leq 1$, and standard Brownian motions $\widetilde{W}$ and Z are independent. Put $\widetilde{Y}_t = Y_t^2$. Then, using the Itô formula, we get

$$d\widetilde{Y}_t = 2Y_t\,dY_t + \sigma^2\,dt = \left(\sigma^2 - 2q\widetilde{Y}_t\right)dt + 2\sigma Y_t\,dZ_t.$$

Define new stochastic processes by

$$d\widehat{W}_t = \operatorname{sign}(Y_t)\,d\widetilde{W}_t \quad \text{and} \quad d\widehat{Z}_t = \operatorname{sign}(Y_t)\,dZ_t.$$

Then we have

$$\begin{cases} dX_t = \mu X_t\,dt + \sqrt{\widetilde{Y}_t}X_t\left(\sqrt{1-\rho^2}\,d\widehat{W}_t + \rho\,d\widehat{Z}_t\right), \\ d\widetilde{Y}_t = \left(\sigma^2 - 2q\widetilde{Y}_t\right)dt + 2\sigma\sqrt{\widetilde{Y}_t}\,d\widehat{Z}_t. \end{cases} \tag{2.18}$$

The processes $\widehat{W}$ and $\widehat{Z}$ are standard Brownian motions (use Lévy's characterization theorem and path properties of the process Y). We will next prove that these Brownian motions are independent. It suffices to prove that $\langle \widehat{W}, \widehat{Z}\rangle_t = 0$ for all $t > 0$. It follows from the definition of the process $\widehat{Z}$ that the processes $\widetilde{W}$ and $\widehat{Z}$ are independent. We will need the following assertion (Theorem 29 in [Pro04]).

Theorem 2.6 *Let G_1 and G_2 be semimartingales, and suppose H_i is an G_i-integrable process for $i = 1, 2$. Put*

$$S_{1,t} = \int_0^t H_{1,s}\,dG_{1,s} \quad \textit{and} \quad S_{2,t} = \int_0^t H_{2,s}\,dG_{2,s}, \quad t \geq 0.$$

Then for every $t > 0$,

$$\langle S_1, S_2\rangle_t = \int_0^t H_{1,s}H_{2,s}\,d\langle G_1, G_2\rangle_s.$$

Applying Theorem 2.6 with $H_{1,t} = \operatorname{sign}(Y_t)$, $H_{2,t} = 1$, $G_{1,t} = \widetilde{W}_t$, and $G_{2,t} = \widehat{Z}_t$, we obtain

$$\langle \widehat{W}, \widehat{Z}\rangle_t = \int_0^t \operatorname{sign}(Y_s)\,d\langle \widetilde{W}, \widehat{Z}\rangle_s = 0. \tag{2.19}$$

The last equality in (2.19) holds since the processes $\widetilde{W}$ and $\widehat{Z}$ are independent. This completes the proof of the independence of the processes $\widehat{W}$ and $\widehat{Z}$.

It follows from the previous reasoning that the model in (2.18) is a special Heston model. The following lemma summarizes what was said above.

Lemma 2.7 *For every $t > 0$ the marginal distribution of the stock price X_t in the correlated Stein–Stein model with $m = 0$, $X_0 = x_0$ $\mathbb{P}$-a.s., and $Y_0 = y_0$ $\mathbb{P}$-a.s. coincides with the marginal distribution of the stock price X_t in the special Heston model given by* (2.18) *with $X_0 = x_0$ and $\widetilde{Y}_0 = y_0^2$ $\mathbb{P}$-a.s.*

2.5 Girsanov's Theorem

Girsanov's theorem explains how stochastic translations of a multi-dimensional Brownian motion are related to absolutely continuous changes of the Wiener measure.

Let $W = (W^{(1)}, \dots, W^{(d)})$ be a d-dimensional standard Brownian motion on $(\Omega, \mathcal{F}, \mathbb{P})$ with respect to a filtration $\{\mathcal{F}_t\}_{t \geq 0}$, satisfying the usual conditions of right-continuity and completeness. Let

$$\lambda_t = \left(\lambda_t^{(1)}, \dots, \lambda_t^{(d)}\right), \quad t \geq 0,$$

be a measurable adapted d-dimensional stochastic process such that

$$\int_0^T \|\lambda_s\|^2 \, ds < \infty \quad \mathbb{P}\text{-a.s.} \tag{2.20}$$

for all $T > 0$. Set

$$W_t^\lambda = W_t + \int_0^t \lambda_s \, ds, \quad t \geq 0, \tag{2.21}$$

and

$$\mathcal{E}_t^\lambda = \exp\left\{-\frac{1}{2}\int_0^t \|\lambda_s\|^2 \, ds - \int_0^t \sum_{i=1}^d \lambda_s^{(i)} \, dW_s^{(i)}\right\}, \quad t \geq 0.$$

Then by Itô's formula we have

$$d\mathcal{E}_t^\lambda = \mathcal{E}_t^\lambda \sum_{i=1}^d \lambda_t^{(i)} \, dW_t^{(i)},$$

and therefore the process $t \mapsto \mathcal{E}_t^\lambda$, $t \geq 0$, is a local martingale. Moreover, this process starts at 1 and is strictly positive. It is a true martingale if and only if

$$\mathbb{E}\left[\mathcal{E}_t^\lambda\right] = 1 \tag{2.22}$$

for all $t \geq 0$. A similar statement is valid on every finite interval $[0, T]$.

Suppose condition (2.22) holds. For every $t \geq 0$ define the probability measure $\mathbb{P}_t^\lambda$ on the σ-algebra $\mathcal{F}_t$ by $d\mathbb{P}_t^\lambda = \mathcal{E}_t^\lambda \, d\mathbb{P}$. Since the process $t \mapsto \mathcal{E}_t^\lambda$ is a martingale, the following consistency condition is satisfied: $\mathbb{P}_t^\lambda(A) = \mathbb{P}_s^\lambda(A)$ for all $0 \leq s \leq t$ and $A \in \mathcal{F}_s$.

We are finally ready to formulate Girsanov's theorem.

Theorem 2.8 *Suppose condition* (2.22) *holds. Then, for every $T > 0$ the process W_t^λ, $0 \leq t \leq T$, where W_t^λ is defined by* (2.21), *is a d-dimensional standard Brownian motion on $(\Omega, \mathcal{F}_T, \{\mathcal{F}_t\}_{0 \leq t \leq T}, \mathbb{P}_T^\lambda)$.*

A useful sufficient condition for the validity of the equality in (2.22) is provided in the next lemma.

Lemma 2.9 *Suppose for some* $T > 0$,

$$\mathbb{E}\left[\exp\left\{\frac{1}{2}\int_0^T \|\lambda_s\|^2 ds\right\}\right] < \infty. \tag{2.23}$$

Then equality (2.22) *holds for all* $0 \le t \le T$.

Condition (2.23) in Lemma 2.9 is called the Novikov condition. The conclusion in Lemma 2.9 is also valid if the Novikov condition is satisfied on small intervals.

Lemma 2.10 *Suppose for some number* $T > 0$ *there exist numbers* t_i, $0 \le i \le n$, *such that* $0 = t_0 < t_1 < \cdots < t_n = T$ *and*

$$\mathbb{E}\left[\exp\left\{\frac{1}{2}\int_{t_{i-1}}^{t_i} \|\lambda_s\|^2 ds\right\}\right] < \infty$$

for all $1 \le i \le n$. *Then equality* (2.22) *holds for all* $0 \le t \le T$.

Corollary 2.11 *Suppose for some number* $T > 0$ *there exists* $\eta > 0$ *such that*

$$\sup_{0 \le s \le T} \mathbb{E}\left[\exp\{\eta \|\lambda_s\|^2\}\right] < \infty. \tag{2.24}$$

Then equality (2.22) *holds for all* $0 \le t \le T$.

Proof Let η be such that (2.24) holds. It is clear that there exist $n > 1$ and numbers t_i, $0 \le i \le n$, such that $0 = t_0 < t_1 < \cdots < t_n = T$ and $t_i - t_{i-1} < 2\eta$ for all $1 \le i \le n$. Using Jensen's inequality and (2.24), we see that for all $1 \le i \le n$,

$$\begin{aligned}\mathbb{E}\left[\exp\left\{\frac{1}{2}\int_{t_{i-1}}^{t_i} \|\lambda_s\|^2 ds\right\}\right] &\le \frac{1}{t_i - t_{i-1}}\int_{t_{i-1}}^{t_i} \mathbb{E}\left[\exp\left\{\frac{t_i - t_{i-1}}{2}\|\lambda_s\|^2\right\}\right] ds \\ &\le \sup_{0 \le s \le T} \mathbb{E}\left[\exp\{\eta \|\lambda_s\|^2\}\right] < \infty.\end{aligned}$$

Now it is clear that Corollary 2.11 follows from Lemma 2.10. □

Under the conditions in Girsanov's theorem, the process W_t^λ, $0 \le t \le T$, is a Brownian motion with respect to the filtration $\{\mathcal{F}_t\}_{0 \le t \le T}$ and the measure $\mathbb{P}_T^\lambda$ defined on the σ-algebra $\mathcal{F}_T$. The measures $\mathbb{P}_T^\lambda$ and $\mathbb{P}$ are mutually absolutely continuous on $\mathcal{F}_T$ (such measures are called equivalent). It would be important to find sufficient conditions for the existence of a single measure $\mathbb{P}^\lambda$ on $\mathcal{F}_\infty = \sigma(\mathcal{F}_t : 0 \le t < \infty)$ such that for every $T > 0$ the measure $\mathbb{P}^\lambda$ coincides with the measure $\mathbb{P}_T^\lambda$ on

$\mathcal{F}_T$. Any two measures satisfying the previous conditions are called locally equivalent. It is known that the measure $\mathbb{P}^\lambda$ exists and is unique if the original filtration $\{\mathcal{F}\}_{t\geq 0}$ coincides with the filtration $\{\mathcal{F}_t^W\}_{t\geq 0}$ generated by Brownian motion W.

The next assertion is a corollary to Girsanov's theorem.

Corollary 2.12 *Under the conditions in Girsavnov's theorem, suppose the process λ is adapted to the filtration $\{\mathcal{F}_t^W\}$, satisfies condition* (2.20), *and is progressively measurable. If the equality in* (2.22) *is satisfied, then the process W^λ defined by* (2.21) *is a d-dimensional standard Brownian motion on $(\Omega, \mathcal{F}_\infty^W, \{\mathcal{F}_t^W\}_{t\geq 0}, \mathbb{P}^\lambda)$.*

Remark 2.13 It is assumed in Corollary 2.12 that the process λ is progressively measurable, since we wish the process W^λ to be adapted to the filtration $\{\mathcal{F}_t^W\}_{t\geq 0}$.

The reader is referred to [KS91], Sect. 3.5, for the proofs of the statements formulated in this section.

2.6 Risk-Neutral Measures

For a stochastic volatility model, a risk-neutral measure is a martingale measure that is equivalent to the "physical" measure governing the dynamics of the stock price. A standard method of constructing risk-neutral measures is to use Girsanov's theorem to replace the drift in the stock price equation by the constant interest rate and also transform the drift in the volatility equation. The acceptable transformations of the volatility drift depend on special stochastic processes called market prices of volatility risk. In a sense, these processes label risk-neutral measures. It is important to bear in mind that Girsanov's theorem can be applied only under certain conditions, which may fail to be true even for classical stochastic volatility models. The latter statement will be explained below. Risk-neutral measures are used in option pricing theory to fairly price options on the underlying stock.

Let us consider a stochastic volatility model described by (2.7), and suppose the restrictions on the functions f, b, and σ formulated in Lemmas 2.3 and 2.4 hold. Denote by $\{\mathcal{F}_t\}$, $\{\mathcal{F}_t^{(1)}\}$, and $\{\mathcal{F}_t^{(2)}\}$ the filtrations generated by the processes $(\widetilde{W}, Z)$, $\widetilde{W}$, and Z, respectively. The volatility process Y is adapted to the filtration $\{\mathcal{F}_t^{(2)}\}$, while the stock price process X is adapted to the filtration $\{\mathcal{F}_t\}$. Given a time horizon $T > 0$, denote by $\mathbb{P}$ the restriction of the original ("physical") measure to the σ-algebra $\mathcal{F}_T$.

Definition 2.14 A probability measure $\mathbb{P}_T^*$ on the σ-algebra $\mathcal{F}_T$ is called a risk-neutral measure for the model described by (2.7) if the measures $\mathbb{P}_T$ and $\mathbb{P}_T^*$ are mutually absolutely continuous (such measures are called equivalent), and the discounted stock price process $t \mapsto e^{-rt}X_t$, $0 \leq t \leq T$, is a martingale with respect to the filtration $\{\mathcal{F}_t\}_{0\leq t\leq T}$ and the measure $\mathbb{P}_T^*$.

We will see below that for a stochastic volatility model, risk-neutral measures (if they exist) can be labeled by certain two-dimensional stochastic processes.

Definition 2.15 A market price of risk is a pair of stochastic processes (ζ, γ) such that the processes ζ and γ are adapted to the filtration $\{\mathcal{F}_t^{(2)}\}_{t\geq 0}$, measurable, and such that for every $t > 0$,

$$\int_0^t \zeta_s^2 \, ds < \infty \quad \text{and} \quad \int_0^t \gamma_s^2 \, ds < \infty$$

$\mathbb{P}$-a.s. The process ζ is called the market price of risk for the stock, while the process γ is called the market price of volatility risk.

It will be assumed in the sequel that the market price of volatility risk satisfies the following condition. For every $t \geq 0$,

$$\mathbb{E}\big[\exp\{-\mathcal{E}_t^{\gamma}\}\big] = 1 \tag{2.25}$$

where

$$\mathcal{E}_t^{\gamma} = \frac{1}{2}\int_0^t \gamma_s^2 \, ds + \int_0^t \gamma_s \, dZ_s. \tag{2.26}$$

Put $W_t^* = \widetilde{W}_t + \int_0^t \zeta_s \, ds$ and $Z_t^* = Z_t + \int_0^t \gamma_s \, ds$. Our next goal is to show that for all $t \geq 0$,

$$\mathbb{E}\big[\exp\{-\mathcal{E}_t^{\zeta,\gamma}\}\big] = 1 \tag{2.27}$$

where

$$\mathcal{E}_t^{\zeta,\gamma} = \frac{1}{2}\int_0^t \zeta_s^2 \, ds + \int_0^t \zeta_s \, d\widetilde{W}_s + \frac{1}{2}\int_0^t \gamma_s^2 \, ds + \int_0^t \gamma_s \, dZ_s.$$

For the sake of simplicity, we will model independent Brownian motions $\widetilde{W}$ and Z on two copies $(\Omega_1, \mathcal{F}^{(1)}, \mathbb{P}^{(1)})$ and $(\Omega_2, \mathcal{F}^{(2)}, \mathbb{P}^{(2)})$ of Wiener spaces. In this interpretation, the stock price process X is defined on the product space of these Wiener spaces, and the physical measure is given by $\mathbb{P} = \mathbb{P}^{(1)} \times \mathbb{P}^{(2)}$. We have

$$\mathbb{E}\big[\exp\{-\mathcal{E}_t^{\zeta,\gamma}\}\big] = \mathbb{E}^{(2)}\big[\exp\{-\mathcal{E}_t^{\gamma}\}\mathbb{E}^{(1)}\big[\exp\{-\mathcal{E}_t^{\zeta}\}\big]\big], \tag{2.28}$$

where $\mathcal{E}_t^{\gamma}$ is defined by (2.26) and

$$\mathcal{E}_t^{\zeta} = \frac{1}{2}\int_0^t \zeta_s^2 \, ds + \int_0^t \zeta_s \, d\widetilde{W}_s.$$

Since the process ζ is adapted to the filtration $\{\mathcal{F}_t^{(2)}\}$, we have

$$\mathbb{E}^{(1)}\big[\exp\{-\mathcal{E}_t^{\zeta}\}\big] = 1 \quad \mathbb{P}^{(2)}\text{-a.s.}$$

Now (2.27) follows from (2.25) and (2.28).

Let us assume that the processes ζ and γ are progressively measurable. Applying Corollary 2.12 with $\lambda = (\zeta, \gamma)$, we see that the processes W^* and Z^* are independent standard Brownian motions under the new measure $\mathbb{P}^{\lambda} = \mathbb{P}^{\zeta,\gamma}$ that is locally equivalent to the measure $\mathbb{P}$ (see Sect. 2.5 for the construction of the measure $\mathbb{P}^{\lambda}$). In addition, under the measure $\mathbb{P}^{\zeta,\gamma}$ the stochastic volatility model in (2.7) takes the following form:

$$\begin{cases} dX_t = X_t\big[\mu - f(Y_t)\big(\sqrt{1-\rho^2}\zeta_t + \rho\gamma_t\big)\big]\,dt \\ \qquad\quad + f(Y_t)X_t\big(\sqrt{1-\rho^2}\,dW_t^* + \rho\,dZ_t^*\big), \\ dY_t = \big[b(Y_t) - \sigma(Y_t)\gamma_t\big]\,dt + \sigma(Y_t)\,dZ_t^*. \end{cases} \tag{2.29}$$

For every $\omega \in \Omega$ define

$$A_\omega = \big\{t \in [0,\infty) : f\big(Y_t(\omega)\big) = 0\big\}.$$

Since the function f is continuous, and Y is a continuous process, the set A_ω is a closed subset of $[0,\infty)$. It will be assumed in the sequel that

$$l(A_\omega) = 0 \quad \text{for all } \omega \in \Omega, \tag{2.30}$$

where l is the Lebesgue measure on $[0,\infty)$. Note that if Y is a measurable process, then the set

$$A = \big\{(t,\omega) \in [0,\infty) \times \Omega : f\big(Y_t(\omega)\big) = 0\big\}$$

belongs to the σ-algebra $\mathbb{B}_{[0,\infty)} \otimes \mathcal{F}$.

Our goal is to find locally equivalent martingale measures among the measures $\mathbb{P}^{\zeta,\gamma}$. From the first equation in (2.29) we see that for every $t \ge 0$,

$$\begin{aligned} X_t = x_0 &+ \int_0^t X_s\big[\mu - f(Y_s)\big(\sqrt{1-\rho^2}\zeta_s + \rho\gamma_s\big)\big]\,ds \\ &+ \sqrt{1-\rho^2}\int_0^t f(Y_s)X_s\,dW_s^* + \rho\int_0^t f(Y_s)X_s\,dZ_s^* \end{aligned}$$

$\mathbb{P}^{\zeta,\gamma}$-a.s. Therefore, if condition (2.30) holds and

$$r = \mu - f(Y_t)\big(\sqrt{1-\rho^2}\zeta_t + \rho\gamma_t\big) \tag{2.31}$$

for all $\omega \in \Omega$ and $t \in [0,\infty)\backslash A_\omega$, then the discounted stock price process $\widetilde{X}_t = e^{-rt}X_t$, $t \ge 0$, is a strictly positive local $\mathbb{P}^{\zeta,\gamma}$-martingale.

Let $\mu \ne r$ and $\rho \in (-1,1)$, and let γ be a market price of volatility risk. Suppose Y is a progressively measurable process and condition (2.31) holds for it. Define the process ζ by the following:

$$\zeta_t = \frac{\mu - r}{\sqrt{1-\rho^2}\,f(Y_t)} - \frac{\rho\gamma_t}{\sqrt{1-\rho^2}} \tag{2.32}$$

for all $(t, \omega) \in ([0, \infty) \times \Omega)\backslash A$, and $\zeta_t = 0$ if $(t, \omega) \in A$. It follows from the progressive measurability of γ and Y and from the definition of the set A that the process ζ is progressively measurable. Let us assume that

$$\int_0^t f(Y_s)^{-2} ds < \infty \quad \mathbb{P}\text{-a.s.} \tag{2.33}$$

for all $t \geq 0$. Then the process ζ is a market price of risk for the stock.

Remark 2.16 In finance, the market price of risk for the stock has an interpretation of an extra compensation (per unit of risk) for taking on risk. Formula (2.32) confirms this interpretation. For instance, if $\rho = 0$, then (2.32) becomes

$$\zeta_t = \frac{\mu - r}{f(Y_t)} \quad \text{if } (t, \omega) \in \big([0, \infty) \times \Omega\big)\backslash A, \tag{2.34}$$

and $\zeta_t = 0$ if $(t, \omega) \in A$. Here μ is equal to the average stock return, $\mu - r$ is the excess return, while $f(Y_t)$ is the variability of the stock (the volatility). The expression on the right-hand side of (2.34) is called the Sharpe ratio.

Let us denote by $\mathbb{P}^{\mu,\gamma}$ the measure $\mathbb{P}^{\zeta,\gamma}$, corresponding to the given market price of volatility risk γ and the process ζ such as in (2.32). It is not hard to see that under the measure $\mathbb{P}^{\mu,\gamma}$ the discounted stock price process $\widetilde{X}$ solves the following system of stochastic differential equations:

$$\begin{cases} d\widetilde{X}_t = f(Y_t)X_t\big(\sqrt{1-\rho^2}\, dW_t^* + \rho\, dZ_t^*\big), \\ dY_t = \big[b(Y_t) - \sigma(Y_t)\gamma_t\big] dt + \sigma(Y_t)\, dZ_t^*. \end{cases} \tag{2.35}$$

Moreover, we have

$$\widetilde{X}_t = x_0 \exp\left\{\int_0^t f(Y_s)\, dW_s - \frac{1}{2}\int_0^t f(Y_s)^2\, ds\right\} \tag{2.36}$$

where $W = \sqrt{1-\rho^2} W^* + \rho Z^*$. The stochastic exponential in (2.36) is a strictly positive local martingale. It is a martingale if and only if

$$\mathbb{E}^{\mu,\gamma}\left[\exp\left\{\int_0^t f(Y_s)\, dW_s - \frac{1}{2}\int_0^t f(Y_s)^2\, ds\right\}\right] = 1$$

for all $t \geq 0$. Since the processes Y and W^* are independent, we can use conditional expectations to show that the previous equality is equivalent to the following:

$$\mathbb{E}^{\mu,\gamma}\left[\exp\left\{\rho\int_0^t f(Y_s)\, dZ_s^* - \frac{1}{2}\rho^2\int_0^t f(Y_s)^2\, ds\right\}\right] = 1 \tag{2.37}$$

for all $t \geq 0$. It follows that if (2.37) holds, then the measure $\mathbb{P}^{\mu,\gamma}$ is a martingale measure locally equivalent to $\mathbb{P}$.

We will next summarize what has been accomplished in this section.

Conclusion 2.17 Consider a stochastic volatility model such as in the beginning of this section, and let γ be a market price of volatility risk. Assume the volatility process Y is progressively measurable and satisfies conditions (2.30), (2.33), and (2.37). Define the market price of risk ζ for the stock by (2.32). Then the measures $\mathbb{P}^{\zeta,\gamma}$ and $\mathbb{P}$ are locally equivalent, and $\mathbb{P}^{\zeta,\gamma}$ is a martingale measure. Moreover, for every time horizon $T > 0$ the measure $\mathbb{P}_T^{\zeta,\gamma}$ is a risk-neutral measure.

Note that in the last statement in Conclusion 2.17, it suffices to assume that the processes ζ and γ are measurable.

Remark 2.18 In the paper [WH06] of B. Wong and C.C. Heide, the equality $r = \mu - f(Y_t)\zeta_t$ appears in the context of no-arbitrage pricing in stochastic volatility models.

In the next three sections, we discuss risk-neutral measures for the uncorrelated Hull–White, Stein–Stein, and Heston models. Conclusion 2.17 will play an important role in this discussion. It will be shown that for the Hull–White model there exists an infinite family of risk-neutral measures labeled by market prices of risk, while for the Stein–Stein model, Girsanov's theorem can be applied only when the drift of the physical measure is equal to the interest rate. For the Heston model the situation is more delicate. Here the existence of risk-neutral measures is determined by the values of the model parameters. The case of correlated models will be addressed in Sects. 2.11 and 2.12.

2.7 Risk-Neutral Measures for Uncorrelated Hull–White Models

If a stochastic volatility model is uncorrelated, then condition (2.37) always holds. For the Hull–White model, we have $f(u) = u$, and since geometric Brownian motions never hit zero, equality (2.30) and condition (2.33) clearly hold. Let γ be a market price of volatility risk satisfying condition (2.25), and let the process ζ be defined by (2.34). It follows from (2.33) that this process satisfies the square integrability condition in Definition 2.15.

Fix the time horizon $T > 0$. Then the results obtained in the previous section imply the following statement.

Conclusion 2.19 Every pair (μ, γ), where μ is the drift coefficient of the stock price and γ is a market price of volatility risk satisfying condition (2.25), generates a risk-neutral measure $\mathbb{P}_T^{\mu,\gamma}$.

The next lemma shows that for the Hull–White model a stronger condition is valid than condition (2.33).

Lemma 2.20 *For every $T > 0$ the following inequality holds*:

$$\mathbb{E}\left[\int_0^T Y_s^{-2}\,ds\right] < \infty.$$

Proof It is clear that

$$\mathbb{E}\left[\int_0^T Y_s^{-2}\,ds\right] = \int_0^T ds \int_0^\infty y^{-2}\rho_s(y)\,dy, \tag{2.38}$$

where ρ_t is the distribution density of Y_t. Since

$$\rho_s(y) = \frac{\sqrt{y_0 e^{\nu s}}}{\sqrt{2\pi s}\xi}\exp\left\{-\frac{\xi^2 s}{8}\right\} y^{-\frac{3}{2}}\exp\left\{-\frac{(\log\frac{y}{y_0 e^{\nu s}})^2}{2s\xi^2}\right\}$$

(see (1.5)), the equality (2.38) implies

$$\mathbb{E}\left[\int_0^T Y_s^{-2}\,ds\right] \le c_1 \int_0^T \frac{ds}{\sqrt{s}} \int_0^\infty y^{-\frac{7}{2}} y^{\frac{c_2}{s}} \exp\left\{-\frac{(\log y)^2}{2s\xi^2}\right\} dy < \infty. \tag{2.39}$$

In (2.39), $c_1 > 0$ and $c_2 > 0$ are positive constants depending only on y_0, T, ν, and ξ. Now it is clear that Lemma 2.20 holds. □

2.8 Local Times for Semimartingales

To study risk-neutral measures in the Stein–Stein or the Heston model, more advanced methods are needed. The present section is devoted to local times for semimartingales. Our presentation follows [KS91], Sect. 3.3.7 (see also [RY04], pp. 209–210).

Let us consider a continuous semimartingale on $(\Omega, \mathcal{F}, \{\mathcal{F}_t\}, \mathbb{P})$ given by

$$X_t = x_0 + M_t + V_t, \quad t \ge 0, \tag{2.40}$$

where M is a continuous local martingale, and V is a continuous adapted process of locally bounded variation with $V_0 = 0$ a.s. It is assumed that the filtration $\{\mathcal{F}_t\}$ satisfies the usual conditions.

Theorem 2.21 *For any continuous semimartingale X given by* (2.40) *there exists a non-negative random field $L = L_t^a(\omega)$, $(t, a) \in [0, \infty) \times \mathbb{R}$, $\omega \in \Omega$ such that*

1. *The mapping $(t, a, \omega) \mapsto L_t^a(\omega)$ is $\mathcal{B}_{[0,\infty)\times\mathbb{R}} \otimes \mathcal{F}/\mathcal{B}_{[0,\infty)}$-measurable.*
2. *For each pair (t, a) the random variable L_t^a is $\mathcal{F}_t$-measurable.*
3. *For each $a \in \mathbb{R}$ the mapping $t \mapsto L_t^a(\omega)$ is continuous and nondecreasing. Moreover, $L_0^a(\omega) = 0$ and*

$$\int_0^\infty \mathbb{1}_{\mathbb{R}\setminus\{a\}}\big(X_t(\omega)\big)\,dL_t^a(\omega) = 0$$

for $\mathbb{P}$-almost all $\omega \in \Omega$.

4. *For every Borel measurable function $f : \mathbb{R} \to [0, \infty)$ the equality*

$$\int_0^t f\big(X_s(\omega)\big)\, d\langle M, M\rangle_s(\omega) = \int_{-\infty}^{\infty} f(a) L_t^a(\omega)\, da, \quad 0 \le t < \infty, \tag{2.41}$$

holds $\mathbb{P}$-a.s.

5. *For all $(t, a) \in [0, \infty) \times \mathbb{R}$, we have $\lim_{\tau \to t, b \downarrow a} L_\tau^a(\omega) = L_t^a(\omega)$. Moreover, the limit $L_t^{a-}(\omega) = \lim_{\tau \to t, b \uparrow a} L_\tau^b(\omega)$ exists $\mathbb{P}$-a.s. on Ω.*

The random field L_t^a is called the local time for X. Formula (2.41) is known as the occupation times formula, while property 5 states that the mapping L is a.s. jointly continuous in t and right-continuous with left limits in a. Note that we use a different normalization in (2.41) than that chosen in [KS91].

We will need the following well-known generalization of the Itô formula.

Theorem 2.22 *Let X be a continuous semimartingale and L_t^a be its family of local times. Suppose that f can be represented as the difference of two convex functions on $(0, \infty)$. Then*

$$f(X_t) = f(X_0) + \int_0^t f'_-(X_s)\, dX_s + \frac{1}{2}\int_0^\infty L_t^a f''(da) \tag{2.42}$$

(f'' is in general a signed Borel measure). In particular, $f(X_t)$ is a semimartingale.

Formula (2.42) is called the Itô–Tanaka formula (see, e.g., [RY04]).

2.9 Risk-Neutral Measures for Uncorrelated Stein–Stein Models

Recall that in the uncorrelated Stein–Stein model, $f(u) = |u|$ and the volatility is modeled by the process $f(Y)$, where Y is an Ornstein–Uhlenbeck process. For the Stein–Stein model condition (2.33) never holds, and hence we cannot change the drift in the stock price equation using Girsanov's theorem. The previous statement can easily be obtained from the following lemma.

Lemma 2.23 *Suppose the initial condition y_0 for the volatility process Y in the Stein–Stein model equals zero. Then*

$$\mathbb{P}\left[\int_0^T Y_s^{-2}\, ds = \infty\right] = 1$$

for all $T > 0$. If $y_0 \neq 0$, then

$$0 < \mathbb{P}\left[\int_0^T Y_s^{-2}\, ds = \infty\right] < 1$$

for all $T > 0$.

Proof Consider the Ornstein–Uhlenbeck process $Y = Y(q, m, \sigma, y_0)$ with $q \geq 0$, $m \in \mathbb{R}$, and $y_0 \in \mathbb{R}$. This process is the solution to the equation

$$dY_t = q(m - Y_t)\,dt + \sigma\,dZ_t, \quad Y_0 = y_0.$$

If $q = 0$, then $Y = \sigma Z$, where Z is a standard Brownian motion. If $q > 0$, then Y_t is given by

$$Y_t(q, m, \sigma, y_0) = e^{-qt}\big(y_0 + \big(e^{qt} - 1\big)m + B_{\phi(t)}\big), \tag{2.43}$$

where $\phi(t) = \frac{\sigma^2}{2q}(e^{2qt} - 1)$, and B is a $\mathcal{F}_{\phi^{-1}(t)}$-standard Brownian motion (see Lemma 1.20).

It is not hard to see using (2.43) that for $q \neq 0$ the convergence properties of the integrals $\int_0^T Y_s(q, m, \sigma, y_0)^{-2}\,ds$ and

$$\int_0^{\frac{\sigma^2}{2q}(e^{2qT}-1)} \left(y_0 + \left(\sqrt{1 + \frac{2q}{\sigma^2}u} - 1\right)m + Z_u\right)^{-2} du$$

are the same. Next, put $\varphi(u) = (\sqrt{1 + \frac{2q}{\sigma^2}u} - 1)m$, $0 \leq u \leq \widetilde{T}$, where $\widetilde{T} = \frac{\sigma^2}{2q}(e^{2qT} - 1)$. It follows from Girsanov's theorem that there exists a probability measure $\mathbb{P}^*$ on $\mathcal{F}$ such that the measures $\mathbb{P}$ and $\mathbb{P}^*$ are mutually absolutely continuous, and the process $\widetilde{Z}_t = \varphi(t) + Z_t$ is a standard Brownian motion under $\mathbb{P}^*$. Since

$$\int_0^{\frac{\sigma^2}{2q}(e^{2qT}-1)} \left(y_0 + \left(\sqrt{1 + \frac{2q}{\sigma^2}u} - 1\right)m + Z_u\right)^{-2} du = \int_0^{\widetilde{T}} (y_0 + \widetilde{Z}_u)^{-2}\,du,$$

the convergence properties of the integrals

$$\int_0^T Y_s(q, m, \sigma, y_0)^{-2}\,ds \quad \text{and} \quad \int_0^{\widetilde{T}} (y_0 + \widetilde{Z}_u)^{-2}\,du$$

are the same. Summarizing what was said above, we see that it suffices to prove Lemma 2.23 for Brownian motion starting at $x \in \mathbb{R}$. With no loss of generality, we may assume that $x \geq 0$, since the process $-Z$ is a standard Brownian motion.

Define the passage time T_b of the standard Brownian motion Z to a level $b \in \mathbb{R}$ by

$$T_b = \inf\{t \geq 0 : Z_t = b\}.$$

Then T_b is a stopping time. The distribution density p_b of T_b is given by

$$p_b(y) = \frac{|b|}{\sqrt{2\pi y^3}} \exp\left\{-\frac{b^2}{2y}\right\} \tag{2.44}$$

(see, e.g., [KS91], p. 80).

Suppose $x > 0$ and $0 < b < x$. It follows from (2.44) that with positive probability, the process $x + Z_t$ never reaches the interval $(-\infty, b)$ before time T. Since Brownian motion is a continuous process, the estimate

$$\int_0^T \frac{1}{(x+Z_s)^2}\, ds < \infty$$

holds with positive probability.

The case where $x = 0$ is special. In this case, we will show that the following equality holds $\mathbb{P}$-a.s.:

$$\int_0^T \frac{1}{Z_s^2}\, ds = \infty. \tag{2.45}$$

To establish (2.45), we use the law of the iterated logarithm for a standard Brownian motion.

Theorem 2.24 *Let Z be a standard Brownian motion. Then*

$$\mathbb{P}\left(\limsup_{t \downarrow 0} \frac{Z_t}{\sqrt{2t \log\log \frac{1}{t}}} = 1\right) = 1.$$

The reader is referred to [RY04], p. 53, for the proof of Theorem 2.24. It follows from Theorem 2.24 that for $\mathbb{P}$-almost all $\omega \in \Omega$, there exists $r(\omega) > 0$ such that

$$\left|Z_s(\omega)\right| \le c\sqrt{s \log\log \frac{1}{s}}, \quad 0 \le s \le r(\omega), \tag{2.46}$$

Here we use the fact that $-Z$ is also a standard Brownian motion. Now it is clear that (2.46) implies (2.45).

It remains to prove that if $x > 0$, then the equality

$$\int_0^T \frac{1}{(x+Z_s)^2}\, ds = \infty \tag{2.47}$$

holds with positive probability. We will derive (2.47) from the following general statement due to G. F. Le Gall (see [LeG83], see also [RY04]).

Lemma 2.25 *Let Y be a continuous semimartingale, and assume that for some $T > 0$ and $E \in \mathcal{F}$ the local time L_T^0 of Y at zero satisfies $L_T^0(\omega) \neq 0$, $\omega \in E$. Then for every positive Borel function f on $(-\infty, \infty)$ with*

$$\int_{0+} \frac{da}{f(a)} = \infty \tag{2.48}$$

the following equality holds:

$$\int_0^T \frac{d\langle Y, Y\rangle_s}{f(Y_s)} = \infty \quad \mathbb{P}\text{-}a.s. \text{ on } E. \tag{2.49}$$

Proof By the occupation times formula,

$$\int_0^T \frac{d\langle Y, Y\rangle_s}{f(Y_s)} = \int_{-\infty}^{\infty} \frac{1}{f(a)} L_T^a \, da.$$

Now let $\omega \in E$. Then, using the right continuity of the local time in a, condition (2.48), and the assumption $L_T^0(\omega) \neq 0$, we see that the integral on the right-hand side of (2.49) diverges.

This completes the proof of Lemma 2.25. □

Let us return to the proof of (2.47). We have

$$d\langle x + Z, x + Z\rangle_s = ds.$$

Set $f(u) = u^{-2}$. Then formula (2.49) applied to the process $x + Z$ shows that (2.47) holds on the set E, where the local time $\widetilde{L}_T^0$ of the process $x + Z$ satisfies $\widetilde{L}_T^0 \neq 0$. This set coincides with the set where the Brownian local time L_T^x satisfies $L_T^x \neq 0$.

It is known that

$$\mathbb{E}\big[L_T^x\big] = \int_0^T \rho_s(x) \, ds \tag{2.50}$$

where ρ_s stands for the distribution density of Z_s (see [MR06]). Since

$$\rho_s(x) = \frac{1}{\sqrt{2\pi s}} \exp\left\{-\frac{x^2}{2s}\right\},$$

we have $\mathbb{E}[L_T^x] > 0$, and hence $L_T^x > 0$ on a set of positive measure. It follows that formula (2.47) is valid on a set of positive measure.

This completes the proof of Lemma 2.23. □

Conclusion 2.26 Girsanov's theorem can be used to find risk-neutral measures in the uncorrelated Stein–Stein model only if $r = \mu$. In this case, any pair (μ, γ), where γ is a market price of volatility risk satisfying condition (2.25), generates a risk-neutral measure $\mathbb{P}_T^{\mu,\zeta}$.

2.10 Risk-Neutral Measures for Uncorrelated Heston Models

In the Heston model, the volatility of the stock is described by the process $\widetilde{Y} = \sqrt{Y}$, where Y is the CIR process satisfying the following stochastic differential equation:

$$dY_t = (a - bY_t) \, dt + c\sqrt{Y_t} \, dZ_t, \quad Y_0 = y_0 \ \mathbb{P}\text{-a.s.} \tag{2.51}$$

It will be assumed that $y_0 > 0$, $a \geq 0$, $b \geq 0$, and $c > 0$. We also assume that at least one of the conditions $a \neq 0$ and $b \neq 0$ holds. This implies the validity of formula (2.30) for the process Y (see Theorem 1.45).

Our next goal is to study the behavior of the random integral

$$\int_0^T \widetilde{Y}_s^{-2} ds = \int_0^T Y_s^{-1} ds.$$

Let $x > 0$, and set $\tau_0^x = \inf\{t \geq 0 : Y_t = 0\}$ where Y is the solution to (2.51) starting at x. Here we assume $\inf\{\emptyset\} = \infty$. Then the random variable τ_0^x is a stopping time. The next statement concerns hitting times of zero by CIR processes.

Theorem 2.27 *Let $x > 0$. Then the following are true*:

1. *If $2a \geq c^2$, then $\mathbb{P}(\tau_0^x = \infty) = 1$.*
2. *If $2a < c^2$, then $\mathbb{P}(\tau_0^x < \infty) = 1$.*

The equality in part 2 of Theorem 2.27 holds under the assumption $b \geq 0$. If $b < 0$, then we have $0 < \mathbb{P}(\tau_0^x < \infty) < 1$. We refer the reader to [LL08], Proposition 6.2.3 and Exercise 37, for a detailed sketch of the proof of Theorem 2.27.

Using part 1 of Theorem 2.27 and the continuity of the process Y, we obtain the following corollary:

Corollary 2.28 *Let Y be a CIR process. If $y_0 > 0$ and $2a \geq c^2$, then*

$$\mathbb{P}\left(\int_0^T Y_s^{-1} ds < \infty\right) = 1,$$

for every $T > 0$.

More results concerning the convergence of the integral $\int_0^T Y_s^{-1} ds$ are contained in the next lemma.

Lemma 2.29 *The following are true for the CIR process Y*:

1. *The inequality*

$$\mathbb{E}\left[\int_0^T Y_s^{-1} ds\right] < \infty$$

   holds if and only if $c^2 < 2a$.
2. *Let $c^2 > 2a$. Then*

$$0 < \mathbb{P}\left(\int_0^T Y_s^{-1} ds = \infty\right) < 1.$$

Proof The structure of the proof of Lemma 2.29 is similar to that of Lemma 2.20. By Theorem 1.40 and Remark 1.41, it suffices to analyze the random integral

$\int_0^T T_s^{-1}\,ds$ where $T = BESQ_{y_0}^{\frac{4a}{c^2}}$, in order to understand the behavior of the integral $\int_0^T Y_s^{-1}\,ds$.

Let $c^2 < 2a$. Then $a \neq 0$ and $\nu > 0$. Denote by ρ_s the distribution density of the random variable $BESQ_{y_0}^{\frac{4a}{c^2}}(s)$. This density is given by

$$\rho_t(y) = \left(\frac{y}{y_0}\right)^{\frac{\nu}{2}} \frac{1}{2t} e^{-\frac{y_0+y}{2t}} I_\nu\left(\frac{\sqrt{y_0 y}}{t}\right), \quad \nu = \frac{2a}{c^2} - 1$$

(see (1.49)). We have

$$\mathbb{E}\left[\int_0^T \frac{ds}{T_s}\right] = \int_0^T ds \int_0^\infty \frac{1}{y} \rho_s(y)\,dy.$$

Moreover,

$$\mathbb{E}\left[\int_0^T \frac{ds}{T_s}\right] = \frac{1}{2} y_0^{-\frac{\nu}{2}} \int_0^T e^{-\frac{y_0}{2s}} \frac{ds}{s} \int_0^\infty y^{\frac{\nu}{2}-1} e^{-\frac{y}{2s}} I_\nu\left(\frac{\sqrt{y_0 y}}{s}\right) dy. \tag{2.52}$$

The modified Bessel function of the first kind $I_\nu(x)$ behaves near zero like the function $\frac{1}{\Gamma(\nu+1)}(\frac{x}{2})^\nu$ and near infinity like the function $\frac{1}{\sqrt{2\pi x}} e^x$ (see 9.6.7 and 9.7.1 in [AS72]). Therefore, the expression on the right-hand side of (2.52) is finite. This proves the sufficiency in part 1 of Lemma 2.29.

Now let $2a = c^2$. Then $\nu = 0$, and since the function I_0 behaves near zero like a constant function, the expression on the right-hand side of (2.52) is infinite. It follows that

$$\mathbb{E}\left[\int_0^T \frac{ds}{T_s}\right] = \infty \tag{2.53}$$

for $c^2 = 2a$. If $c^2 > 2a$, then (2.53) follows from the second statement in Lemma 2.29, which will be established next.

Suppose $c^2 > 2a$ and put $\delta = \frac{4a}{c^2}$. Then we have $0 \leq \delta < 2$. Denote by $T_0^{y_0}$ the hitting time of zero by the process $BESQ_{y_0}^\delta$, that is,

$$T_0^{y_0} = \inf\{t > 0 : BESQ_{y_0}^\delta(t) = 0\}.$$

The distribution density p of $T_0^{y_0}$ is given by

$$p(y) = \frac{1}{\Gamma(1 - \frac{\delta}{2}) y} \left(\frac{y_0}{2y}\right)^{1-\frac{\delta}{2}} \exp\left\{-\frac{y_0}{2y}\right\} \tag{2.54}$$

for all $y > 0$ (see formula (15) in [G-JY03]). It follows from (2.54) that

$$\mathbb{P}(T_0^{y_0} > T) > 0$$

for all $T > 0$. Since T is a continuous process, the following inequality holds:

$$\mathbb{P}\left(\int_0^T \frac{ds}{T_s} < \infty\right) > 0.$$

We will next prove the inequality

$$\mathbb{P}\left(\int_0^T \frac{ds}{T_s} = \infty\right) > 0.$$

Note that for $\nu = \frac{\delta}{2} - 1$, we have $-1 \le \nu < 0$. It is clear from the stochastic differential equation for the process T that this process is a non-negative semimartingale.

Consider the δ-dimensional Bessel process starting at $\sqrt{y_0}$. This process is defined by

$$BES^{\delta}_{\sqrt{y_0}} = \sqrt{BESQ^{\delta}_{y_0}} = \sqrt{T}.$$

It follows from Theorem 2.22 applied to the process T and the function

$$f(u) = -\sqrt{u}, \quad u > 0$$

that

$$\sqrt{T_t} = \sqrt{y_0} + \frac{1}{2}\int_0^t \frac{dT_s}{\sqrt{T_s}} - \frac{1}{8}\int_0^{\infty} \widetilde{L}_t^a a^{-\frac{3}{2}}\, da,$$

where $\widetilde{L}_t^a$ is the family of local times for the process T. Therefore, the process $\sqrt{T}$ is a continuous semimartingale, for which the quadratic variation is equal to t. Now, applying Lemma 2.25 to the process $\sqrt{T}$ and the function $f(u) = u^{-2}$, we see that

$$\int_0^T \frac{ds}{T_s} = \infty \tag{2.55}$$

on the set where the local time L_T^0 for the process $\sqrt{T}$ satisfies $L_T^0 > 0$.

The distribution density $\tilde{\rho}_s$ of the random variable $\sqrt{T_s}$ is given by

$$\tilde{\rho}_s(y) = \frac{1}{s} y_0^{\frac{1}{4}} y^{\frac{1}{2}} \exp\left\{-\frac{y_0 + y^2}{2s}\right\} I_{-\frac{1}{2}}\left(\frac{\sqrt{y_0}y}{s}\right). \tag{2.56}$$

Formula (2.56) can be derived from (1.49). It is not hard to see, using the occupation times formula (formula (2.41)), that the following equality holds:

$$\mathbb{E}\left[L_T^0\right] = \int_0^T \tilde{\rho}_s(0)\, ds. \tag{2.57}$$

Since the function $I_\nu(x)$ behaves near zero like the function $const \times x^{-\frac{1}{2}}$ (see 9.6.7 in [AS72]), formulas (2.56) and (2.57) imply that $\mathbb{E}[L_T^0] > 0$. It follows that $L_T^0 > 0$

on a set of positive probability, and hence (2.55) holds on a set of positive probability.

This completes the proof of part 2 of Lemma 2.29. □

Conclusion 2.30 For the Heston model with $c^2 \le 2a$, any pair (μ, γ), where μ is the drift of the stock price and γ is a market price of volatility risk satisfying condition (2.25), generates a risk-neutral measure $\mathbb{P}_T^{\mu,\gamma}$. On the other hand, for the Heston model with $c^2 > 2a$, the same is true only if $r = \mu$.

2.11 Hull–White Models. Complications with Correlations

Consider the following correlated Hull–White model:

$$\begin{cases} dX_t = Y_t X_t \left(\sqrt{1-\rho^2}\, d\widetilde{W}_t + \rho\, dZ_t\right), \\ dY_t = \nu Y_t\, dt + \xi Y_t\, dZ_t, \end{cases} \tag{2.58}$$

and suppose $\mu, \nu \in \mathbb{R}$, $\xi > 0$ and $\rho \in (-1, 0) \cup (0, 1)$. Our goal in this section is to study risk-neutral measures in the model described by (2.58).

In an important paper [Sin98], titled "Complications with stochastic volatility models", C. Sin established that the existence of risk-neutral measures for the correlated Hull–White model and for similar models is determined by the possibility of explosions in finite time for solutions of certain auxiliary stochastic differential equations. Let us consider the following stochastic model:

$$\begin{cases} dX_t = Y_t^{\alpha} X_t (\sigma_1\, dW_t + \sigma_2\, dZ_t), \\ dY_t = q(m - Y_t)\, dt + Y_t (a_1\, dW_t + a_2\, dZ_t). \end{cases} \tag{2.59}$$

In (2.59), (W, Z) is a two-dimensional standard Brownian motion on the filtered probability space $(\Omega, \mathcal{F}, \{\mathcal{F}_t\}, \mathbb{P})$, and $\sigma_1, \sigma_2, a_1, a_2, q, m$, and $\alpha > 0$ are real constants. The initial conditions for X and Y are denoted by x_0 and y_0, respectively. The next assertions were obtained by Sin (see Lemma 4.2 and Lemma 4.3 in [Sin98]).

Lemma 2.31 *Suppose (X, Y) is a solution to* (2.59). *Then X is a supermartingale, and for every $T > 0$,*

$$\mathbb{E}[X_T] = x_0 \mathbb{P}(H \text{ does not explode on } [0, T]),$$

where H is the unique solution up to explosion time to the following stochastic differential equation:

$$dH_t = q(m - H_t)\, dt + H_t^{\alpha+1} (\sigma_1 a_1 + \sigma_2 a_2)\, dt + H_t (a_1\, dW_t + a_2\, dZ_t). \tag{2.60}$$

Lemma 2.32 *The unique solution H to* (2.60) *explodes in finite time with positive probability if and only if $\sigma_1 a_1 + \sigma_2 a_2 > 0$.*

The next theorem is the main result obtained in [Sin98].

Theorem 2.33 *The process X in* (2.59) *is a martingale if and only if*

$$\sigma_1 a_1 + \sigma_2 a_2 \le 0.$$

It is easy to see that Theorem 2.33 follows from Lemmas 2.31 and 2.32. We omit the proofs of these lemmas, and refer the reader to [Sin98] for more details.

The model in (2.58) is a special case of (2.59) where $\sigma_1 = \sqrt{1-\rho^2}$, $\sigma_2 = \rho$, $m = 0$, $q = -\nu$, $a_1 = 0$, $a_2 = \xi$, and $\alpha = 1$. The auxiliary equation in (2.60), corresponding to (2.58), has the following form:

$$\begin{cases} dH_t = \left(\nu H_t + \rho\xi H_t^2\right) dt + \xi H_t \, dZ_t, \\ H_0 = y_0. \end{cases}$$

Recall that $\rho \neq 0$ and $\xi > 0$. Applying Lemmas 2.31 and 2.32, we see that if $\rho > 0$, then the process H explodes in finite time with positive probability, and hence for some $T > 0$, we have $\mathbb{E}[X_T] < x_0$. It follows that for the Hull–White model in (2.58) the stock price process X is not a martingale on $[0, T]$. On the other hand, if $\rho < 0$, then the process X is a positive supermartingale such that $\mathbb{E}[X_T] = x_0$ for all $T \ge 0$. Therefore the process X_t, $t \ge 0$, is a martingale.

Conclusion 2.34 For the Hull–White model given by (2.58), the "physical" measure $\mathbb{P}$ is a martingale measure if and only if the standard Brownian motions W and Z, driving the stock price equation and the volatility equation, respectively, are negatively correlated.

We will next give special examples of market prices of volatility risk, which generate risk-neutral measures for the Hull–White model in the case of positive correlation. We will need the following assertion obtained by A. Lewis (see Theorem 9.2 in [Lew00]). This assertion is a generalization of Sin's results.

Theorem 2.35 *Suppose a stochastic model is of the following form under the measure $\mathbb{P}$*:

$$\begin{cases} dX_t = \sqrt{Y_t} X_t \left(\sqrt{1-\rho^2}\, dW_t + \rho \, dZ_t\right), \\ dY_t = b(Y_t)\, dt + \sigma(Y_t)\, dZ_t. \end{cases} \tag{2.61}$$

Suppose also that the second equation in (2.61) *has a unique strong non-exploding solution Y, which is a non-negative process. Then for every $T > 0$,*

$$\mathbb{E}[X_T] = x_0 \mathbb{P}(H \text{ does not explode on } [0, T]),$$

where H is the unique solution up to explosion time to the following stochastic differential equation:

$$dH_t = \left[b(H_t) + \rho\sqrt{H_t}\sigma(H_t)\right] dt + \sigma(H_t)\, dZ_t. \tag{2.62}$$

Corollary 2.36 *The process X_t, $t \geq 0$, in* (2.61) *is a martingale with respect to the measure $\mathbb{P}$ and the filtration generated by the processes W and Z if and only if the solution H to the auxiliary equation* (2.62) *does not explode in finite time with positive probability.*

Let us fix $T > 0$, and suppose the market price of volatility risk is defined by $\gamma_t = bY_t$, $t \geq 0$, for some $b > 0$. Then under the measure $P_T^{0,\gamma}$ the stock price process X in the Hull–White model described by (2.58) satisfies the following equation:

$$\begin{cases} dX_t = Y_t X_t \left(\sqrt{1-\rho^2}\, dW_t^* + \rho\, dZ_t^*\right), \\ dY_t = \left(\nu Y_t - \xi b Y_t^2\right) dt + \xi Y_t\, dZ_t^* \end{cases} \tag{2.63}$$

(see (2.35)). Put $V = Y^2$. Then, using Itô's lemma, we see that (2.63) can be rewritten as follows:

$$\begin{cases} dX_t = \sqrt{V_t} X_t \left(\sqrt{1-\rho^2}\, dW_t^* + \rho\, dZ_t^*\right), \\ dV_t = \left[(2\nu + \xi^2) V_t - 2\xi b V_t^{\frac{3}{2}}\right] dt + 2\xi V_t\, dZ_t^*. \end{cases} \tag{2.64}$$

In (2.64), we assume $V_0 = y_0^2$ $P_T^{0,\gamma}$-a.s. It is clear that the process V is the unique strong solution to the second equation in (2.64). Moreover, since the process Y does not explode in finite time, the same is true for the process V.

Our next goal is to apply Theorem 2.35 to the model in (2.64). In this case, the auxiliary equation in (2.62) is as follows:

$$dH_t = \left[(2\nu + \xi^2) H_t + 2\xi(\rho - b) H_t^{\frac{3}{2}}\right] dt + 2\xi H_t\, dZ_t^*,$$

which is a special case of the auxiliary equation in (2.60) with $m = 0$, $\alpha = \frac{1}{2}$, $q = -(2\nu + \xi^2)$, $a_1 = 0$, $a_2 = 2\xi$, and $\sigma_2 = \rho - b$. Note that no restrictions are needed on the parameter σ_1. Next, using Lemma 2.32, we see that the process H explodes in finite time with positive probability if and only if $\rho > b$. Finally, applying Theorem 2.35 to (2.64), we arrive at the following assertion.

Conclusion 2.37 For the Hull–White model with $\rho > 0$, let the market price of volatility risk be given by $\gamma = bY$ with $b > 0$. Fix the time horizon $T > 0$. Then the measure $\mathbb{P}_T^{0,\gamma}$ is a risk-neutral measure if and only if $b \geq \rho$.

2.12 Heston Models and Stein–Stein Models. No Complications with Correlations

For the Heston model and the Stein–Stein model, we can rely on the techniques developed in Sect. 2.6 to find risk-neutral measures.

Let us begin with the correlated Heston model such that $c^2 \le 2a$ and $-1 < \rho < 1$. Recall that the volatility process $\sqrt{Y}$ in this model does not reach zero almost surely (see Theorem 2.27), and hence condition (2.33) holds. Let γ be a market price of volatility risk satisfying condition (2.25), and let ζ be the market price of risk for the stock defined by formula (2.32).

Our next goal is to prove that for any $T > 0$ and all t with $0 \le t \le T$,

$$\mathbb{E}_T^{\mu,\gamma}\left[\exp\left\{-\frac{1}{2}\rho^2 \int_0^t Y_s\,ds + \rho \int_0^t \sqrt{Y_s}\,dZ_s^*\right\}\right] = 1. \tag{2.65}$$

Note that the equality in (2.65) is nothing else but condition (2.37) formulated for the Heston model.

Equality (2.65) will be derived from Corollary 2.11. Let us first show that there exists $\eta > 0$ for which

$$\sup_{0\le s\le T} \mathbb{E}\left[\exp\{\eta Y_s\}\right] < \infty. \tag{2.66}$$

To prove (2.66), we notice that for every $\eta > 0$,

$$\sup_{0\le s\le T} \mathbb{E}\left[\exp\{\eta Y_s\}\right] = \sup_{0\le s\le T} \int_0^\infty e^{\eta y}\,d\rho_s(y), \tag{2.67}$$

where the symbol ρ_s stands for the distribution of the random variable Y_s. Explicit formulas for the distributions ρ_s are given in Theorem 1.45. Using these formulas in (2.67) and taking into account the asymptotics of the I-Bessel function, we can prove that there exists a number η for which (2.66) holds.

It is not hard to see that the process $\gamma = \eta\sqrt{Y}$ with $\eta \in \mathbb{R}$ is a special example of a market price of volatility risk in the Heston model, satisfying condition (2.25). For the process γ defined above this condition is as follows:

$$\mathbb{E}\left[\exp\left\{-\frac{1}{2}\eta^2 \int_0^t Y_s\,ds - \eta \int_0^t \sqrt{Y_s}\,dZ_s\right\}\right] = 1, \tag{2.68}$$

for all $0 \le t \le T$. It is not hard to see that equality (2.68) can be established exactly as equality (2.65). Here we use formula (2.67).

For the Heston model with $c^2 > 2a$, or the Stein–Stein model, the volatility processes reach zero with positive probability, and condition (2.33) does not hold (see Sects. 2.9 and 2.10). Therefore, it is necessary to assume that $\mu = r$ from the very beginning in order to find risk-neutral measures. It is not hard to see that under the previous assumption all the techniques, used in this section for the Heston model with $c^2 \le 2a$, are still applicable.

Conclusion 2.38 For the correlated Heston model with $c^2 \le 2a$ and $-1 < \rho < 1$, any pair (μ, γ), where μ is the drift of the stock price and γ is a market price of volatility risk satisfying condition (2.25), generates a risk-neutral measure $\mathbb{P}_T^{\mu,\gamma}$. On the other hand, for the correlated Heston model with $c^2 > 2a$ and $-1 < \rho < 1$, the same is true only if $r = \mu$.

The reason why we cannot use the methods, employed in the present section, in the case of the correlated Hull–White model is that the marginal distribution densities of a geometric Brownian motion have log-normal decay, which does not guarantee the validity of inequality (2.66).

2.13 Notes and References

- Geometric Brownain motion is used as the volatility process in the models developed by J. Hull and A. White in [HW87] and J. Wiggins in [Wig87]. The Ornstein–Uhlenbeck process plays the role of the volatility process in the models suggested and studied by L.O. Scott in [Sco87] and E.M. Stein and J. Stein in [SS91]. L.O. Scott also studied a model where the volatility is described by the exponential Orstein–Uhlenbeck process (see [Sco87]). The CIR process is the volatility process in the Heston model (see [Hes93]). The Heston model with time-dependent parameters is studied in [BGM10]. We refer the interested reader to the following books and surveys on stochastic volatility models and related topics: [GHR96, Hob98, FPS00, Lew00, Lip01, RW02, B-NNS02, Jäc03, Tau04, She05, She06, SA09, FPSS11, Gob11].
- It is a common practice to assume that the stock returns and the volatility are negatively correlated. This property is called the leverage effect. For a discussion of the leverage effect, see [FW01] and [BLT06].
- The notion of a risk-neutral (martingale) measure arises in the theory of no-arbitrage pricing. For more information on this theory the reader should refer to [Bjö04, MR05, DS06]. Girsanov's theorem is an important tool in the study of risk-neutral measures. It explains how the behavior of a stochastic process changes when one passes from the physical measure to a special equivalent measure (a risk-neutral measure). More about Girsanov's theorem can be found in [RY04, KS91]
- The work of C. Sin on risk-neutral measures in correlated Hull–White type models (see [Sin98]) was continued in [HLW07, WH04], and [WH06].
- In [K-R11], M. Keller-Ressel studied stock price processes in general affine models. He found necessary and sufficient conditions under which the discounted stock price process in an affine model is a martingale (see Theorem 2.5 in [K-R11]). The results of Keller-Ressel can be used to give an alternative proof of the statement in Conclusion 2.38 formulated in the previous section.
- The following are selected publications on affine processes and affine models: [DPS00, DFS03, FM09, CFT10, K-R11, K-RST11].

Chapter 3
Realized Volatility and Mixing Distributions

Realized volatility, or historical volatility, is obtained by averaging the volatility process over time intervals. Historical volatility is not an instantaneous feature of the stock price, since it depends on the behavior of the volatility in the past. Note that averages similar to the realized volatility play an important role in the theory of Asian style options. A marginal distribution of the realized volatility process in an uncorrelated stochastic volatility model is called a mixing distribution. Such distributions play the role of mixing factors in representations of stock price densities in terms of Black–Scholes densities. For a correlated stochastic volatility model, mixing distributions are higher-dimensional. They coincide with joint distributions of various combinations of the integrated volatility, the integrated variance, and the volatility.

3.1 Asymptotic Relations Between Functions

Asymptotic formulas play an important role in this book. The following definitions introduce certain asymptotic relations between positive functions.

1. Let φ_1 and φ_2 be positive functions defined on the interval (a,∞). If there exist $\alpha_1 > 0$, $\alpha_2 > 0$, and $y_0 > 0$ such that

$$\alpha_1\varphi_1(y) \le \varphi_2(y) \le \alpha_2\varphi_1(y)$$

   for all $y > y_0$, then we write $\varphi_1(y) \approx \varphi_2(y)$ as $y \to \infty$.
2. If the condition

$$\lim_{y\to\infty}\left[\varphi_2(y)\right]^{-1}\varphi_1(y) = 1$$

   holds, then we write $\varphi_1(y) \sim \varphi_2(y)$ as $y \to \infty$.
3. Let ρ be a positive function on $(0,\infty)$. We use the notation

$$\varphi_1(y) = \varphi_2(y) + O\big(\rho(y)\big) \quad \text{as } y \to \infty,$$

A. Gulisashvili, *Analytically Tractable Stochastic Stock Price Models*,
Springer Finance, DOI 10.1007/978-3-642-31214-4_3,
© Springer-Verlag Berlin Heidelberg 2012

if there exist $\alpha > 0$ and $y_0 > 0$ such that

$$\left|\varphi_1(y) - \varphi_2(y)\right| \le \alpha\rho(y)$$

for all $y > y_0$.

Similar relations can be introduced in the case where $y \downarrow 0$ for positive functions defined on the interval $(0, b)$.

The following statement holds.

Lemma 3.1

1. *Let $a \ge 0$, and let f be a positive function on (a, ∞). Suppose*

$$f(x) = O\big(\psi(x)\big) \quad \text{as } x \to \infty,$$

for any positive increasing function ψ on (a, ∞) satisfying $\psi(x) \to \infty$ as $x \to \infty$. Then $f(x) = O(1)$ as $x \to \infty$.

2. *Let $b > 0$, and let g be a positive function on $(0, b)$. Suppose*

$$g(x) = O\big(\tau(x)\big) \quad \text{as } x \to 0,$$

for any positive decreasing function τ on $(0, b)$ satisfying $\tau(x) \to \infty$ as $x \to 0$. Then $g(x) = O(1)$ as $x \to 0$.

The proof of Lemma 3.1 is left as an exercise for the reader.

3.2 Mixing Distributions and Stock Price Distributions

For an uncorrelated stochastic volatility model described by (2.3), the realized volatility process α is defined by

$$\alpha_t = \left\{\frac{1}{t}\int_0^t f(Y_s)^2\,ds\right\}^{\frac{1}{2}}, \quad t \ge 0.$$

The distribution μ_t of the realized volatility α_t is called the mixing distribution. If μ_t admits a density m_t, then m_t is called the mixing distribution density. Let us denote by ρ_t the distribution of the stock price X_t. If ρ_t admits a density D_t, then D_t is called the stock price distribution density. Note that the stock price distribution depends on the initial conditions x_0 and y_0, while the mixing distribution depends on the initial condition y_0. The next lemma shows that for every uncorrelated stochastic volatility model the stock price density exists, and this density can be represented as a log-normal transformation of the mixing distribution.

Lemma 3.2 *Consider an uncorrelated stochastic volatility model defined by* (2.3). *Then for every $t > 0$ there exists the stock price density D_t. Moreover, the following equality holds*:

$$D_t(x) = \frac{1}{x_0 e^{\mu t}} \int_0^{\infty} L\left(t, y, \frac{x}{x_0 e^{\mu t}}\right) d\mu_t(y), \tag{3.1}$$

where L is the log-normal density defined by

$$L(t, y, v) = \frac{1}{\sqrt{2\pi t} y v} \exp\left\{-\left(2ty^2\right)^{-1}\left(\log v + \frac{ty^2}{2}\right)^2\right\}.$$

Proof There exists a weak solution (X, Y, W, Z) to the system in (2.3) such that standard Brownian motions W and Z are defined on the Wiener spaces $(\Omega_1, \mathcal{F}_1, \mathbb{P}^{(1)})$ and $(\Omega_2, \mathcal{F}_2, \mathbb{P}^{(2)})$, respectively. The process Y is defined on the space $(\Omega_2, \mathcal{F}_2, \mathbb{P}^{(2)})$, while the stock price process X is defined on the product space $(\Omega, \mathcal{F}, \mathbb{P})$ of the spaces $(\Omega_i, \mathcal{F}_i, \mathbb{P}^{(i)})$, $i = 1, 2$. Moreover, the process X is given by the exponential formula (2.10). The independence of the processes W and Z plays an important role in the reasoning above.

It is not hard to see that (2.10) implies the following:

$$\begin{aligned}
&\mathbb{P}(X_t \le \lambda) \\
&\quad = \mathbb{E}^{(2)}\left[\mathbb{P}^{(1)}\left(\int_0^t f(Y_s)(\omega_2)\, dW_s(w_1) \le \log \frac{\lambda}{x_0 e^{\mu t}} + \frac{t(\alpha_t(\omega_2))^2}{2}\right)\right].
\end{aligned}$$

Next, using elementary properties of stochastic integrals, we obtain

$$\begin{aligned}
\mathbb{P}(X_t \le \lambda) &= \mathbb{E}^{(2)}\left[\int_{-\infty}^{\eta_t(\alpha_t(\omega_2))} \frac{1}{\sqrt{2\pi t}\alpha_t(\omega_2)} \exp\left\{-\frac{z^2}{2t(\alpha_t(\omega_2))^2}\right\} dz\right] \\
&= \int_0^{\infty} d\mu_t(y) \int_0^{\eta_t(y)} \frac{1}{\sqrt{2\pi t} y} \exp\left\{-\frac{z^2}{2ty^2}\right\} dz,
\end{aligned} \tag{3.2}$$

where

$$\eta_t(\sigma) = \log \frac{\lambda}{x_0 e^{\mu t}} + \frac{t\sigma^2}{2}, \quad \sigma > 0.$$

Making the substitution

$$z = \log \frac{x}{x_0 e^{\mu t}} + \frac{ty^2}{2}$$

in formula (3.2), we get

$$\mathbb{P}(X_t \le \lambda) = \int_0^{\infty} d\mu_t(y) \int_0^{\lambda} \frac{1}{\sqrt{2\pi t} y} \exp\left\{-\frac{(\log \frac{x}{x_0 e^{\mu t}} + \frac{ty^2}{2})^2}{2ty^2}\right\} \frac{dx}{x}. \tag{3.3}$$

Now it is clear that Lemma 3.2 follows from (3.3). □

The next statement is a generalization of Lemma 3.2. Let us denote by ζ_t the joint distribution of the random variable Y_t and the stock price X_t in an uncorrelated stochastic volatility model, and by τ_t the joint distribution of Y_t and the realized volatility α_t.

Lemma 3.3 *Suppose the distribution τ_t has a density $m_t^{(2)}$. Then the distribution ζ_t also has a density $D_t^{(2)}$ given by*

$$D_t^{(2)}(x, y) = \frac{1}{x_0 e^{\mu t}} \int_0^\infty L\left(t, z, \frac{x}{x_0 e^{\mu t}}\right) m_t^{(2)}(z, y)\, dz.$$

The role of the mixing density in Lemma 3.3 is played by the density $m_t^{(2)}$. The proof of Lemma 3.3 is similar to that of Lemma 3.2, and we leave it as an exercise for the interested reader. Note that formula (3.1) in Lemma 3.2 can be obtained from Lemma 3.3 by integrating out the variable y.

3.3 Stock Price Densities in Uncorrelated Models as Mixtures of Black–Scholes Densities

Let us recall that the stock price distribution density $\rho_t^{(\sigma)}$ in the Black–Scholes model with the volatility parameter σ and with the initial condition x_0 for the stock price process is given by the formula in (2.2). It follows from (3.1) and (2.2) that

$$D_t(x) = \int_0^\infty \rho_t^{(\sigma)}(x)\, d\mu_t(\sigma) \tag{3.4}$$

for all $x > 0$. Formulas similar to (3.4) also hold for many other quantities associated with uncorrelated stochastic volatility models. One can say that, in a sense, any uncorrelated stochastic volatility model is a mixture of Black–Scholes models.

Equality (3.1) can be rewritten as follows:

$$D_t\left(x_0 e^{\mu t} x\right) = \frac{1}{x_0 e^{\mu t} \sqrt{2\pi t}} x^{-\frac{3}{2}} \int_0^\infty y^{-1} \exp\left\{-\left[\frac{\log^2 x}{2ty^2} + \frac{ty^2}{8}\right]\right\} d\mu_t(y). \tag{3.5}$$

The following symmetry property of the stock price distribution density D_t in an uncorrelated stochastic volatility model can be established using formula (3.5):

$$D_t\left(x_0 e^{\mu t} x\right) = x^{-3} D_t\left(x_0 e^{\mu t} \frac{1}{x}\right) \tag{3.6}$$

for all $x > 0$. This property allows one to study the behavior of the stock price density D_t near zero, knowing its behavior near infinity.

Formula (3.1) explains how the stock price density in an uncorrelated stochastic volatility model depends on the distribution of the integrated variance. However, if the stock price and the volatility are correlated, then higher-dimensional joint distributions appear in representation formulas for the stock price density. For instance, various joint distributions of the integrated volatility, the integrated variance, and the volatility are used in such representation formulas. We call these joint distributions mixing distributions by analogy with the case of zero correlation. In the next sections, we provide several results concerning higher-dimensional mixing distributions arising in classical stochastic volatility models.

3.4 Mixing Distributions and Heston Models

Let us consider the correlated Heston model defined by

$$\begin{cases} dX_t = \mu X_t\,dt + \sqrt{Y_t}X_t\left(\sqrt{1-\rho^2}\,d\widetilde{W}_t + \rho\,dZ_t\right), \\ dY_t = (a - bY_t)\,dt + c\sqrt{Y_t}\,dZ_t, \end{cases} \tag{3.7}$$

where $\mu \in \mathbb{R}$, $a \geq 0$, $b \geq 0$, $c > 0$, and $\rho \in (-1, 1)$ (see Sect. 2.3). Then the exponential formula for the stock price process gives

$$X_t = x_0 \exp\left\{\mu t - \frac{1}{2}\int_0^t Y_s\,ds + \int_0^t \sqrt{Y_s}\,dW_s\right\} \tag{3.8}$$

where

$$W_t = \sqrt{1-\rho^2}\int_0^t \sqrt{Y_s}\,d\widetilde{W}_s + \rho\int_0^t \sqrt{Y_s}\,dZ_s$$

(see (2.10)). Our next goal is to get rid of the stochastic integral $\int_0^t \sqrt{Y_s}\,dZ_s$ in formula (3.8). To achieve this goal, we rewrite the second equation in (3.7) in the integrated form and solve for the above-mentioned integral. As a consequence, we get the following:

$$\int_0^t \sqrt{Y_s}\,dZ_s = \frac{1}{c}\left(Y_t - y_0 - at + b\int_0^t Y_s\,ds\right).$$

Therefore,

$$\begin{aligned} X_t &= x_0 \exp\left\{\left(\mu - \frac{\rho a}{c}\right)t - \frac{\rho y_0}{c}\right\} \\ &\quad \times \exp\left\{\frac{\rho}{c}Y_t + \left(\frac{\rho b}{c} - \frac{1}{2}\right)\int_0^t Y_s\,ds + \sqrt{1-\rho^2}\int_0^t \sqrt{Y_s}\,d\widetilde{W}_t\right\}. \end{aligned} \tag{3.9}$$

Let us denote by $\mu_t^{(2)}$ the joint distribution of the integrated variance $\beta_t = \int_0^t Y_s\,ds$ and the variance Y_t. We will next reason as in the proof of Lemma 3.2.

It follows from (3.9) that

$$\mathbb{P}(X_t \le \lambda) = \mathbb{E}^{(2)}\left[\mathbb{P}^{(1)}\left(\int_0^t \sqrt{Y}_s \, d\widetilde{W}_s \le \frac{A_t(\lambda, \beta_t, Y_t)}{\sqrt{1-\rho^2}}\right)\right],$$

where

$$A_t(\lambda, y, z) = \log \frac{\lambda}{x_0 e^{\mu t}} + \frac{\rho y_0 + \rho a t - \rho z}{c} - \left(\frac{\rho b}{c} - \frac{1}{2}\right) y. \tag{3.10}$$

Therefore,

$$\begin{aligned}
\mathbb{P}(X_t \le \lambda) &= \mathbb{E}^{(2)}\left[\int_{-\infty}^{\frac{1}{\sqrt{1-\rho^2}} A_t(\lambda, \beta_t, Y_t)} \frac{1}{\sqrt{2\pi \beta_t}} \exp\left\{-\frac{w^2}{2\beta_t}\right\} dw\right] \\
&= \int_{[0,\infty)^2} d\mu_t^{(2)}(y, z) \\
&\quad \times \int_{-\infty}^{\frac{1}{\sqrt{1-\rho^2}} A_t(\lambda, y, z)} \frac{1}{\sqrt{2\pi y}} \exp\left\{-\frac{w^2}{2y}\right\} dw.
\end{aligned} \tag{3.11}$$

Making the substitution $w = \frac{1}{\sqrt{1-\rho^2}} A_t(u, y, z)$ in (3.11), we obtain

$$\begin{aligned}
\mathbb{P}(X_t \le \lambda) &= \frac{1}{\sqrt{2\pi(1-\rho^2)}} \int_{[0,\infty)^2} d\mu_t^{(2)}(y, z) \\
&\quad \times \int_0^\lambda \frac{1}{\sqrt{y}} \exp\left\{-\frac{A_t(u, y, z)^2}{2(1-\rho^2) y}\right\} \frac{du}{u}.
\end{aligned}$$

Hence, the following assertion holds for the stock price distribution density D_t in the Heston model.

Lemma 3.4 *For any $t > 0$ and $x > 0$,*

$$D_t(x) = \frac{1}{\sqrt{2\pi(1-\rho^2)} x} \int_{[0,\infty)^2} \frac{1}{\sqrt{y}} \exp\left\{-\frac{A_t(x, y, z)^2}{2(1-\rho^2) y}\right\} d\mu_t^{(2)}(y, z), \tag{3.12}$$

where the function A_t is defined by (3.10).

Lemma 3.4 provides a representation of the stock price density D_t in the Heston model in the form of a certain integral transformation of the mixing density $\mu_t^{(2)}$.

3.5 Mixing Distributions and Hull–White Models with Driftless Volatility

In this section, we consider the correlated Hull–White model with driftless volatility. For such a model we will find a formula for the stock price density, similar to that in (3.12).

Consider the Hull–White model of the following form:

$$\begin{cases} dX_t = \mu X_t\,dt + Y_t X_t\left(\sqrt{1-\rho^2}\,d\widetilde{W}_t + \rho\,dZ_t\right), \\ dY_t = \xi Y_t\,dZ_t, \end{cases} \tag{3.13}$$

where $\mu \in \mathbb{R}$, $\xi > 0$, and $-1 < \rho < 1$. In (3.13), $(\widetilde{W}, Z)$ is a two-dimensional standard Brownian motion. Reasoning as in the proof of formula (3.9), we obtain

$$X_t = x_0 \exp\left\{\mu t - \frac{1}{2}\int_0^t Y_s^2\,ds + \sqrt{1-\rho^2}\int_0^t Y_s\,d\widetilde{W}_s + \frac{\rho}{\xi}(Y_t - y_0)\right\}.$$

As before, denote by D_t the distribution density of the stock price X_t, and let $\mu_t^{(2)}$ be the joint distribution of the integrated variance $\beta_t = \int_0^t Y_s^2\,ds$ and the volatility Y_t. The next assertion provides an explicit formula for the stock price distribution density D_t in the correlated Hull–White model with driftless volatility in terms of the two-dimensional mixing distribution $\mu_t^{(2)}$.

Lemma 3.5 *For all $t > 0$ and $x > 0$,*

$$D_t(x) = \frac{1}{\sqrt{2\pi(1-\rho^2)}\,x}\int_{[0,\infty)^2} \frac{1}{\sqrt{y}} \exp\left\{-\frac{A_t(x,y,z)^2}{2(1-\rho^2)y}\right\} d\mu_t^{(2)}(y,z) \tag{3.14}$$

where

$$A_t(x,y,z) = \log\frac{x}{x_0 e^{\mu t}} + \frac{1}{2}y - \frac{\rho}{\xi}(z - y_0).$$

The proof of Lemma 3.5 is similar to that of Lemma 3.4, and we omit it.

3.6 Mixing Distributions and Hull–White Models

In this section, we turn our attention to the general Hull–White model given by

$$\begin{cases} dX_t = \mu X_t\,dt + Y_t X_t\left(\sqrt{1-\rho^2}\,d\widetilde{W}_t + \rho\,dZ_t\right), \\ dY_t = \nu Y_t\,dt + \xi Y_t\,dZ_t, \end{cases} \tag{3.15}$$

where $\mu \in \mathbb{R}$, $\nu \in \mathbb{R}$, $\xi > 0$, and $-1 < \rho < 1$. In (3.15), $(\widetilde{W}, Z)$ is a two-dimensional standard Brownian motion. The mixing distribution $\mu_t^{(3)}$ in the Hull–White model described by (3.15) is three-dimensional. It coincides with the joint distribution of the integrated variance $\beta_t = \int_0^t Y_s^2\, ds$, the integrated volatility $\gamma_t = \int_0^t Y_s\, ds$, and the volatility Y_t. The next assertion provides a representation formula for the stock price density D_t in the correlated Heston model.

Lemma 3.6 *For all $t > 0$ and $x > 0$,*

$$D_t(x) = \frac{1}{\sqrt{2\pi(1-\rho^2)}x} \int_{[0,\infty)^3} \frac{1}{\sqrt{y}} \exp\left\{ -\frac{A_t(x,y,z,w)^2}{2(1-\rho^2)y} \right\} d\mu_t^{(3)}(y,z,w) \tag{3.16}$$

where

$$A_t(x,y,z,w) = \log \frac{x}{x_0 e^{\mu t}} + \frac{1}{2} y - \frac{\rho}{\xi}(z - y_0 - \nu w).$$

The proof of Lemma 3.6 is similar to that of Lemma 3.4. Here we take into account the exponential formula for the stock price (formula (3.8)) and the equality

$$\int_0^t Y_s\, dZ_s = \frac{1}{\xi}\left(Y_t - y_0 - \nu \int_0^t Y_s\, ds \right).$$

3.7 Mixing Distributions and Stein–Stein Models

The methods developed in the previous sections also apply to the correlated Stein–Stein model given by

$$\begin{cases} dX_t = \mu X_t\, dt + Y_t X_t \left(\sqrt{1-\rho^2}\, d\widetilde{W}_t + \rho\, dZ_t\right), \\ dY_t = q(m - Y_t)\, dt + \sigma\, dZ_t, \end{cases} \tag{3.17}$$

where $\mu \in \mathbb{R}$, $q \geq 0$, $m \geq 0$, and $\rho \in (-1, 1)$. However, in this case we cannot simplify the stochastic integral $\int_0^t Y_s\, dZ_s$ by using the second equation in (3.17). In order to overcome this difficulty we introduce a new stochastic variable R defined by $R_t = Y_t^2$, $t \geq 0$. The dynamics of R are described by

$$dR_t = \left(\sigma^2 + 2qmY_t - 2qR_t\right) dt + 2\sigma Y_t\, dZ_t \tag{3.18}$$

(use the Itô formula). Combining (3.17) and (3.18), we get a representation of the Stein–Stein model as a three-dimensional affine model.

It follows from (3.18) that

$$\int_0^t Y_s\, dZ_s = \frac{1}{2\sigma}\left(Y_t^2 - y_0^2 - \sigma^2 t - 2qm \int_0^t Y_s\, ds + 2q \int_0^t Y_s^2\, ds \right).$$

Denote by $\mu_t^{(3)}$ the distribution of the integrated variance $\int_0^t Y_s^2\,ds$, the variance Y_t^2, and the integrated volatility $\int_0^t Y_s\,ds$. The measure $\mu_t^{(3)}$ is the mixing distribution in the Stein–Stein model.

The next lemma provides an integral representation of the stock price density in the Stein–Stein model.

Lemma 3.7 *The following formula holds for the stock price density in the Stein–Stein model*:

$$D_t(x) = \frac{1}{\sqrt{2\pi(1-\rho^2)}x}\int_{[0,\infty)^2\times\mathbb{R}} \frac{1}{\sqrt{y}}\exp\left\{-\frac{A_t(x,y,z,w)^2}{2(1-\rho^2)y}\right\}d\mu_t^{(3)}(y,z,w), \tag{3.19}$$

where

$$A_t(x,y,z,w) = \log\frac{x}{x_0e^{\mu t}} + \frac{\rho\sigma}{2}t + \frac{\rho y_0^2}{2\sigma} - \frac{\rho}{2\sigma}z + \left(\frac{1}{2} - \frac{\rho q}{\sigma}\right)y + \frac{\rho q m}{\sigma}w. \tag{3.20}$$

One can establish Lemma 3.7 by reasoning as in the proofs of similar lemmas in the previous sections.

3.8 Notes and References

- Marginal distribution densities of stock price processes in stochastic volatility models can be often represented as integral transforms of special distributions, which are called mixing distributions. For an uncorrelated stochastic volatility model, the integral transform has a log-normal kernel, and the mixing distribution is the distribution of the realized volatility. In the presence of correlation between the stock price and the volatility, mixing distributions may be higher-dimensional. They are usually joint distributions of various combinations of the variance, the integrated volatility, and the integrated variance.
- Lemma 3.1 was brought to our attention by R. Lee.

Chapter 4
Integral Transforms of Distribution Densities

Analytical tractability is a desirable property of a stochastic stock price model. Informally speaking, a stochastic model has this property if explicit or asymptotic formulas are available for certain important functions associated with the model. Note that the Hull–White, Stein–Stein, and Heston models belong to the class of analytically tractable models. To illustrate the previous statement we find in this chapter explicit formulas for Laplace transforms of mixing densities and Mellin transforms of stock price densities in these models.

4.1 Geometric Brownian Motions and Laplace Transforms of Mixing Distributions

In this section, we compute the Laplace transform of the mixing distribution density m_t, associated with a special geometric Brownian motion Y satisfying the equation

$$dY_t = \frac{1}{2} Y_t \, dt + Y_t \, dZ_t$$

with $Y_0 = y_0$ $\mathbb{P}$-a.s. Note that the process Y is the volatility process in the Hull–White model with $\nu = \frac{1}{2}$ and $\xi = 1$. Note that $Y_t = y_0 \exp\{Z_t\}$ for all $t \geq 0$.

The next assertion provides an explicit formula for the mathematical expectation of the exponential functional of the process Y. This assertion was established in [AG97], see also [GS10a].

Theorem 4.1 *The following equality holds for the geometric Brownian motion with* $\nu = \frac{1}{2}$ *and* $\xi = 1$:

$$\mathbb{E}\left[\exp\left\{-\lambda \int_0^t Y_s^2 \, ds\right\}\right] = \frac{1}{\sqrt{2\pi t}} \int_{-\infty}^{\infty} \exp\{i\sqrt{2\lambda} y_0 \sinh y\} e^{-\frac{y^2}{2t}} \, dy \tag{4.1}$$

for all $\lambda > 0$ *and* $t > 0$.

A. Gulisashvili, *Analytically Tractable Stochastic Stock Price Models*,
Springer Finance, DOI 10.1007/978-3-642-31214-4_4,
© Springer-Verlag Berlin Heidelberg 2012

Proof One of the main ideas used in the proof is to establish a relation between the solutions to the classical heat equation on the one hand and the solutions to the equation

$$\frac{\partial^2 w}{\partial y^2} = x^2 \frac{\partial^2 w}{\partial x^2} + x \frac{\partial w}{\partial x} - x^2 w \tag{4.2}$$

on the other hand. We will use this idea in a special case.

Lemma 4.2 *Let the function w be defined on $\mathbb{R}^2$ by*

$$w(x, y) = \exp\{ix \sinh y\}. \tag{4.3}$$

Then the function

$$u(x,t) = \frac{1}{2\sqrt{\pi t}} \int_{-\infty}^{\infty} w(x, y) e^{-\frac{y^2}{4t}}\, dy, \tag{4.4}$$

where $x \in \mathbb{R}$ and $t > 0$, solves the following initial value problem

$$\begin{cases} \dfrac{\partial u}{\partial t} = x^2 \dfrac{\partial^2 u}{\partial x^2} + x \dfrac{\partial u}{\partial x} - x^2 u, \\ u(0) = 1. \end{cases} \tag{4.5}$$

Proof of Lemma 4.2 It is not hard to check that the function w defined by (4.3) is a solution to the equation in (4.2). Let u be defined by (4.4). It is well known that the function

$$s(y,t) = \frac{1}{2\sqrt{\pi t}} e^{-\frac{y^2}{4t}}, \quad y \in \mathbb{R},\ t > 0,$$

is a solution to the one-dimensional heat equation

$$\frac{\partial s(y,t)}{\partial t} = \frac{\partial^2 s(y,t)}{\partial y^2}$$

with $s(0) = \delta_0$. Therefore, the function u satisfies the initial condition in (4.5). We have

$$\begin{aligned}
\frac{\partial u(x,t)}{\partial t} &= \int_{-\infty}^{\infty} w(x,y) \frac{\partial s(y,t)}{\partial t}\, dy \\
&= \int_{-\infty}^{\infty} w(x,y) \frac{\partial^2 s(y,t)}{\partial y^2}\, dy = \int_{-\infty}^{\infty} \frac{\partial^2 w(x,y)}{\partial y^2} s(y,t)\, dy \\
&= \int_{-\infty}^{\infty} \left(x^2 \frac{\partial^2 w(x,y)}{\partial x^2} + x \frac{\partial w(x,y)}{\partial x} - x^2 w(x,y) \right) s(y,t)\, dy \\
&= x^2 \frac{\partial^2 u(x,t)}{\partial x^2} + x \frac{\partial u(x,t)}{\partial x} - x^2 u(x,t).
\end{aligned}$$

It follows that the function u satisfies (4.5).

This completes the proof of Lemma 4.2. □

It is not hard to see using Lemma 4.2 that for every $\lambda > 0$ the function

$$v_\lambda(x,t) = u\left(\sqrt{2\lambda}x, \frac{t}{2}\right),$$

where u is defined by (4.4), is a classical solution to the following initial value problem:

$$\begin{cases} \dfrac{\partial v_\lambda}{\partial t} = \dfrac{1}{2}x^2\dfrac{\partial^2 v_\lambda}{\partial x^2} + \dfrac{1}{2}x\dfrac{\partial v_\lambda}{\partial x} - \lambda x^2 v_\lambda, \\ v_\lambda(0) = 1. \end{cases} \tag{4.6}$$

Therefore, the uniqueness theorem for diffusion equations implies that the function v_λ coincides with the solution to (4.6) given by the Feynman–Kac formula, that is,

$$\begin{aligned} &\mathbb{E}\left[\exp\left\{-\lambda y_0^2 \int_0^t \exp\{2Z_s\}\,ds\right\}\right] \\ &\quad = \frac{1}{\sqrt{2\pi t}} \int_{-\infty}^{\infty} \exp\{i\sqrt{2\lambda}y_0 \sinh y\} e^{-\frac{y^2}{2t}}\,dy \end{aligned} \tag{4.7}$$

for all y_0. This establishes (4.1). □

Remark 4.3 The reader is referred to [Jef96, Øks03, RY04, GvC06] for more information on the Feynman–Kac formula.

Formula (4.7) is equivalent to the following formula:

$$\begin{aligned} &\mathbb{E}\left[\exp\left\{-\frac{x^2}{2} \int_0^t \exp\{2Z_s\}\,ds\right\}\right] \\ &\quad = \frac{1}{\sqrt{2\pi t}} \int_{-\infty}^{\infty} \exp\{ix \sinh y\} e^{-\frac{y^2}{2t}}\,dy, \end{aligned} \tag{4.8}$$

for all $x \in \mathbb{R}$.

Remark 4.4 The uniqueness theorem mentioned above can be found, e.g., in [Øks03], Theorem 8.2.1. Note that in the book by Øksendal the uniqueness theorem is proved for initial conditions from the space C_0^2 of twice differentiable functions vanishing at infinity. However, the theorem also holds in the case where the initial condition is a bounded continuous function. The proof of this fact is exactly the same as that in [Øks03].

4.2 Bougerol's Identity in Law

Consider the following family of random variables:

$$A_t^\nu = \int_0^t \exp\{2(\nu s + Z_s)\}\, ds, \quad t \geq 0,$$

where $\nu \in \mathbb{R}$ and Z is a standard Brownian motion. If $\nu = 0$, then we will write A_t instead of A_t^0. In [Bou83], the following formula was obtained for the exponential functional A_t:

$$\sinh(Z_t) = \gamma_{A_t} \quad \text{in law}, \tag{4.9}$$

where γ is a standard Brownian motion independent of Z. Bougerol's formula states that the random process $\sinh(Z)$ is a time-changed Brownian motion with the stochastic clock characterized by the process A. It is interesting that Bougerol's identity (4.9) is equivalent to formula (4.8) (this will be shown below). Hence, formula (4.8) can be considered as an analytical counterpart of Bougerol's identity.

Proof of Bougerol's identity The characteristic function of the random variable on the right-hand side of (4.9) is given by

$$\mathbb{E}\big[\exp\{ix \sinh(Z_t)\}\big] = \frac{1}{\sqrt{2\pi t}} \int \exp\{ix \sinh y\} \exp\left\{-\frac{y^2}{2t}\right\} dy. \tag{4.10}$$

By (4.8) and (4.10), we have

$$\mathbb{E}\big[\exp\{ix \sinh(Z_t)\}\big] = \mathbb{E}\left[\exp\left\{-\frac{x^2}{2} A_t\right\}\right] = \int \exp\left\{-\frac{x^2}{2} y\right\} \tilde{m}_t(y)\, dy, \tag{4.11}$$

where $\tilde{m}_t$ is the distribution density of A_t.

It is known that the following formula holds for the Fourier transform of the Gaussian density:

$$\int \exp\{-\alpha \xi^2\} \exp\{ix\xi\}\, d\xi = \sqrt{\frac{\pi}{\alpha}} \exp\left\{-\frac{x^2}{4\alpha}\right\}.$$

Put $\alpha = \frac{1}{2y}$ in the previous formula. Then we get

$$\exp\left\{-\frac{x^2}{2} y\right\} = \frac{1}{\sqrt{2\pi y}} \int \exp\left\{-\frac{\xi^2}{2y}\right\} \exp\{ix\xi\}\, d\xi.$$

It follows from (4.11) that

$$\mathbb{E}\big[\exp\{ix \sinh(Z_t)\}\big] = \int \exp\{ix\xi\}\, d\xi \int \frac{1}{\sqrt{2\pi y}} \exp\left\{-\frac{\xi^2}{2y}\right\} \tilde{m}_t(y)\, dy \tag{4.12}$$

for all $x \in \mathbb{R}$. Now it is not hard to see, using the independence of γ and Z, that the distribution density n_t of γ_{A_t} is given by

$$n_t(\xi) = \int \frac{1}{\sqrt{2\pi y}} \exp\left\{-\frac{\xi^2}{2y}\right\} \tilde{m}_t(y)\,dy.$$

It follows from (4.12) that

$$\mathbb{E}\left[\exp\left\{ix \sinh(Z_t)\right\}\right] = \mathbb{E}\left[\exp\{ix\gamma_{A_t}\}\right]$$

for all $x \in \mathbb{R}$, which establishes Bougerol's formula. It is also clear from the proof above that formulas (4.9) and (4.8) are equivalent. □

4.3 Squared Bessel Processes and Laplace Transforms of Mixing Distributions

In this section, we turn our attention to exponential functionals of squared Bessel processes.

Theorem 4.5 *Let $\lambda \in \mathbb{R}$, $t > 0$, $\delta \geq 0$, and $x \geq 0$. Then*

$$\mathbb{E}\left[\exp\left\{-\frac{\lambda^2}{2}\int_0^t BESQ_x^\delta(u)\,du\right\}\right] = \left[\cosh(\lambda t)\right]^{-\frac{\delta}{2}} \exp\left\{-\frac{x\lambda}{2}\tanh(\lambda t)\right\}. \tag{4.13}$$

Since $BESQ_0^1 = W^2$, Theorem 4.5 contains a celebrated Cameron–Martin formula as a special case.

Corollary 4.6 (Cameron–Martin Formula) *Let W be a standard Brownian motion. Then for all $\lambda > 0$ and $t > 0$,*

$$\mathbb{E}\left[\exp\left\{-\frac{\lambda^2}{2}\int_0^t W_u^2\,du\right\}\right] = \left[\cosh(\lambda t)\right]^{-\frac{1}{2}}.$$

Proof of Theorem 4.5 We will first solve a certain Sturm–Liouville problem. Fix $t > 0$ and $\lambda > 0$. Our goal is to find a function φ on the interval $[0, t]$ such that

1. $\varphi(0) = 1$,
2. φ is twice continuously differentiable on $[0, t]$,
3. φ is nonincreasing on $[0, t]$,
4. $\varphi'(t) = 0$,

5. φ satisfies the equation

$$\varphi''(u) = \lambda^2 \varphi(u) \tag{4.14}$$

for all $u \in [0, t]$.

The next lemma provides an explicit formula for the solution to the Sturm–Liouville problem formulated above.

Lemma 4.7 *Define a function by*

$$\varphi(u) = \cosh(\lambda u) - \tanh(\lambda t)\sinh(\lambda u), \quad u \in [0, t]. \tag{4.15}$$

Then the function φ satisfies conditions 1–5 *listed above.*

Proof It is easy to see that the functions $u \mapsto \sinh(\lambda u)$ and $u \mapsto \cosh(\lambda u)$ satisfy the equation in (4.14). Hence, the function defined by $\varphi(u) = c_1 \cosh(\lambda u) + c_2 \sinh(\lambda u)$ with $c_1, c_2 \in \mathbb{R}$ also satisfies (4.14). It remains to choose the constants c_1 and c_2 so that conditions 1–4 hold for the resulting function.

It is clear that condition 1 implies $c_1 = 1$. On the other hand, condition 4 implies $\lambda \sinh(\lambda t) + \lambda c_2 \cosh(\lambda t) = 0$. Therefore, $c_2 = -\tanh(\lambda t)$, and hence

$$\varphi(u) = \cosh(\lambda u) - \tanh(\lambda t)\sinh(\lambda u). \tag{4.16}$$

It follows from (4.16) that

$$\begin{aligned}\varphi'(u) &= \lambda \sinh(\lambda u) - \lambda \tanh(\lambda t)\cosh(\lambda u)\\ &= \lambda \cosh(\lambda u)\big(\tanh(\lambda u) - \tanh(\lambda t)\big).\end{aligned}$$

Since the function $y \mapsto \tanh y$ increases on $[0, \infty)$, we have $\varphi'(u) \le 0$. Therefore, the function φ is nonincreasing.

This completes the proof of Lemma 4.7. □

Let us continue the proof of Theorem 4.5. Put $X = BESQ_x^\delta$ and let φ be the function given by (4.15). Define

$$F(u) = \frac{\varphi'(u)}{\varphi(u)}. \tag{4.17}$$

Then the following equality holds:

$$F(0) = -\lambda \tanh(\lambda t) \quad \text{and} \quad F(t) = 0. \tag{4.18}$$

Since $\varphi'(u) \le 0$ and $\varphi(u) \ge 0$, the function $F(u)$ is negative on the interval $[0, t]$. Moreover,

$$F'(u) = \frac{\varphi''(u)}{\varphi(u)} - \frac{\varphi'(u)}{\varphi(u)^2} = \lambda^2 - \frac{\varphi'(u)^2}{\varphi(u)^2} = \lambda^2 - F(u)^2. \tag{4.19}$$

Using (4.18), (4.19), and the integration by parts formula, we see that for all $0 \le s \le t$,

$$F(s)X_s = -\lambda \tanh(\lambda t)x + \int_0^s F(u)\,dX_u + \int_0^s X_u dF(u)$$
$$= -\lambda \tanh(\lambda t)x + \int_0^s F(u)\,dX_u + \lambda^2 \int_0^s X_u\,du$$
$$- \int_0^s X_u F(u)^2\,du. \tag{4.20}$$

It follows from the equality $d(X_s - \delta s) = 2\sqrt{X_s}\,dZ_s$ and from Remark 2.5 that the process

$$M_s = X_s - \delta s, \quad 0 \le s \le t,$$

is a continuous martingale. Moreover, the process

$$s \mapsto \frac{1}{2}\int_0^s F(u)\,dM_u = \int_0^s F(u)\sqrt{X_u}\,dZ_u, \quad 0 \le s \le t,$$

is also a continuous martingale. Its quadratic variation process is

$$s \mapsto \int_0^s X_u F(u)^2\,du, \quad 0 \le s \le t.$$

Hence, the corresponding exponential martingale has the following form:

$$\mathcal{E}_s = \exp\left\{\frac{1}{2}\int_0^s F(u)\,dM_u - \frac{1}{2}\int_0^s X_u F(u)^2\,du\right\}, \quad 0 \le s \le t.$$

Using (4.17), (4.20), and the equality $\varphi(0) = 1$, we see that

$$\frac{1}{2}\int_0^s F(u)\,dM_u - \frac{1}{2}\int_0^s X_u F(u)^2\,du$$
$$= \frac{1}{2}\int_0^s F(u)\,dX_u - \frac{\delta}{2}\int_0^s F(u)\,du - \frac{1}{2}\int_0^s X_u F(u)^2\,du$$
$$= \frac{1}{2}\int_0^s F(u)\,dX_u - \frac{\delta}{2}\log\varphi(s) - \frac{1}{2}\int_0^s X_u F(u)^2\,du$$
$$= \frac{1}{2}F(s)X_s + \frac{1}{2}\lambda \tanh(\lambda t)x - \frac{\delta}{2}\log\varphi(s) - \frac{\lambda^2}{2}\int_0^t X_s\,ds.$$

Therefore,

$$\mathcal{E}_s = \exp\left\{\frac{1}{2}F(s)X_s + \frac{1}{2}\lambda \tanh(\lambda t)x - \frac{\delta}{2}\log\varphi(s) - \frac{\lambda^2}{2}\int_0^s X_u\,du\right\}$$

for $0 \le s \le t$.

The function φ is positive, bounded, and decreasing. In addition, the function F is negative and the process X is a positive process. It follows that each random variable $\mathcal{E}_s$ with $0 \le s \le t$ is bounded, and hence the expectation of this random variable is finite. Since the process $\mathcal{E}_s$, $0 \le s \le t$, is a martingale, we have $\mathbb{E}[\mathcal{E}_t] = \mathbb{E}[\mathcal{E}_0]$. Next, using (4.18) and the equality $\varphi(0) = 1$, we obtain

$$\mathbb{E}\left[\exp\left\{\frac{1}{2}\big(F(t)X_t + \lambda\tanh(\lambda t)x - \delta\log\varphi(t)\big) - \frac{\lambda^2}{2}\int_0^t X_u\,du\right\}\right] = 1.$$

Since $F(t) = 0$, we see that

$$\mathbb{E}\left[\exp\left\{-\frac{\lambda^2}{2}\int_0^t X_u\,du\right\}\right] = \varphi(t)^{\frac{\delta}{2}}\exp\left\{-\frac{x\lambda}{2}\tanh(\lambda t)\right\}. \tag{4.21}$$

Finally, using (4.16), we get

$$\varphi(t) = \cosh(\lambda t) - \tanh(\lambda t)\sinh(\lambda t) = \frac{1}{\cosh(\lambda t)}. \tag{4.22}$$

Now, (4.21) and (4.22) give (4.13).

This completes the proof of Theorem 4.5. □

The next assertion is a generalization of Theorem 4.5.

Theorem 4.8 *Let τ be a continuous function on $[0, t]$. Then for all $x \ge 0$ and $\delta \ge 0$,*

$$\mathbb{E}\left[\exp\left\{-\frac{1}{2}\int_0^t BESQ_x^\delta(u)\tau(u)\,du\right\}\right] = \varphi(t)^{-\frac{\delta}{2}}\exp\left\{\frac{x}{2}\varphi'(0)\right\}.$$

The proof of Theorem 4.8 is similar to that of Theorem 4.5. We would only like to mention that the Sturm–Liouville problem appearing in this proof is as follows. Find a function φ on $[0, t]$ such that

1. $\varphi(0) = 1$,
2. φ is twice continuously differentiable on $[0, t]$,
3. φ is nonincreasing on $[0, t]$,
4. $\varphi'(t) = 0$,
5. the function φ satisfies the equation

$$\varphi''(u) = \tau(u)\varphi(u) \tag{4.23}$$

for all $u \in [0, t]$.

4.4 CIR Processes and Laplace Transforms of Mixing Distributions

Consider the CIR process Y satisfying the equation

$$dY_t = (a - bY_t)\,dt + c\sqrt{Y_t}\,dZ_t, \quad Y_0 = y_0 \ \mathbb{P}\text{-a.s.} \tag{4.24}$$

This process is related to a time-changed squared Bessel process (see Theorem 1.40). More precisely,

$$Y_t = e^{-bt} BESQ_{y_0}^{\frac{4a}{c^2}}\left(\frac{c^2}{4b}\left(e^{bt} - 1\right)\right), \quad t > 0. \tag{4.25}$$

Therefore, we can use Theorem 4.8 to find an explicit formula for the mathematical expectation of the exponential functional of a CIR process.

Theorem 4.9 *Let $b \neq 0$, and let Y be the CIR process determined from* (4.24). *Then for every $\eta > \frac{1}{2}$,*

$$\begin{aligned}
&\mathbb{E}\left[\exp\left\{-\frac{b^2(4\eta^2 - 1)}{2c^2}\int_0^t Y_s\,ds\right\}\right] \\
&= \exp\left\{\frac{abt}{c^2}\right\}\left(\frac{2\eta}{2\eta\cosh(bt\eta) + \sinh(bt\eta)}\right)^{\frac{2a}{c^2}} \\
&\quad\times \exp\left\{-\frac{by_0(4\eta^2 - 1)\sinh(bt\eta)}{2c^2\eta\cosh(bt\eta) + c^2\sinh(bt\eta)}\right\}.
\end{aligned} \tag{4.26}$$

Proof We will prove Theorem 4.9 for $b > 0$. The proof in the case $b < 0$ is similar. It follows from (4.25) that for every $\lambda > 0$,

$$\begin{aligned}
&\mathbb{E}\left[\exp\left\{-\frac{\lambda}{2}\int_0^t Y_s\,ds\right\}\right] \\
&= \mathbb{E}\left[\exp\left\{-\frac{\lambda}{2}\int_0^t e^{-bs} BESQ_{y_0}^{\frac{4a}{c^2}}\left(\frac{c^2}{4b}\left(e^{bs} - 1\right)\right)ds\right\}\right].
\end{aligned} \tag{4.27}$$

By the scaling property of Bessel processes,

$$BESQ_{y_0}^{\frac{4a}{c^2}}\left(\frac{c^2}{4b}\left(e^{bs} - 1\right)\right) = \frac{c^2}{4b} BESQ_{\frac{4by_0}{c^2}}^{\frac{4a}{c^2}}\left(e^{bs} - 1\right) \quad \text{in law.} \tag{4.28}$$

Next, using (4.27), (4.28), and making the substitution $u = e^{bs} - 1$, we obtain

$$\mathbb{E}\left[\exp\left\{-\frac{\lambda}{2}\int_0^t Y_s\,ds\right\}\right]$$

$$= \mathbb{E}\left[\exp\left\{-\frac{c^2\lambda}{8b^2}\int_0^{e^{bt}-1}\frac{1}{(1+u)^2}BESQ^{\frac{4a}{c^2}}_{\frac{4by_0}{c^2}}(u)\,du\right\}\right].$$

Therefore, for every $\lambda > 0$,

$$\mathbb{E}\left[\exp\left\{-\frac{2b^2\lambda}{c^2}\int_0^t Y_s\,ds\right\}\right]$$
$$= \mathbb{E}\left[\exp\left\{-\frac{\lambda}{2}\int_0^{e^{bt}-1}\frac{1}{(1+u)^2}BESQ^{\frac{4a}{c^2}}_{\frac{4by_0}{c^2}}(u)\,du\right\}\right]. \tag{4.29}$$

We will next apply Theorem 4.8 with $x = 4by_0c^{-2}$, $\delta = 4ac^{-2}$, and with the measure μ defined on the interval $[0, e^{bt} - 1]$ by

$$d\mu(u) = \frac{\lambda}{(1+u)^2}\,du.$$

Let us note first that equation (4.23) can be rewritten as follows:

$$\varphi''(u) = \varphi(u)\frac{\lambda}{(1+u)^2}, \quad 0 \le u \le e^{bt} - 1. \tag{4.30}$$

The function φ in (4.30) should satisfy the conditions

$$\varphi(0) = 1, \qquad \phi'\left(e^{bt} - 1\right) = 0, \tag{4.31}$$

and be monotonically decreasing. We will look for a solution to (4.30) having the following form:

$$\varphi(u) = \beta_1(1+u)^{\alpha} + \beta_2(1+u)^{1-\alpha}. \tag{4.32}$$

The constants α, β_1, and β_2 in (4.32) will be determined later.

Differentiating the function φ twice, we get

$$\varphi'(u) = \alpha\beta_1(1+u)^{\alpha-1} + (1-\alpha)\beta_2(1+u)^{-\alpha} \tag{4.33}$$

and

$$\varphi''(u) = \alpha(\alpha-1)\beta_1(1+u)^{\alpha-2} + \alpha(\alpha-1)\beta_2(1+u)^{-\alpha-1}. \tag{4.34}$$

It is not hard to see that the conditions in (4.31) and equation (4.30) are satisfied if

$$\beta_1 + \beta_2 = 1, \tag{4.35}$$

$$\alpha\beta_1 e^{bt(\alpha-1)} + (1-\alpha)\beta_2 e^{-bt\alpha} = 0, \tag{4.36}$$

and

$$\alpha(\alpha - 1) = \lambda \tag{4.37}$$

(we use (4.34) in the proof of (4.37)). Next, we choose α satisfying (4.37), i.e.

$$\alpha = \frac{1+\sqrt{1+4\lambda}}{2}.$$

The constants β_1 and β_2 can be computed using (4.35) and (4.36). This gives

$$\beta_1 = \frac{(\alpha-1)e^{-bt\alpha}}{\alpha e^{bt(\alpha-1)} + (\alpha-1)e^{-bt\alpha}} \tag{4.38}$$

and

$$\beta_2 = \frac{\alpha e^{bt(\alpha-1)}}{\alpha e^{bt(\alpha-1)} + (\alpha-1)e^{-bt\alpha}}. \tag{4.39}$$

We also need to compute the numbers $\varphi(e^{bt}-1)$ and $\varphi'(0)$. It follows from (4.32), (4.38), and (4.39) that

$$\varphi\left(e^{bt}-1\right) = \beta_1 e^{bt\alpha} + \beta_2 e^{bt(1-\alpha)} = \frac{2\alpha-1}{\alpha e^{bt(\alpha-1)} + (\alpha-1)e^{-bt\alpha}}. \tag{4.40}$$

Using (4.33), (4.37), (4.38), and (4.40), we get

$$\varphi'(0) = \alpha\beta_1 + (1-\alpha)\beta_2 = \frac{\lambda(-e^{bt(\alpha-1)} + e^{-bt\alpha})}{\alpha e^{bt(\alpha-1)} + (\alpha-1)e^{-bt\alpha}}. \tag{4.41}$$

It is not hard to prove that

$$\varphi'(u) < 0 \quad \text{if } 0 < u < e^{bt} - 1. \tag{4.42}$$

Indeed, using (4.33), (4.38), (4.39), and the fact that $\alpha > 1$, we see that the number $\varphi'(u)$ has the same sign as the expression

$$e^{-bt\alpha}(1+u)^{\alpha-1} - e^{bt(\alpha-1)}(1+u)^{-\alpha}. \tag{4.43}$$

Since the inequality

$$e^{-bt\alpha}(1+u)^{\alpha-1} < e^{bt(\alpha-1)}(1+u)^{-\alpha}$$

is equivalent to the inequality

$$e^{-bt(2\alpha-1)}(1+u)^{2\alpha-1} < 1,$$

and in addition $\alpha > 1$ and $1+u < e^{bt}$, the expression in (4.43) is negative. This establishes (4.42).

It is now clear that we can apply Theorem 4.8, taking into account (4.29), (4.40), and (4.41). This gives

$$\mathbb{E}\left[\exp\left\{-\frac{2b^2\lambda}{c^2}\int_0^t Y_s\,ds\right\}\right]$$

$$= \left(\frac{2\alpha - 1}{\alpha e^{bt(\alpha-1)} + (\alpha - 1)e^{-bt\alpha}} \right)^{\frac{2a}{c^2}}$$
$$\times \exp\left\{ -\frac{2by_0}{c^2} \frac{\lambda(e^{bt(\alpha-1)} - e^{-bt\alpha})}{\alpha e^{bt(\alpha-1)} + (\alpha - 1)e^{-bt\alpha}} \right\}. \tag{4.44}$$

We will next simplify formula (4.44). Let us put $\lambda = \eta^2 - \frac{1}{4}$ where $\eta > \frac{1}{2}$. Then we have $\eta = \alpha - \frac{1}{2}$. Moreover,

$$\alpha e^{bt(\alpha-1)} + (\alpha - 1)e^{-bt\alpha} = 2e^{-\frac{bt}{2}} \left(\eta \cosh(bt\eta) + \frac{1}{2} \sinh(bt\eta) \right)$$

and

$$e^{bt(\alpha-1)} - e^{-bt\alpha} = 2e^{-\frac{bt}{2}} \sinh(bt\eta).$$

Using the previous equality, we see that formula (4.26) follows from formula (4.44).
This completes the proof of Theorem 4.9. □

Let Y be the CIR process determined from (4.24). Denote by $\bar{m}_t$ the distribution density of the random variable $\int_0^t Y_s\,ds$ and by m_t the mixing distribution density associated with the process $\sqrt{Y_t}$, that is, the distribution density of the random variable α_t defined by

$$\alpha_t = \left\{ \frac{1}{t} \int_0^t Y_s\,ds \right\}^{\frac{1}{2}}.$$

Applying Theorem 4.9, we see that

$$\int_0^\infty \exp\left\{ -\frac{b^2(4\eta^2 - 1)y}{2c^2} \right\} \bar{m}_t(y)\,dy$$
$$= \exp\left\{ \frac{abt}{c^2} \right\} \left(\frac{2\eta}{2\eta \cosh(bt\eta) + \sinh(bt\eta)} \right)^{\frac{2a}{c^2}}$$
$$\times \exp\left\{ -\frac{by_0(4\eta^2 - 1)\sinh(bt\eta)}{2c^2\eta \cosh(bt\eta) + c^2 \sinh(bt\eta)} \right\} \tag{4.45}$$

for all $\eta > \frac{1}{2}$. Since

$$\bar{m}_t(y) = \frac{1}{2\sqrt{ty}} m_t\left(t^{-\frac{1}{2}} y^{\frac{1}{2}}\right),$$

formula (4.45) implies the following equality:

$$\int_0^\infty \exp\left\{ -\frac{b^2(4\eta^2 - 1)y}{2c^2} \right\} y^{-\frac{1}{2}} m_t\left(t^{-\frac{1}{2}} y^{\frac{1}{2}}\right) dy$$

$$= 2\sqrt{t}\exp\left\{\frac{abt}{c^2}\right\}\left(\frac{2\eta}{2\eta\cosh(bt\eta)+\sinh(bt\eta)}\right)^{\frac{2a}{c^2}}$$
$$\times\exp\left\{-\frac{by_0(4\eta^2-1)\sinh(bt\eta)}{2c^2\eta\cosh(bt\eta)+c^2\sinh(bt\eta)}\right\} \tag{4.46}$$

for all $\eta > \frac{1}{2}$.

The next statement provides an explicit formula for the Laplace transform of a certain function depending on the mixing distribution density m_t in the Heston model.

Theorem 4.10 *For every* $\lambda > 0$,

$$\int_0^\infty \exp\{-\lambda y\}y^{-\frac{1}{2}}m_t\big(t^{-\frac{1}{2}}y^{\frac{1}{2}}\big)\,dy$$
$$= 2\sqrt{t}\exp\left\{\frac{abt}{c^2}\right\}$$
$$\times\left(\frac{\sqrt{b^2+2c^2\lambda}}{\sqrt{b^2+2c^2\lambda}\cosh(\frac{1}{2}t\sqrt{b^2+2c^2\lambda})+b\sinh(\frac{1}{2}t\sqrt{b^2+2c^2\lambda})}\right)^{\frac{2a}{c^2}}$$
$$\times\exp\left\{-\frac{2y_0\lambda\sinh(\frac{1}{2}t\sqrt{b^2+2c^2\lambda})}{\sqrt{b^2+2c^2\lambda}\cosh(\frac{1}{2}t\sqrt{b^2+2c^2\lambda})+b\sinh(\frac{1}{2}t\sqrt{b^2+2c^2\lambda})}\right\}. \tag{4.47}$$

Proof If $b \neq 0$, then we obtain formula (4.47) by making the substitution $\lambda = (2c^2)^{-1}b^2(4\eta^2-1)$ in formula (4.46).

Next, suppose $b = 0$. Then

$$Y_t = BESQ_{y_0}^{\frac{4a}{c^2}}\left(\frac{c^2}{4}t\right) = \frac{c^2}{4}BESQ_{\frac{4y_0}{c^2}}^{\frac{4a}{c^2}}(t) \quad \text{in law,}$$

by Remark 1.41 and by the scaling property of a squared Bessel process. It follows that, for every $\lambda > 0$,

$$\mathbb{E}\left[\exp\left\{-\frac{\lambda}{2}\int_0^t Y_s\,ds\right\}\right] = \mathbb{E}\left[\exp\left\{-\frac{\lambda c^2}{8}\int_0^t BESQ_{\frac{4y_0}{c^2}}^{\frac{4a}{c^2}}(u)\,du\right\}\right].$$

Using Theorem 4.5, we obtain

$$\mathbb{E}\left[\exp\left\{-\frac{\lambda}{2}\int_0^t Y_s\,ds\right\}\right]$$
$$= \left[\cosh\left(\frac{ct\sqrt{\lambda}}{2}\right)\right]^{-\frac{2a}{c^2}}\exp\left\{-\frac{y_0\sqrt{\lambda}}{c}\tanh\frac{ct\sqrt{\lambda}}{2}\right\}. \tag{4.48}$$

Now it is not hard to see that formula (4.47) with $b=0$, i.e.

$$\int_0^\infty e^{-\lambda y} y^{-\frac{1}{2}} m_t\left(t^{-\frac{1}{2}} y^{\frac{1}{2}}\right) dy$$

$$= 2\sqrt{t}\left[\cosh\left(\frac{ct\sqrt{\lambda}}{\sqrt{2}}\right)\right]^{-\frac{2a}{c^2}} \exp\left\{-\frac{y_0\sqrt{2}\sqrt{\lambda}}{c}\tanh\frac{ct\sqrt{\lambda}}{\sqrt{2}}\right\},$$

follows from (4.48).

This completes the proof of Theorem 4.10. □

4.5 Ornstein–Uhlenbeck Processes and Laplace Transforms of Mixing Distributions

This section deals with the Laplace transform of the mixing distribution density associated with an Ornstein–Uhlenbeck process. We first assume that the long-run mean m of the process is equal to zero. This case is similar to that of a CIR process, while the case where $m \neq 0$ is more challenging.

Let $\widetilde{Y}$ be the Ornstein–Uhlenbeck process for which

$$d\widetilde{Y}_t = -q\widetilde{Y}_t\,dt + \sigma\,dZ_t, \quad \widetilde{Y}_0 = y_0 \ \mathbb{P}\text{-a.s.}$$

It was shown in Sect. 1.16 that the process Y defined by $Y = \widetilde{Y}^2$ has the same law as the CIR process H satisfying

$$dH_t = \left(\sigma^2 - 2qH_t\right)dt + 2\sigma\sqrt{H_t}\,dZ_t, \quad H_0 = y_0^2 \ \mathbb{P}\text{-a.s.}$$

Theorem 4.11 *For every $\lambda > 0$ the following formula holds for the mixing distribution density m_t associated with the process $\widetilde{Y}$:*

$$\int_0^\infty \exp\{-\lambda y\} y^{-\frac{1}{2}} m_t\left(t^{-\frac{1}{2}} y^{\frac{1}{2}}\right) dy$$

$$= 2\sqrt{t}\exp\left\{\frac{qt}{2}\right\}$$

$$\times\left(\frac{\sqrt{q^2+2\sigma^2\lambda}}{\sqrt{q^2+2\sigma^2\lambda}\cosh(t\sqrt{q^2+2\sigma^2\lambda}) + q\sinh(t\sqrt{q^2+2\sigma^2\lambda})}\right)^{\frac{1}{2}}$$

$$\times\exp\left\{-\frac{y_0^2\lambda\sinh(t\sqrt{q^2+2\sigma^2\lambda})}{\sqrt{q^2+2\sigma^2\lambda}\cosh(t\sqrt{q^2+2\sigma^2\lambda}) + q\sinh(t\sqrt{q^2+2\sigma^2\lambda})}\right\}. \tag{4.49}$$

Proof Using the fact that $Y = H$ (in law) and Theorem 4.10, we get

$$\int_0^\infty \exp\{-\lambda y\} y^{-\frac{1}{2}} m_t\big(t^{-\frac{1}{2}} y^{\frac{1}{2}}\big)\,dy$$

$$= 2\sqrt{t}\exp\left\{\frac{2q\sigma^2 t}{4\sigma^2}\right\}$$

$$\times\left(\frac{\sqrt{4q^2+8\sigma^2\lambda}}{\sqrt{4q^2+8\sigma^2\lambda}\cosh(\frac{1}{2}t\sqrt{4q^2+8\sigma^2\lambda})+2q\sinh(\frac{1}{2}t\sqrt{4q^2+8\sigma^2\lambda})}\right)^{\frac{2\sigma^2}{4\sigma^2}}$$

$$\times\exp\left\{-\frac{2y_0^2\lambda\sinh(\frac{1}{2}t\sqrt{4q^2+8\sigma^2\lambda})}{\sqrt{4q^2+8\sigma^2\lambda}\cosh(\frac{1}{2}t\sqrt{4q^2+8\sigma^2\lambda})+2q\sinh(\frac{1}{2}t\sqrt{4q^2+8\sigma^2\lambda})}\right\}$$

$$= 2\sqrt{t}\exp\left\{\frac{qt}{2}\right\}$$

$$\times\left(\frac{2\sqrt{q^2+2\sigma^2\lambda}}{2\sqrt{q^2+2\sigma^2\lambda}\cosh(t\sqrt{q^2+2\sigma^2\lambda})+2q\sinh(t\sqrt{q^2+2\sigma^2\lambda})}\right)^{\frac{1}{2}}$$

$$\times\exp\left\{-\frac{2y_0^2\lambda\sinh(t\sqrt{q^2+2\sigma^2\lambda})}{2\sqrt{q^2+2\sigma^2\lambda}\cosh(t\sqrt{q^2+2\sigma^2\lambda})+2q\sinh(t\sqrt{q^2+2\sigma^2\lambda})}\right\}. \tag{4.50}$$

Now it is clear that (4.50) implies (4.49).

This completes the proof of Theorem 4.11. □

If the long-run mean m of the Ornstein–Uhlenbeck process is different from zero, then explicit formulas representing the Laplace transform on the left-hand side of (4.49) are more complicated. Such formulas were obtained independently in [Wen90] and [SS91]. Our presentation in this section is based on [SS91] (see the lemma on p. 745). However, the paper [SS91] includes only a very short sketch of the proof of this lemma.

We will first formulate the result due to E.M. Stein and J. Stein exactly as it is given in their paper, and then give an equivalent and more transparent formulation. Finally, we will provide a detailed proof.

Let σ be the Ornstein–Uhlenbeck process satisfying the equation

$$d\sigma_t = \delta(\theta - \sigma_t)\,dt + k\,dz_2$$

where z_2 is a standard Brownian motion, and $\sigma_0 = \sigma_0$ a.s. Suppose $\lambda > 0$ and define new parameters by

$$A = -\frac{\delta}{k^2},\quad B = \frac{\theta\delta}{k^2},\quad \text{and}\quad C = -\frac{\lambda}{k^2 t}.$$

Let us also set

$$a = \left(A^2 - 2C\right)^{\frac{1}{2}}, \quad b = -\frac{A}{a}, \quad L = -A - \left(\frac{\sinh(ak^2t) + b\cosh(ak^2t)}{\cosh(ak^2t) + b\sinh(ak^2t)}\right),$$

$$M = B\left\{\frac{b\sinh(ak^2t) + b^2\cosh(ak^2t) + 1 - b^2}{\cosh(ak^2t) + b\sinh(ak^2t)} - 1\right\},$$

$$\begin{aligned} N = {} & \frac{a-A}{2a^2}\left[a^2 - AB^2 - B^2a\right]k^2t \\ & + \frac{B^2(A^2-a^2)}{2a^3}\left\{\frac{(2A+a) + (2A-a)e^{2ak^2t}}{A + a + (a-A)e^{2ak^2t}}\right\} \\ & + \frac{2AB^2(a^2-A^2)e^{ak^2t}}{a^3(A + a + (a-A)e^{2ak^2t})} - \frac{1}{2}\log\left\{\frac{1}{2}\left(\frac{A}{a} + 1\right) + \frac{1}{2}\left(1 - \frac{A}{a}\right)e^{2ak^2t}\right\}. \end{aligned}$$

The next statement was obtained in [SS91].

Theorem 4.12 *For every $\lambda > 0$ the mixing distribution density m_t associated with the process $\widetilde{Y}$ satisfies the following condition:*

$$\int_0^\infty e^{-\lambda\sigma^2} m_t(\sigma)\,d\sigma = \exp\left\{\frac{L\sigma_0^2}{2} + M\sigma_0 + N\right\}.$$

Theorem 4.12 can be reformulated as follows.

Theorem 4.13 *For every $\lambda > 0$,*

$$\begin{aligned} & \int_0^\infty e^{-\lambda t y^2} m_t(y)\,dy \\ & \quad = e^{\frac{qt}{2}}\left(\frac{\sqrt{q^2 + 2\sigma^2\lambda}}{\sqrt{q^2 + 2\sigma^2\lambda}\cosh(t\sqrt{q^2 + 2\sigma^2\lambda}) + q\sinh(t\sqrt{q^2 + 2\sigma^2\lambda})}\right)^{\frac{1}{2}} \\ & \quad \times \exp\left\{-\frac{y_0^2\lambda\sinh(t\sqrt{q^2 + 2\sigma^2\lambda})}{\sqrt{q^2 + 2\sigma^2\lambda}\cosh(t\sqrt{q^2 + 2\sigma^2\lambda}) + q\sinh(t\sqrt{q^2 + 2\sigma^2\lambda})}\right\} \\ & \quad \times \exp\left\{-\Lambda_1(\lambda)\right\}\exp\left\{\Lambda_2(\lambda)\right\}\exp\left\{\Lambda_3(\lambda)\right\}, \end{aligned} \tag{4.51}$$

where the functions $\Lambda_1(\lambda)$, $\Lambda_2(\lambda)$, and $\Lambda_3(\lambda)$ are defined by

$$\frac{2mqy_0\lambda(\cosh(t\sqrt{q^2 + 2\sigma^2\lambda}) - 1)}{\sqrt{q^2 + 2\sigma^2\lambda}[\sqrt{q^2 + 2\sigma^2\lambda}\cosh(t\sqrt{q^2 + 2\sigma^2\lambda}) + q\sinh(t\sqrt{q^2 + 2\sigma^2\lambda})]},$$

$$\frac{m^2q^2\lambda(\sinh(t\sqrt{q^2 + 2\sigma^2\lambda}) - t\sqrt{q^2 + 2\sigma^2\lambda}\cosh(t\sqrt{q^2 + 2\sigma^2\lambda}))}{(q^2 + 2\sigma^2\lambda)[\sqrt{q^2 + 2\sigma^2\lambda}\cosh(t\sqrt{q^2 + 2\sigma^2\lambda}) + q\sinh(t\sqrt{q^2 + 2\sigma^2\lambda})]},$$

and

$$\frac{m^2q^3\lambda[2(\cosh(t\sqrt{q^2+2\sigma^2\lambda})-1)-t\sqrt{q^2+2\sigma^2\lambda}\sinh(t\sqrt{q^2+2\sigma^2\lambda})]}{(q^2+2\sigma^2\lambda)^{\frac{3}{2}}[\sqrt{q^2+2\sigma^2\lambda}\cosh(t\sqrt{q^2+2\sigma^2\lambda})+q\sinh(t\sqrt{q^2+2\sigma^2\lambda})]},$$

respectively.

To prove the equivalence of Theorems 4.12 and 4.13, we replace the symbols δ, θ, k, σ_0, and λ in the paper of Stein and Stein by the symbols q, m, σ, y_0, and $t\lambda$, respectively, and take into account that

$$A=-\frac{q}{\sigma^2},\quad B=\frac{mq}{\sigma^2},\quad C=-\frac{\lambda}{\sigma^2},$$
$$a=\frac{1}{\sigma^2}\sqrt{q^2+2\sigma^2\lambda},\quad b=\frac{q}{\sqrt{q^2+2\sigma^2\lambda}},\quad \text{and}\quad ak^2t=t\sqrt{q^2+2\sigma^2\lambda}.$$

Combining the terms

$$\frac{a-A}{2a^2}a^2k^2t \quad\text{and}\quad -\frac{1}{2}\log\left\{\frac{1}{2}\left(\frac{A}{a}+1\right)+\frac{1}{2}\left(1-\frac{A}{a}\right)e^{2ak^2t}\right\}$$

in the definition of the parameter N given above and performing long and tedious computations, we can show that Theorems 4.12 and 4.13 are equivalent. Note that if we take $m=0$ in (4.51) and change variables, we recover formula (4.49).

Proof of Theorem 4.13 Suppose Y is the CIR process satisfying the volatility equation in (2.14) with $Y_0=y$ $\mathbb{P}$-a.s. By the Feynman–Kac formula, the solution u_λ to the initial value problem,

$$\begin{cases}\dfrac{\partial u_\lambda}{\partial t}=\dfrac{1}{2}\sigma^2\dfrac{\partial^2u_\lambda}{\partial y^2}+q(m-y)\dfrac{\partial u_\lambda}{\partial y}-\lambda y^2u_\lambda,\\ u(0,y)=1,\end{cases}\tag{4.52}$$

is given by

$$u_\lambda(t,y)=\mathbb{E}\left[\exp\left\{-\lambda\int_0^t Y_s^2\,ds\right\}\right].$$

It follows that the mixing distribution density m_t corresponding to the process Y (note that this process depends on y) satisfies

$$\int_0^\infty e^{-\lambda t\sigma^2}m_t(\sigma)\,d\sigma=u_\lambda(t,y).\tag{4.53}$$

We will look for the solution of (4.52) having the following form:

$$u_\lambda(t,y)=\exp\{\phi_1(t)y^2+\phi_2(t)y+\phi_3(t)\},\tag{4.54}$$

where ϕ_1, ϕ_2, and ϕ_3 are functions of t such that

$$\phi_1(0) = \phi_2(0) = \phi_3(0) = 0.$$

Note that these functions depend on λ. It is clear that the initial condition in (4.52) is satisfied for the solution given by (4.54). Moreover,

$$\begin{aligned}\phi_1'(t)y^2 + \phi_2'(t)y + \phi_3'(t) &= \frac{\sigma^2}{2}\left[2\phi_1(t) + \left(2y\phi_1(t) + \phi_2(t)\right)^2\right] \\ &\quad + q(m-y)\left(2y\phi_1(t) + \phi_2(t)\right) - \lambda y^2.\end{aligned}$$

Therefore,

$$\begin{cases} \phi_1'(t) = 2\sigma^2\phi_1(t)^2 - 2q\phi_1(t) - \lambda, \\ \phi_2'(t) = 2qm\phi_1(t) + 2\sigma^2\phi_1(t)\phi_2(t) - q\phi_2(t), \\ \phi_3'(t) = \sigma^2\phi_1(t) + \dfrac{\sigma^2}{2}\phi_2(t)^2 + qm\phi_2(t). \end{cases} \tag{4.55}$$

Now it is clear that in order to solve the system of equations in (4.55), we can first solve the first equation with the initial condition $\phi_1(0) = 0$, then plug ϕ_1 into the second equation, solve it taking into account that $\phi_2(0) = 0$, and finally plug ϕ_1 and ϕ_2 into the third equation, and solve this equation using the equality $\phi_3(0) = 0$.

Solution of the first equation in (4.55). Let us consider the following initial value problem:

$$\begin{cases} \phi'(t) = s_1\phi(t)^2 + s_2\phi(t) + s_3, \\ \phi(0) = 0, \end{cases} \tag{4.56}$$

where s_1, s_2, and s_3 are real numbers. Put

$$\phi(t) = \frac{\sinh(\gamma t)}{\alpha\cosh(\gamma t) + \beta\sinh(\gamma t)}, \quad \alpha \neq 0,\ \gamma \neq 0. \tag{4.57}$$

Our goal is to choose the constants α, β, and γ so that the equation in (4.56) holds. We have

$$\phi'(t) = \frac{\alpha\gamma}{(\alpha\cosh(\gamma t) + \beta\sinh(\gamma t))^2}.$$

Hence the function ϕ satisfies the equation in (4.56) if and only if

$$\begin{aligned}\alpha\gamma &= s_1\sinh^2(\gamma t) + s_2\sinh(\gamma t)\left(\alpha\cosh(\gamma t) + \beta\sinh(\gamma t)\right) \\ &\quad + s_3\left(\alpha\cosh(\gamma t) + \beta\sinh(\gamma t)\right)^2\end{aligned}$$

for all $t > 0$. The previous equality is equivalent to the equality

$$\begin{aligned}&\left(s_1 + \beta s_2 + \beta^2 s_3 + \alpha^2 s_3\right)\sinh^2(\gamma t) + (\alpha s_2 + 2\alpha\beta s_3)\sinh(\gamma t)\cosh(\gamma t) \\ &\quad + \left(\alpha^2 s_3 - \alpha\gamma\right) = 0.\end{aligned}$$

It follows that

$$s_1 + \beta s_2 + \beta^2 s_3 + \alpha^2 s_3 = 0, \qquad \alpha s_2 + 2\alpha\beta s_3 = 0, \qquad \alpha^2 s_3 - \alpha\gamma = 0,$$

and therefore

$$\alpha = \frac{\sqrt{s_2^2 - 4s_1 s_3}}{2s_3}, \quad \beta = -\frac{s_2}{2s_3}, \quad \text{and} \quad \gamma = \frac{\sqrt{s_2^2 - 4s_1 s_3}}{2}. \tag{4.58}$$

Since $s_1 = 2\sigma^2$, $s_2 = -2q$, and $s_3 = -\lambda$ in the first equation in (4.55), and the equalities in (4.58) hold, we have

$$\alpha = -\frac{\sqrt{q^2 + 2\sigma^2\lambda}}{\lambda}, \quad \beta = -\frac{q}{\lambda}, \quad \text{and} \quad \gamma = \sqrt{q^2 + 2\sigma^2\lambda}. \tag{4.59}$$

Next using (4.57) and (4.59) we obtain

$$\phi_1(t) = -\frac{\lambda \sinh(t\sqrt{q^2 + 2\sigma^2\lambda})}{\sqrt{q^2 + 2\sigma^2\lambda}\cosh(t\sqrt{q^2 + 2\sigma^2\lambda}) + q\sinh(t\sqrt{q^2 + 2\sigma^2\lambda})}. \tag{4.60}$$

Solution of the second equation in (4.55). Consider the following function:

$$\phi_2(t) = \frac{\delta(\cosh(\gamma t) - 1)}{\alpha\cosh(\gamma t) + \beta\sinh(\gamma t)}, \tag{4.61}$$

where α, β, and γ are the same constants as in (4.57), and $\delta \neq 0$. We have

$$\phi_2'(t) = \delta\gamma \frac{\alpha\sinh(\gamma t) + \beta\cosh(\gamma t) - \beta}{(\alpha\cosh(\gamma t) + \beta\sinh(\gamma t))^2}.$$

Our next goal is to show that there exists δ such that

$$\phi_2'(t) = 2qm\phi(t) + 2\sigma^2\phi(t)\phi_2(t) - q\phi_2(t). \tag{4.62}$$

In (4.62), ϕ and ϕ_2 are given by (4.57) and (4.61), respectively, with α, β, and γ defined by (4.59). It follows that

$$\begin{aligned} &\delta\gamma\alpha\sinh(\gamma t) + \delta\gamma\beta\cosh(\gamma t) - \delta\gamma\beta \\ &\quad = 2qm\sinh(\gamma t)\big(\alpha\cosh(\gamma t) + \beta\sinh(\gamma t)\big) + 2\sigma^2\delta\sinh(\gamma t)\big(\cosh(\gamma t) - 1\big) \\ &\quad - q\delta\big(\cosh(\gamma t) - 1\big)\big(\alpha\cosh(\gamma t) + \beta\sinh(\gamma t)\big). \end{aligned} \tag{4.63}$$

Moreover, it is not hard to see that the equality in (4.63) holds if

$$\delta = 2mq\big(q^2 + 2\sigma^2\lambda\big)^{-\frac{1}{2}}.$$

Therefore the second equation in (4.55) is satisfied if the function ϕ_2 is given by the following expression:

$$-\frac{2mq\lambda(\cosh(t\sqrt{q^2+2\sigma^2\lambda})-1)}{\sqrt{q^2+2\sigma^2\lambda}(\sqrt{q^2+2\sigma^2\lambda}\cosh(t\sqrt{q^2+2\sigma^2\lambda})+q\sinh(t\sqrt{q^2+2\sigma^2\lambda}))}. \tag{4.64}$$

Solution of the third equation in (4.55). Let ϕ be the function defined by (4.57). It is not hard to see that

$$\phi(t) = -\frac{\beta}{\alpha^2-\beta^2} + \frac{\alpha}{\gamma(\alpha^2-\beta^2)}\frac{(\alpha\cosh(\gamma t)+\beta\sinh(\gamma t))'}{\alpha\cosh(\gamma t)+\beta\sinh(\gamma t)}.$$

Hence

$$\int \phi(t)\,dt = -\frac{\beta}{\alpha^2-\beta^2}t + \frac{\alpha}{\gamma(\alpha^2-\beta^2)}\log\bigl[\alpha\cosh(\gamma t)+\beta\sinh(\gamma t)\bigr]+C. \tag{4.65}$$

Next, (4.60) and (4.65) with $\alpha=\gamma=\sqrt{q^2+2\sigma^2\lambda}$ and $\beta=q$ give

$$\int \sigma^2\phi_1(t)\,dt$$
$$= \frac{qt}{2} - \frac{1}{2}\log\bigl[\sqrt{q^2+2\sigma^2\lambda}\cosh\bigl(t\sqrt{q^2+2\sigma^2\lambda}\bigr)+q\sinh\bigl(t\sqrt{q^2+2\sigma^2\lambda}\bigr)\bigr]+C.$$

Since we expect that $\phi_3(0)=0$, it is convenient to choose

$$C = \frac{1}{2}\log\sqrt{q^2+2\sigma^2\lambda}.$$

Then

$$\int \sigma^2\phi_1(t)\,dt$$
$$= \frac{qt}{2} + \frac{1}{2}\log\frac{\sqrt{q^2+2\sigma^2\lambda}}{\sqrt{q^2+2\sigma^2\lambda}\cosh(t\sqrt{q^2+2\sigma^2\lambda})+q\sinh(t\sqrt{q^2+2\sigma^2\lambda})}. \tag{4.66}$$

We will next integrate the function

$$f(t) = \frac{\sigma^2}{2}\phi_2(t)^2 + qm\phi_2(t)$$

where ϕ_2 is given by (4.64). First note that

$$f(t) = A_1\frac{(\cosh(\gamma t)-1)^2}{(\alpha\cosh(\gamma t)+\beta\sinh(\gamma t))^2} + A_2\frac{\cosh(\gamma t)-1}{\alpha\cosh(\gamma t)+\beta\sinh(\gamma t)}, \tag{4.67}$$

where

$$\alpha = \gamma = \sqrt{q^2 + 2\sigma^2\lambda}, \qquad \beta = q,$$

$$A_1 = \frac{2\sigma^2 m^2 q^2 \lambda^2}{q^2 + 2\sigma^2\lambda}, \qquad A_2 = -\frac{2m^2 q^2 \lambda}{\sqrt{q^2 + 2\sigma^2\lambda}}.$$

We will look for a solution of the equation $f(t) = g(t)'$ having the following form:

$$g(t) = \frac{B_1 t \cosh(\gamma t) + B_2 t \sinh(\gamma t) + B_3 \cosh(\gamma t) + B_4 \sinh(\gamma t) + B_5}{\alpha \cosh(\gamma t) + \beta \sinh(\gamma t)},$$

where B_i, $1 \le i \le 5$, are real constants. Differentiating the function g, we see that

$$g'(t) = \frac{h(t)}{(\alpha \cosh(\gamma t) + \beta \sinh(\gamma t))^2},$$

where

$$\begin{aligned} h(t) = {} & (B_1\alpha + B_2\beta)\cosh^2(\gamma t) + (B_4\gamma\alpha - B_3\beta\gamma - B_2\beta) \\ & + (B_2\gamma\alpha - B_1\beta\gamma)t + (B_1\beta + B_2\alpha)\sinh(\gamma t)\cosh(\gamma t) \\ & - B_5\alpha\gamma\sinh(\gamma t) - B_5\beta\gamma\cosh(\gamma t). \end{aligned}$$

Since (4.67) implies that

$$f(t) = Z\big(\alpha\cosh(\gamma t) + \beta\sinh(\gamma t)\big)^{-2},$$

where

$$\begin{aligned} Z = {} & (A_1 + A_2\alpha)\cosh^2(\gamma t) - (2A_1 + A_2\alpha)\cosh(\gamma t) \\ & + A_2\beta\sinh(\gamma t)\cosh(\gamma t) - A_2\beta\sinh(\gamma t) + A_1, \end{aligned}$$

the constants B_1 and B_2 should be chosen so that

$$B_2\alpha - B_1\beta = 0.$$

In addition, the following conditions should be satisfied:

$$\begin{aligned} & B_1\alpha + B_2\beta = A_1 + A_2\alpha, \quad B_4\gamma\alpha - B_3\beta\gamma - B_2\beta = A_1, \\ & B_1\beta + B_2\alpha = A_2\beta, \quad B_5\alpha\gamma = A_2\beta, \quad \text{and} \quad B_5\beta\gamma = 2A_1 + A_2\alpha. \end{aligned}$$

It follows that

$$B_1 = \frac{A_2}{2} = -\frac{m^2 q^2 \lambda}{\sqrt{q^2 + 2\sigma^2\lambda}}, \tag{4.68}$$

$$B_2 = \frac{A_2 \beta}{2\alpha} = -\frac{m^2 q^3 \lambda}{q^2 + 2\sigma^2 \lambda}, \tag{4.69}$$

and

$$B_5 = \frac{A_2 \beta}{\alpha \gamma} = -\frac{2m^2 q^3 \lambda}{(q^2 + 2\sigma^2 \lambda)^{\frac{3}{2}}}.$$

Moreover, since we expect that $\phi_3(0) = 0$, the constant B_3 has to be chosen as follows:

$$B_3 = -B_5 = \frac{2m^2 q^3 \lambda}{(q^2 + 2\sigma^2 \lambda)^{\frac{3}{2}}}. \tag{4.70}$$

This implies that

$$B_4 = \frac{A_1 + B_2 \beta + B_3 \beta \gamma}{\alpha \gamma} = \frac{2\sigma^2 m^2 q^2 \lambda^2 + m^2 q^4 \lambda}{(q^2 + 2\sigma^2 \lambda)^2} = \frac{m^2 q^2 \lambda}{q^2 + 2\sigma^2 \lambda}, \tag{4.71}$$

and hence

$$\int \left(\frac{\sigma^2}{2} \phi_2(t)^2 + qm\phi_2(t) \right) dt$$

$$= \frac{B_1 t \cosh(\gamma t) + B_2 t \sinh(\gamma t) + B_3 \cosh(\gamma t) + B_4 \sinh(\gamma t) - B_3}{\alpha \cosh(\gamma t) + \beta \sinh(\gamma t)}. \tag{4.72}$$

Finally, taking into account (4.66) and (4.72), we obtain

$$\phi_3(t) = \frac{qt}{2} + \frac{1}{2} \log \frac{\sqrt{q^2 + 2\sigma^2 \lambda}}{\sqrt{q^2 + 2\sigma^2 \lambda} \cosh(t\sqrt{q^2 + 2\sigma^2 \lambda}) + q \sinh(t\sqrt{q^2 + 2\sigma^2 \lambda})}$$

$$+ \frac{E(t)}{\sqrt{q^2 + 2\sigma^2 \lambda} \cosh(t\sqrt{q^2 + 2\sigma^2 \lambda}) + q \sinh(t\sqrt{q^2 + 2\sigma^2 \lambda})}, \tag{4.73}$$

where

$$E(t) = B_1 t \cosh\left(t\sqrt{q^2 + 2\sigma^2 \lambda}\right) + B_2 t \sinh\left(t\sqrt{q^2 + 2\sigma^2 \lambda}\right)$$

$$+ B_3 \cosh\left(t\sqrt{q^2 + 2\sigma^2 \lambda}\right) + B_4 \sinh\left(t\sqrt{q^2 + 2\sigma^2 \lambda}\right) - B_3, \tag{4.74}$$

and the constants B_1, B_2, B_3, and B_4 are given by (4.68), (4.69), (4.70), and (4.71), respectively. Now it is not hard to see that (4.51) follows from (4.53), (4.54), (4.60), (4.64), (4.73), and (4.74).

This completes the proof of Theorem 4.13. □

4.6 Hull–White Models with Driftless Volatility and Hartman–Watson Distributions

In this section, we return to the discussion of the special Hull–White model given by (3.13). The Hartman–Watson distribution will play an important role in this discussion.

For every $s \in \mathbb{R}$ consider the measure η_s defined implicitly as follows:

$$\int_0^\infty e^{-\nu^2 t/2} \, d\eta_s(t) = \frac{I_{|\nu|}(s)}{I_0(s)}, \quad \nu \in \mathbb{R}, \tag{4.75}$$

where $I_{|\nu|}$ and I_0 are the modified Bessel functions of the first kind. P. Hartman and G.S. Watson proved in [HW74] that the measure η_s satisfying (4.75) exists. Moreover, η_s is a probability distribution and has a density γ_s. More information concerning Hartman–Watson distributions can be found in [BRY04, MY05a, MY05b, Ger11].

The next statement provides an explicit formula for the Hartman–Watson density. This formula was obtained in [Yor80].

Lemma 4.14 *For every $s > 0$ the measure η_s is absolutely continuous with respect to the Lebesgue measure on $(0, \infty)$. The density γ_s of the measure η_s is given by*

$$\gamma_s(t) = \theta(s,t) I_0(s)^{-1},$$

where

$$\theta(s,t) = \frac{s}{(2t)^{\frac{1}{2}} \pi^{\frac{3}{2}}} \exp\left\{\frac{\pi^2}{2t}\right\} \int_0^\infty \exp\left\{-\frac{y^2}{2t}\right\} \exp\{-s\cosh y\} \times \sinh y \sin\left(\frac{\pi y}{t}\right) dy. \tag{4.76}$$

Proof We follow [MY05a], Appendix A, and [Yor80] in the proof of Lemma 4.14. It is known that the following integral representation formula holds for the modified Bessel function of the first kind:

$$I_\nu(z) = \frac{1}{2\pi i} \int_C \exp\{z\cosh\omega - \nu\omega\} \, d\omega, \quad \mathrm{Re}(z) > 0, \ \mathrm{Re}(\nu) > -\frac{1}{2},$$

where the contour C is taken to be three sides of a rectangle with corners at $\infty - \pi i$, $-\pi i$, πi, and $\infty + \pi i$. It is clear that we may assume that $\nu > 0$ in the definition of η_s.

The next equality, which can be checked using the Fourier transform, is called the subordination formula:

$$e^{-\beta} = \int_0^\infty \exp\left\{-\frac{\beta^2}{4z}\right\} \frac{e^{-z}}{\sqrt{\pi z}} \, dz, \quad \beta \ge 0. \tag{4.77}$$

The functions appearing in formula (4.77) can be analytically extended into the right half-plane. Using the transformation $z = \frac{\omega^2}{2t}$, we see that

$$e^{-\nu\omega} = \int_0^\infty e^{-\frac{\nu^2 t}{2}} \frac{\omega}{\sqrt{2\pi t^3}} e^{-\frac{\omega^2}{2t}} \, dt.$$

Therefore

$$I_\nu(s) = \int_0^\infty e^{-\nu^2 t/2} \bar{\theta}(s,t) \, dt \tag{4.78}$$

where

$$\bar{\theta}(s,t) = \frac{1}{2\pi i} \int_C e^{s\cosh\omega} \frac{\omega}{\sqrt{2\pi t^3}} e^{-\frac{\omega^2}{2t}} \, d\omega. \tag{4.79}$$

It follows from (4.75) and (4.78) that the density γ_s exists and

$$I_0(s)\gamma_s(t) = \bar{\theta}(s,t).$$

Our next goal is to simplify the contour integral in (4.79). Recalling the definition of the contour C, we see that

$$\begin{aligned}
\sqrt{2\pi t^3}\bar{\theta}(s,t) = &-\frac{1}{2\pi i} \int_0^\infty e^{-s\cosh\eta} (\eta - \pi i) e^{-\frac{(\eta-\pi i)^2}{2t}} \, d\eta \\
&- \frac{1}{2\pi i} \int_{-\pi}^{\pi} e^{s\cos\eta} e^{\frac{\eta^2}{2t}} \eta \, d\eta \\
&+ \frac{1}{2\pi i} \int_0^\infty e^{-s\cosh\eta} (\eta + \pi i) e^{-\frac{(\eta+\pi i)^2}{2t}} \, d\eta.
\end{aligned} \tag{4.80}$$

The second integral on the right-hand side of (4.80) vanishes, since the integrand is an odd function. Combining the first and the third integral, and taking into account that the function on the left-hand side of (4.80) is real, we obtain

$$\sqrt{2\pi t^3}\bar{\theta}(s,t) = \frac{1}{\pi} e^{\frac{\pi^2}{2t}} \int_0^\infty e^{-s\cosh\eta} e^{-\frac{\eta^2}{2t}} \left(\pi \cos\frac{\pi\eta}{t} - \eta \sin\frac{\pi\eta}{t} \right) d\eta.$$

Next, using the integration by parts formula, we see that

$$\sqrt{2\pi t^3}\bar{\theta}(s,t) = \frac{st}{\pi} e^{\frac{\pi^2}{2t}} \int_0^\infty e^{-s\cosh\eta} \sinh\eta \, e^{-\frac{\eta^2}{2t}} \sin\frac{\pi\eta}{t} \, d\eta.$$

It follows that $\theta(s,t) = \bar{\theta}(s,t)$.

The proof of Lemma 4.14 is thus completed. □

4.7 Mixing Density and Stock Price Density in the Correlated Hull–White Model

Recall that by $\mu_t^{(2)}$ was denoted the joint distribution of the random variables β_t and Y_t in the correlated Hull–White model with driftless volatility described in (3.13). It is easy to see that

$$\beta_t = y_0^2 \int_0^t \exp\left\{2\left(-\frac{1}{2}\xi^2 s + \xi Z_s\right)\right\} ds$$

and

$$Y_t = y_0 \exp\left\{-\frac{1}{2}\xi^2 t + \xi Z_t\right\}.$$

Put

$$\widehat{B}_t = \int_0^t \exp\left\{2\left(-\frac{1}{2}s + Z_s\right)\right\} ds$$

and $B_t = -\frac{1}{2}t + Z_t$. It was established in [Yor92b] (see also [Yor92a, Yor01] and [MY05a]) that the joint distribution of $\widehat{B}_t$ and B_t has a density $\tilde{p}_t$ given by

$$\tilde{p}_t(y, z) = \frac{1}{y} \exp\left\{-\frac{z}{2} - \frac{t}{8}\right\} \exp\left\{-\frac{1 + e^{2z}}{2y}\right\} \theta\left(\frac{e^z}{y}, t\right) \tag{4.81}$$

where θ is defined in (4.76).

Our next goal is to prove that the distribution $\mu_t^{(2)}$ has a density $m_t^{(2)}$, and provide an explicit expression for this density. Using formula (4.81), we see that the function $\tilde{p}_{\xi^2 t}(y, z)$ is the joint density of the random variables $\widehat{B}_{\xi^2 t}$ and $B_{\xi^2 t}$. Since the process $Z_{\xi^2 t}$ is indistinguishable from the process ξZ_t (use the scaling property of Brownian motion), the function $\tilde{p}_{\xi^2 t}(y, z)$ is the joint density of the random variables

$$\xi^2 \int_0^t \exp\left\{2\left(-\frac{1}{2}\xi^2 s + \xi Z_s\right)\right\} ds \quad \text{and} \quad -\frac{1}{2}\xi^2 t + \xi Z_t.$$

Therefore, the function

$$m_t^{(2)}(y, z) = \frac{\xi^2}{y_0^2 z} \tilde{p}_{\xi^2 t}\left(\frac{\xi^2}{y_0^2} y, \log \frac{z}{y_0}\right) \tag{4.82}$$

is the joint density of the random variables β_t and Y_t.

It follows from (4.81) and (4.82) that

$$m_t^{(2)}(y, z) = \frac{y_0^{\frac{1}{2}}}{y z^{\frac{3}{2}}} \exp\left\{-\frac{y_0^2 + z^2}{2\xi^2 y}\right\} \theta\left(\frac{y_0 z}{\xi^2 y}, \xi^2 t\right). \tag{4.83}$$

Next, using (4.76) and (4.83) and making simplifications, we see that

$$m_t^{(2)}(y,z) = \frac{y_0^{\frac{3}{2}}}{(2t)^{\frac{1}{2}}\pi^{\frac{3}{2}}\xi^3 y^2 z^{\frac{1}{2}}} \exp\left\{-\frac{\xi^2 t}{8}\right\} \exp\left\{\frac{\pi^2}{2\xi^2 t}\right\} \exp\left\{-\frac{y_0^2+z^2}{2\xi^2 y}\right\}$$
$$\times \int_0^\infty \exp\left\{-\frac{u^2}{2\xi^2 t}\right\} \exp\left\{-\frac{y_0 z}{\xi^2 y}\cosh u\right\} \sinh u \sin\left(\frac{\pi u}{\xi^2 t}\right) du. \tag{4.84}$$

Formula (4.84) provides an explicit expression for the two-dimensional mixing distribution density $m_t^{(2)}$ in the special Hull–White model defined by (3.13).

The results presented in Sect. 4.6 and in the present section can be combined to establish an explicit formula for the stock price density D_t.

Theorem 4.15 *Let D_t be the stock price density in the correlated Hull–White model with driftless volatility. Then*

$$D_t(x) = \frac{y_0^{\frac{3}{2}}}{2\pi^2\sqrt{t(1-\rho^2)}\xi^3 x} \exp\left\{-\frac{\xi^2 t}{8} + \frac{\pi^2}{2\xi^2 t}\right\}$$
$$\times \int_0^\infty \int_0^\infty \frac{1}{y^{\frac{5}{2}} z^{\frac{1}{2}}} \exp\left\{-\frac{y_0^2+z^2}{2\xi^2 y}\right\}$$
$$\times \exp\left\{-\frac{(\log\frac{x}{x_0 e^{rt}} + \frac{1}{2}y - \frac{\rho}{\xi}(z-y_0))^2}{2(1-\rho^2)y}\right\} dy\,dz$$
$$\times \int_0^\infty \exp\left\{-\frac{u^2}{2\xi^2 t}\right\} \exp\left\{-\frac{y_0 z}{\xi^2 y}\cosh u\right\} \sinh u \sin\left(\frac{\pi u}{\xi^2 t}\right) du.$$

It is not hard to see that Theorem 4.15 follows from (3.14) and (4.84).

4.8 Mellin Transform of the Stock Price Density in the Correlated Heston Model

Let X be the stock price process in a stochastic stock price model, and denote by $X^{\log}$ the log-price process given by $X^{\log} = \log X$. Let the symbols D_t and $D_t^{\log}$ stand for the distribution density of the random variables X_t and $X_t^{\log}$, respectively. The Mellin transform of the density D_t is defined as follows:

$$MD_t(u) = \mathbb{E}[X_t^{u-1}] = \int_0^\infty x^{u-1} D_t(x)\,dx \tag{4.85}$$

where u is a complex number. The domain $\mathrm{Dom}(MD_t)$ of MD_t is the set of all complex numbers, for which the integral in (4.85) is finite. It will be assumed in the sequel that the function D_t is continuous on $(0,\infty)$. For real numbers $\alpha_1<\alpha_2$, denote the open strip $\{z\in\mathbb{C}:\alpha_1<\Re(z)<\alpha_2\}$ by $\mathbb{C}_{\alpha_1,\alpha_2}$. It is not hard to see that if for some real numbers r_1 and r_2 with $r_2<r_1$, the conditions $D_t(x)=O(x^{r_1})$ as $x\downarrow 0$ and $D_t(x)=O(x^{r_2})$ as $x\to\infty$ hold, then $\mathbb{C}_{-r_1,-r_2}\subset \mathrm{Dom}(MD_t)$. The Mellin inversion theorem states that under the restrictions imposed above,

$$D_t(x)=\frac{1}{2\pi i}\int_{r-i\infty}^{r+i\infty}x^{-u}MD_t(u)\,du,\quad x>0, \tag{4.86}$$

where $-r_1<r<-r_2$. It is clear that $MD_t(u)=G_t(u-1)$, where G_t is the moment generating function of the log-price $X_t^{\log}$ defined by

$$G_t(u)=\int_{-\infty}^{\infty}e^{uy}D_t^{\log}(y)\,dy.$$

In the remaining part of the present section, we provide an explicit formula for the Mellin transform of the stock price distribution density in the correlated Heston model. The reasoning below is based on affine principles.

Consider the log-price process $X_t^{\log}=\log X_t$ in the Heston model. From basic principles of affine diffusions, we know that

$$\log\mathbb{E}\left[\exp\left\{sX_t^{\log}\right\}\right]=\phi(s,t)+y_0\psi(s,t), \tag{4.87}$$

where the functions ϕ and ψ satisfy the following Riccati equations:

$$\dot{\phi}=F(s,\psi),\qquad \phi(0)=0, \tag{4.88}$$

$$\dot{\psi}=R(s,\psi),\qquad \psi(0)=0, \tag{4.89}$$

with

$$F(s,v)=av\quad\text{and}\quad R(s,v)=\frac{1}{2}\left(s^2-s\right)+\frac{1}{2}c^2v^2-bv+s\rho cv.$$

In (4.89), $\dot{\phi}$ and $\dot{\psi}$ are the partial derivatives with respect to t of the functions ϕ and ψ, respectively. The symbol s in (4.87) denotes a real parameter. However, the Riccati ODEs in (4.88) and (4.89) are also valid when s is replaced by a complex parameter $u=s+iy$. The solution to the system in (4.88) and (4.89) is given by

$$\psi(u,t)=\frac{(u^2-u)\sinh\frac{P(u)t}{2}}{P(u)\cosh\frac{P(u)t}{2}+(b-c\rho u)\sinh\frac{P(u)t}{2}}$$

and

$$\phi(u,t)=\frac{2a}{c^2}\log\left\{\frac{P(u)\exp\{(b-c\rho u)\frac{t}{2}\}}{P(u)\cosh\frac{P(u)t}{2}+(b-c\rho u)\sinh\frac{P(u)t}{2}}\right\}, \tag{4.90}$$

where

$$P(u)=\sqrt{(b-\rho cu)^2+c^2\left(u-u^2\right)}. \tag{4.91}$$

The following statement concerning the Mellin transform of the stock price density (the moment generating function of the log-price density) in the Heston model follows from the previous formulas.

Lemma 4.16 *For every $t>0$ and $u\in\mathbb{C}$ with $u+1\in\operatorname{Dom}(MD_t)$,*

$$\begin{aligned}MD_t(u+1)&=G_t(u)\\&=\left\{\frac{P(u)\exp\{(b-c\rho u)\frac{t}{2}\}}{P(u)\cosh\frac{P(u)t}{2}+(b-c\rho u)\sinh\frac{P(u)t}{2}}\right\}^{\frac{2a}{c^2}}\\&\quad\times\exp\left\{-y_0\frac{(u-u^2)\sinh\frac{P(u)t}{2}}{P(u)\cosh\frac{P(u)t}{2}+(b-c\rho u)\sinh\frac{P(u)t}{2}}\right\}\end{aligned} \tag{4.92}$$

where the function P is defined by (4.91).

We refer the reader to [Hes93, dBRF-CU10], and [K-R11] for more details.

4.9 Mellin Transform of the Stock Price Density in the Correlated Stein–Stein Model

In this section, we use Lemma 3.7 to compute the Mellin transform of the stock price density D_t in the correlated Stein–Stein model defined by (3.17). It follows from (3.19) and (4.85) that

$$\begin{aligned}MD_t(u)&=\frac{1}{\sqrt{2\pi(1-\rho^2)}}\int_{[0,\infty)^2\times\mathbb{R}}\frac{1}{\sqrt{y}}\,d\mu_t^{(3)}(y,z,w)\\&\quad\times\int_0^\infty x^{u-2}\exp\left\{-\frac{A_t(x,y,z,w)^2}{2(1-\rho^2)y}\right\}dx.\end{aligned}$$

Next, making the substitution $s=\log\frac{x}{x_0e^{\mu t}}$, we obtain

$$\begin{aligned}MD_t(u)&=\frac{(x_0e^{\mu t})^{u-1}}{\sqrt{2\pi(1-\rho^2)}}\int_{[0,\infty)^2\times\mathbb{R}}\frac{1}{\sqrt{y}}\,d\mu_t^{(3)}(y,z,w)\\&\quad\times\int_{-\infty}^{\infty}e^{(u-1)s}\exp\left\{-\frac{A_t(x_0e^{\mu t}e^s,y,z,w)^2}{2(1-\rho^2)y}\right\}ds.\end{aligned}$$

Now, using (3.20) in the previous equation, we see that

$$
\begin{aligned}
MD_t(u) &= \frac{(x_0 e^{\mu t})^{u-1}}{\sqrt{2\pi(1-\rho^2)}} \int_{[0,\infty)^2\times\mathbb{R}} \frac{1}{\sqrt{y}} \\
&\quad \times \exp\left\{-(u-1)\left[\frac{\rho\sigma}{2}t + \frac{\rho y_0^2}{2\sigma} - \frac{\rho}{2\sigma}z + \left(\frac{1}{2} - \frac{\rho q}{\sigma}\right)y + \frac{\rho q m}{\sigma}w\right]\right\} \\
&\quad \times d\mu_t^{(3)}(y,z,w) \int_{-\infty}^{\infty} e^{(u-1)\xi} \exp\left\{-\frac{\xi^2}{2(1-\rho^2)y}\right\} d\xi.
\end{aligned}
$$

Making the substitution $\eta = \frac{\xi}{\sqrt{2(1-\rho^2)y}}$ in the last integral and simplifying the resulting expression, we get

$$
\begin{aligned}
MD_t(u) &= \left(x_0 e^{\mu t}\right)^{u-1} \int_{[0,\infty)^2\times\mathbb{R}} \exp\left\{-(u-1)\left[\frac{\rho\sigma}{2}t + \frac{\rho y_0^2}{2\sigma} - \frac{\rho}{2\sigma}z \right.\right. \\
&\quad \left.\left. + \left(\frac{1}{2} - \frac{\rho q}{\sigma}\right)y + \frac{\rho q m}{\sigma}w\right]\right\} \\
&\quad \times \exp\left\{\frac{(u-1)^2(1-\rho^2)y}{2}\right\} d\mu_t^{(3)}(y,z,w) \\
&= \left(x_0 e^{\mu t}\right)^{u-1} \exp\left\{-(u-1)\left(\frac{\rho\sigma}{2}t + \frac{\rho y_0^2}{2\sigma}\right)\right\} \\
&\quad \times \mathbb{E}\left[\exp\{\mathcal{E}_t\} \mid Y_0 = y_0\right], \qquad (4.93)
\end{aligned}
$$

where

$$
\begin{aligned}
\mathcal{E}_t &= \frac{(u-1)\rho}{2\sigma} Y_t^2 + \left[(u-1)\left(\frac{\rho q}{\sigma} - \frac{1}{2}\right) + \frac{(u-1)^2(1-\rho^2)}{2}\right] \int_0^t Y_s^2\, ds \\
&\quad - (u-1)\frac{\rho q m}{\sigma} \int_0^t Y_s\, ds. \qquad (4.94)
\end{aligned}
$$

According to (4.94) and the Feynman–Kac formula, the function

$$
\varphi(t,y) = \mathbb{E}\left[\exp\{\mathcal{E}_t\} \mid Y_0 = y\right], \quad y > 0,
$$

satisfies the following partial differential equation:

$$
\frac{\partial \varphi}{\partial t} = \frac{1}{2}\sigma^2 \frac{\partial^2 \varphi}{\partial y^2} + q(m-y)\frac{\partial \varphi}{\partial y} + V(u,y)\varphi \qquad (4.95)
$$

with the initial condition given by

$$
\varphi(0,y) = \exp\left\{\frac{(u-1)\rho}{2\sigma} y^2\right\}, \qquad (4.96)
$$

where

$$V(u,y)=\left[(u-1)\left(\frac{\rho q}{\sigma}-\frac{1}{2}\right)+\frac{(u-1)^2(1-\rho^2)}{2}\right]y^2$$
$$-(u-1)\frac{\rho q m}{\sigma}y.$$

In [SZ99], R. Schöbel and J. Zhu found an explicit expression for the characteristic function of the log-price in the Stein–Stein model. We will next continue our computation of the Mellin transform of the stock price density in the Stein–Stein model, using the results obtained in [SZ99].

Let us assume the following ansatz for the solution of (4.95)–(4.96):

$$\varphi(t,y)=\exp\left\{\frac{1}{2}D(t)y^2+B(t)y+C(t)\right\}, \tag{4.97}$$

where the functions D, B, and C do not depend on y. Put

$$v_1=(u-1)\left(\frac{\rho q}{\sigma}-\frac{1}{2}\right)+\frac{(u-1)^2(1-\rho^2)}{2}$$

and

$$v_2=-(u-1)\frac{\rho q m}{\sigma}.$$

Then $V(u,y)=v_1y^2+v_2y$. Substituting (4.97) into (4.95) and (4.96), we obtain the following system of ordinary differential equations for D, B, and C:

$$\begin{cases}\dot{D}=\sigma^2D^2-2qD+2v_1,\\ \dot{B}=(\sigma^2D-q)B+qmD+v_2,\\ \dot{C}=\frac{1}{2}\sigma^2B^2+qmB+\frac{1}{2}\sigma^2D\end{cases} \tag{4.98}$$

with $D(0)=\frac{(u-1)\rho}{\sigma}$ and $B(0)=C(0)=0$. As in the previous section, the dots over the symbols in (4.98) denote the partial derivatives with respect to t. The solution to the system in (4.98) is given by

$$D(t)=\frac{1}{\sigma^2}\left[q-\gamma_1\frac{\sinh(t\gamma_1)+\gamma_2\cosh(t\gamma_1)}{\cosh(t\gamma_1)+\gamma_2\sinh(t\gamma_1)}\right], \tag{4.99}$$

$$B(t)=\frac{1}{\sigma^2\gamma_1}\left[\frac{qm\gamma_1-\gamma_2\gamma_3+\gamma_3(\sinh(t\gamma_1)+\gamma_2\cosh(t\gamma_1))}{\cosh(t\gamma_1)+\gamma_2\sinh(t\gamma_1)}-qm\gamma_1\right], \tag{4.100}$$

and

$$C(t)=-\frac{1}{2}\log\left[\cosh(t\gamma_1)+\gamma_2\sinh(t\gamma_1)\right]+\frac{1}{2}qt$$

$$+\frac{q^2m^2\gamma_1^2-\gamma_3^2}{2\sigma^2\gamma_1^3}\left[\frac{\sinh(t\gamma_1)}{\cosh(t\gamma_1)+\gamma_2\sinh(t\gamma_1)}-\gamma_1 t\right]$$
$$+\frac{(qm\gamma_1-\gamma_2\gamma_3)\gamma_3}{\sigma^2\gamma_1^3}\left[\frac{\cosh(t\gamma_1)-1}{\cosh(t\gamma_1)+\gamma_2\sinh(t\gamma_1)}\right] \tag{4.101}$$

where

$$\gamma_1=\sqrt{-\sigma^2(u-1)^2\big(1-\rho^2\big)+\sigma^2(u-1)\left(1-\frac{2q\rho}{\sigma}\right)+q^2}, \tag{4.102}$$

$$\gamma_2=\frac{q-(u-1)\rho\sigma}{\sqrt{-\sigma^2(u-1)^2(1-\rho^2)+\sigma^2(u-1)(1-\frac{2q\rho}{\sigma})+q^2}}, \tag{4.103}$$

and

$$\gamma_3=q^2m-(u-1)qm\rho\sigma. \tag{4.104}$$

Remark 4.17 The formulas in (4.99)–(4.104) should be compared with the similar formulas in the appendix of [SZ99]. Note that the correspondence between our notation and the notation used in [SZ99] is as follows: $u-1=i\phi$, $q=\kappa$, $m=\theta$, $\sigma=\sigma$, $\rho=\rho$, $v_1=-\hat{s}_1$, $v_2=-\hat{s}_2$, and $(u-1)\rho 2^{-1}\sigma^{-1}=\hat{s}_3$.

The next statement provides a closed form expression for the Mellin transform of the stock price density in the correlated Stein–Stein model.

Lemma 4.18 *For all $t>0$ and $u\in \mathrm{Dom}(MD_t)$,*

$$MD_t(u)=\big(x_0e^{\mu t}\big)^{u-1}\exp\left\{-(u-1)\left(\frac{\rho\sigma}{2}t+\frac{\rho y_0^2}{2\sigma}\right)\right\}$$
$$\times\exp\left\{\frac{1}{2}D(t)y_0^2+B(t)y_0+C(t)\right\}$$

where the functions D, B, and C are defined by (4.99), (4.100), *and* (4.101), *respectively.*

Lemma 4.18 follows from (4.93), (4.97), (4.99), (4.100), and (4.101). The numbers γ_1, γ_2, and γ_3, appearing in (4.99), (4.100), and (4.101), are given by (4.102), (4.103), and (4.104).

4.10 Notes and References

- Formula (4.1) is due to L. Alili and J.C. Gruet (see [AG97]). The proof of this result in Sect. 4.1 is taken from ([GS10a]). Formula (4.13) was obtained in [PY82].

Closed form expressions for the expectation of the exponential functional of the Ornstein–Uhlenbeck process with the long-run mean m different from zero were found independently in [Wen90] and [SS91]. Our presentation of these results in Sect. 4.5 is different from that in [Wen90] and [SS91].

- The distribution η_s defined by formula (4.75) (the Hartman–Watson distribution) was introduced in the paper [HW74] devoted to Brownian motion on the n-dimensional sphere S_n. For more applications of the Hartman–Watson distribution in financial mathematics, see [BRY04, Ger11].
- Most of the methods used in Sect. 4.7 are borrowed from [Mag07]. However, our presentation is different in details.
- For the theory of the Mellin transform, the reader can consult the books [PK01] and [FS09].

Chapter 5
Asymptotic Analysis of Mixing Distributions

In this chapter, we obtain asymptotic formulas with relative error estimates for mixing distribution densities associated with classical volatility processes. For geometric Brownian motions the asymptotic behavior of mixing densities is characterized by formula (5.45), while for CIR processes and Ornstein–Uhlenbeck processes sharp asymptotic formulas for the corresponding mixing densities are given in (5.133) and (5.134), respectively.

We use Abelian and Tauberian theorems in the analysis of the asymptotic behavior of mixing distribution densities and stock price densities. An informal explanation of the difference between Abelian and Tauberian theorems is as follows. Let $f \mapsto Uf$ be an integral transform. Abelian theorems for this transform connect the asymptotic behavior of the function f with that of the function Uf, while the inverse results, or Tauberian theorems, link the behavior of the integral transform with that of the original function. We refer the reader to the book [Kor04] by J. Korevaar for more details. In Sect. 5.1 of this chapter, we prove a Tauberian theorem for the two-sided Laplace transform (see Theorem 5.1). In our opinion, this theorem has an independent interest. In the present chapter, Theorem 5.1 is used to construct asymptotic inverses of Laplace transforms of certain functions associated with mixing densities in the Heston and Stein–Stein models, and to obtain sharp asymptotic formulas for these densities.

Two special Abelian theorems included in the present chapter concern fractional integral operators and integral operators with log-normal kernels (see Theorem 5.5 and Theorem 5.3, respectively). One of the reasons why fractional integrals are useful tools in financial mathematics is that the call pricing function in a stochastic asset price model can be represented as the fractional integral operator of second order applied to the stock price distribution density. Fractional integrals also appear in Sect. 5.4.2, where an equivalent formulation of Dufresne's recurrence formula is given. Dufresne's formula helps to navigate from one Hull–White model to the other. Integral operators with log-normal kernels will be used in the next chapter to study the asymptotic behavior of stock price densities in classical stochastic volatility models.

A. Gulisashvili, *Analytically Tractable Stochastic Stock Price Models*,
Springer Finance, DOI 10.1007/978-3-642-31214-4_5,
© Springer-Verlag Berlin Heidelberg 2012

5.1 Asymptotic Inversion of the Laplace Transform

Our objective for this section is to prove a special Tauberian theorem for the two-sided Laplace transform. This theorem will be a useful tool in the asymptotic analysis of the Stein–Stein and the Heston model.

Recall that for $-\infty \le \alpha_1 < \alpha_2 \le \infty$, we denoted by $\mathbb{C}_{\alpha_1,\alpha_2}$ the open strip $\{z \in \mathbb{C} : \alpha_1 < \Re(z) < \alpha_2\}$. Let M be a non-negative locally integrable function on $(-\infty, \infty)$. Note that in many cases of interest in the theory of stochastic volatility models, the function M is either the distribution density of the log-price, or the mixing distribution density in a stochastic volatility model.

The two-sided Laplace transform of the function M is defined by the following formula:

$$\mathcal{L}M(\lambda) = \int_{-\infty}^{\infty} e^{-\lambda y} M(y)\, dy. \tag{5.1}$$

The domain $D(\mathcal{L}M)$ of $\mathcal{L}M$ is the set of all complex numbers λ, for which the integral in (5.1) converges absolutely. It will be assumed in the sequel that there exist α_1 and α_2 with $-\infty < \alpha_1 < \alpha_2 \le \infty$ such that $\mathbb{C}_{\alpha_1,\alpha_2} \subset D(\mathcal{L}M)$. Then the function $\mathcal{L}M$ is analytic in the strip $\mathbb{C}_{\alpha_1,\alpha_2}$.

Suppose there exist a number $\tau > 0$ and a function I, analytic in the strip $\mathbb{C}_{0,\alpha_2-\alpha_1+\tau}$ and such that

$$\mathcal{L}M(\lambda) = I(\lambda - \alpha_1 + \tau), \quad \lambda \in \mathbb{C}_{\alpha_1,\alpha_2}. \tag{5.2}$$

Fix a small number $\kappa > 0$, and define an auxiliary function ψ of a real variable z_2 by

$$\psi(z_2) = \sup_{z_1 \in [0, \alpha_2 - \alpha_1 + \tau - \kappa]} \left| I(z_1 + i z_2) \right|. \tag{5.3}$$

It is assumed in (5.3) that $|z_2| > 1$.

We now list several Tauberian conditions, under which Theorem 5.1 formulated below holds.

1. The function I admits the following factorization in $\mathbb{C}_{0,\alpha_2-\alpha_1+\tau}$:

$$I(z) = z^{\gamma_1} G_1(z)^{\gamma_2} G_2(z) \exp\{F(z)\}, \tag{5.4}$$

where γ_1 and γ_2 are non-negative constants.

2. For some $\delta > 0$, the functions $\lambda \mapsto G_1(\lambda)$, $\lambda \mapsto G_2(\lambda)$, and $\lambda \mapsto F(\lambda)$ are real-valued on the interval $(0, \delta)$. In addition, $G_1(\lambda) \to \infty$ and $F(\lambda) \to \infty$ as $\lambda \downarrow 0$.
3. The function G_1 never vanishes in $\mathbb{C}_{0,\alpha_2-\alpha_1+\tau}$.
4. There exists $\varepsilon > 0$ such that the functions G_1 and F are analytic in $\mathbb{C}_{-\varepsilon,\alpha_2-\alpha_1+\tau}$ except for a simple pole at $z = 0$.
5. The function G_2 is analytic in $\mathbb{C}_{-\varepsilon,\alpha_2-\alpha_1+\tau}$ and satisfies $G_2(0) \ne 0$.
6. For some $N > 1$, the function ψ defined by (5.3) is Lebesgue integrable over the set $\{z_2 : |z_2| > N\}$. In addition $\psi(z_2) \to 0$ as $|z_2| \to \infty$.

It is known that condition 2 implies the existence of a continuous branch of the function $\log G_1$ in the strip $\mathbb{C}_{0,\alpha_2-\alpha_1+\tau}$. In other words, for some function g analytic in this strip, we have

$$G_1(z) = \exp\{g(z)\} \quad \text{for all } z \in \mathbb{C}_{0,\alpha_2-\alpha_1+\tau}$$

(see Corollary 6.17 in [Con78]). Let us fix such a function g. Then the function $(G_1)^{\gamma_2}(z)$ can be defined for all $z \in \mathbb{C}_{0,\alpha_2-\alpha_1+\tau}$ as follows:

$$(G_1)^{\gamma_2}(z) = \exp\{\gamma_2 g(z)\}.$$

According to condition 3, we have

$$G_1(z) = \frac{\beta}{z} + \widetilde{G}(z) \tag{5.5}$$

and

$$F(z) = \frac{\alpha}{z} + \widetilde{F}(z) \tag{5.6}$$

for all $z \in \mathbb{C}_{-\varepsilon,\alpha_2-\alpha_1+\tau}\backslash 0$. In (5.5) and (5.6), β and α are nonzero complex numbers, while $\widetilde{G}$ and $\widetilde{F}$ are analytic functions in $\mathbb{C}_{-\varepsilon,\alpha_2-\alpha_1+\tau}$.

The next statement is a Tauberian theorem for the two-sided Laplace transform.

Theorem 5.1 *Suppose the function $\mathcal{L}M$ satisfies* (5.2), *and suppose also that conditions* 1–6 *hold for the function I. Then the numbers β and α in* (5.5) *and* (5.6) *are real and positive, and the following asymptotic formula holds*:

$$\begin{aligned} M(y) = {} & \frac{1}{2\sqrt{\pi}} \alpha^{\frac{1}{4}+\frac{\gamma_1-\gamma_2}{2}} \beta^{\gamma_2} G_2(0) \exp\{\widetilde{F}(0)\} y^{-\frac{3}{4}+\frac{\gamma_2-\gamma_1}{2}} e^{(\alpha_1-\tau)y} e^{2\sqrt{\alpha}\sqrt{y}} \\ & \times \left(1 + O\left(y^{-\frac{1}{2}}\right)\right) \end{aligned} \tag{5.7}$$

as $y \to \infty$.

Proof Let us first note that the inequalities $\beta > 0$ and $\alpha > 0$ follow from (5.5), (5.6), and condition 2.

It is not hard to prove that with no loss of generality we can assume $\beta = 1$. Using the Laplace inversion formula, we see that for every ξ with $\alpha_1 < \xi < \alpha_2$,

$$M(y) = \frac{1}{2\pi i} \int_{\xi-\infty i}^{\xi+\infty i} \mathcal{L}M(\lambda) e^{y\lambda} \, d\lambda.$$

It follows from (5.2) that

$$e^{(\tau-\alpha_1)\alpha y} M(\alpha y) = \frac{1}{2\pi i} \int_{\tau-\alpha_1+\xi-\infty i}^{\tau-\alpha_1+\xi+\infty i} I(z) e^{yz} \, dz.$$

By Cauchy's formula, condition 6, and the dominated convergence theorem, we can deform the contour of integration into a new contour η, consisting of the following three parts: the half-line $(-\infty i, -y^{-\frac{1}{2}}i]$, the half-circle Γ in the right half-plane of radius $y^{-\frac{1}{2}}$ centered at 0 (it is oriented counterclockwise), and finally the half-line $[y^{-\frac{1}{2}}i, \infty i)$. It follows that

$$\begin{aligned}
e^{(\tau-\alpha_1)\alpha y} M(\alpha y) &= \frac{1}{2\pi i} \int_\eta I(z) e^{\alpha y z}\, dz \\
&= \frac{1}{2\pi} \int_{-\infty}^{-y^{-\frac{1}{2}}} (ir)^{\gamma_1} G_1(ir)^{\gamma_2} G_2(ir) e^{\widetilde{F}(ir)} e^{-\frac{i\alpha}{r}} e^{i\alpha y r}\, dr \\
&\quad + \frac{1}{2\pi i} \int_\Gamma z^{\gamma_1} G_1(z)^{\gamma_2} G_2(z) e^{\widetilde{F}(z)} e^{\frac{\alpha}{z}} e^{\alpha y z}\, dz \\
&\quad + \frac{1}{2\pi} \int_{y^{-\frac{1}{2}}}^{\infty} (ir)^{\gamma_1} G_1(ir)^{\gamma_2} G_2(ir) e^{\widetilde{F}(ir)} e^{-\frac{i\alpha}{r}} e^{i\alpha y r}\, dr \\
&= I_1(y) + I_2(y) + I_3(y). \qquad (5.8)
\end{aligned}$$

We will first estimate $I_2(y)$. This will give the main contribution to the asymptotics. By making a substitution $z = y^{-\frac{1}{2}} e^{i\theta}$, $-\frac{\pi}{2} \le \theta \le \frac{\pi}{2}$, we see that

$$\begin{aligned}
I_2(y) &= \frac{1}{2\pi} y^{-\frac{1+\gamma_1}{2}} \int_{-\frac{\pi}{2}}^{\frac{\pi}{2}} e^{i\theta\gamma_1} G_1\big(y^{-\frac{1}{2}} e^{i\theta}\big)^{\gamma_2} G_2\big(y^{-\frac{1}{2}} e^{i\theta}\big) \\
&\quad \times \exp\big\{\widetilde{F}\big(y^{-\frac{1}{2}} e^{i\theta}\big)\big\} \exp\big\{\alpha\sqrt{y} e^{-i\theta}\big\} \exp\big\{\alpha\sqrt{y} e^{i\theta}\big\} e^{i\theta}\, d\theta.
\end{aligned}$$

Next, taking into account that

$$\sqrt{y}\big(e^{i\theta} + e^{-i\theta}\big) = 2\sqrt{y}\cos\theta = 2\sqrt{y} + 2\sqrt{y}(\cos\theta - 1),$$

we obtain

$$\begin{aligned}
I_2(y) &= \frac{1}{2\pi} y^{-\frac{1+\gamma_1}{2}} e^{2\alpha\sqrt{y}} \int_{-\frac{\pi}{2}}^{\frac{\pi}{2}} e^{i\theta(1+\gamma_1)} G_1\big(y^{-\frac{1}{2}} e^{i\theta}\big)^{\gamma_2} G_2\big(y^{-\frac{1}{2}} e^{i\theta}\big) \\
&\quad \times \exp\big\{\widetilde{F}\big(y^{-\frac{1}{2}} e^{i\theta}\big)\big\} \exp\big\{2\alpha\sqrt{y}(\cos\theta - 1)\big\}\, d\theta. \qquad (5.9)
\end{aligned}$$

It is easy to see that

$$e^{i\theta(1+\gamma_1)} = 1 + O(|\theta|), \qquad (5.10)$$

$$G_2\big(y^{-\frac{1}{2}} e^{i\theta}\big) - G_2(0) = O\big(y^{-\frac{1}{2}}\big), \qquad (5.11)$$

and

$$\exp\big\{\widetilde{F}\big(y^{-\frac{1}{2}} e^{i\theta}\big)\big\} - e^{\widetilde{F}(0)} = O\big(y^{-\frac{1}{2}}\big) \qquad (5.12)$$

on the contour Γ. Moreover, using (5.5) and the Taylor series of the function $z \mapsto (1+z)^{\gamma_2}$, we conclude that

$$G_1\big(y^{-\frac{1}{2}}e^{i\theta}\big)^{\gamma_2} - \big(\sqrt{y}e^{-i\theta}\big)^{\gamma_2} = \big[\sqrt{y}e^{-i\theta} + \widetilde{G}\big(y^{-\frac{1}{2}}e^{i\theta}\big)\big]^{\gamma_2} - \big(\sqrt{y}e^{-i\theta}\big)^{\gamma_2}$$
$$= O\big(y^{\frac{\gamma_2-1}{2}}\big)$$

on Γ. Note that the constant in the previous O-large estimate depends on $\widetilde{G}(0)$. We also have

$$G_1\big(y^{-\frac{1}{2}}e^{i\theta}\big)^{\gamma_2} - y^{\frac{\gamma_2}{2}} = O\big(y^{\frac{\gamma_2}{2}}|\theta|\big) + O\big(y^{\frac{\gamma_2-1}{2}}\big) \tag{5.13}$$

on Γ. It follows from (5.9)–(5.13) that

$$I_2(y) = \frac{1}{2\pi}G_2(0)e^{\widetilde{F}(0)}y^{-\frac{1+\gamma_1-\gamma_2}{2}}e^{2\alpha\sqrt{y}}\int_{-\frac{\pi}{2}}^{\frac{\pi}{2}}\exp\big\{2\alpha\sqrt{y}(\cos\theta-1)\big\}$$
$$\times\big(1+O\big(y^{-\frac{1}{2}}\big)+O\big(|\theta|\big)\big)\,d\theta. \tag{5.14}$$

We will next employ Laplace's method to estimate the integral appearing in (5.14). Consider the integral $\int_a^b e^{-s\Phi(x)}\psi(x)\,dx$, where $\Phi \in C^\infty[a,b]$ and $\psi \in C^\infty[a,b]$ (much less is needed from the functions Φ and ψ), and assume that there is an $x_0 \in (a,b)$ such that $\Phi'(x_0) = 0$, and $\Phi''(x_0) > 0$ throughout $[a,b]$. Then the following assertion holds:

Lemma 5.2 *Under the above assumptions, with $s > 0$ and $s \to \infty$,*

$$\int_a^b e^{-s\Phi(x)}\psi(x)\,dx = e^{-s\Phi(x_0)}\bigg[\frac{A}{\sqrt{s}} + O\big(s^{-\frac{3}{2}}\big)\bigg], \tag{5.15}$$

where $A = \sqrt{2\pi}\,\psi(x_0)(\Phi''(x_0))^{-\frac{1}{2}}$.

The proof of Lemma 5.2 can be found, e.g., in [BH95], formula (5.1.21). Next, using (5.15) with $a = -\frac{\pi}{2}$, $b = \frac{\pi}{2}$, $\Phi(x) = 1 - \cos x$, $\psi(x) = 1$, $x_0 = 0$, and $s = 2\alpha\sqrt{y}$, we see that

$$\int_{-\frac{\pi}{2}}^{\frac{\pi}{2}} e^{2\alpha\sqrt{y}(\cos\theta-1)}\,d\theta = \frac{\sqrt{\pi}}{\sqrt{\alpha}}y^{-\frac{1}{4}} + O\big(y^{-\frac{3}{4}}\big) \tag{5.16}$$

as $y \to \infty$. Similarly

$$\int_{-\frac{\pi}{2}}^{\frac{\pi}{2}} |\theta|e^{2\alpha\sqrt{y}(\cos\theta-1)}\,d\theta = O\big(y^{-\frac{3}{4}}\big) \tag{5.17}$$

as $y \to \infty$. Therefore, (5.14), (5.16), and (5.17) give

$$I_2(y) = \frac{1}{2\sqrt{\pi\alpha}}G_2(0)e^{\widetilde{F}(0)}y^{-\frac{3}{4}+\frac{\gamma_2-\gamma_1}{2}}e^{2\alpha\sqrt{y}}\big(1+O\big(y^{-\frac{1}{2}}\big)\big) \tag{5.18}$$

as $y \to \infty$.

It remains to show that the integrals $I_1(y)$ and $I_3(y)$ can be included in the error term in formula (5.7). The proof of the previous statement uses condition 6 formulated above. We have

$$I_3(y) = \frac{1}{2\pi} \int_{y^{-\frac{1}{2}}}^{\infty} I(ir) e^{i\alpha y r} \, dr.$$

Next using (5.4) and condition 6, we see that for large values of y,

$$\begin{aligned} |I_3(y)| &\le \frac{1}{2\pi} \int_{y^{-\frac{1}{2}}}^{\infty} |I(ir)| \, dr \le \frac{1}{2\pi} \int_{y^{-\frac{1}{2}}}^{N} |I(ir)| \, dr + \frac{1}{2\pi} \int_N^{\infty} \psi(r) \, dr \\ &\le c \int_{y^{-\frac{1}{2}}}^{N} r^{\gamma_1 - \gamma_2} \, dr + \frac{1}{2\pi} \int_N^{\infty} \psi(r) \, dr, \end{aligned}$$

where the constant c does not depend on y. Similar reasoning can be used in the case of the integral I_1. Combining the resulting estimates, we obtain

$$I_1(y) + I_3(y) = O\big(y^{\frac{\gamma_2 - \gamma_1 - 1}{2}}\big) + O(1) \tag{5.19}$$

as $y \to \infty$. It follows from (5.8), (5.18), and (5.19) that

$$e^{(\tau - \alpha_1)\alpha y} M(\alpha y) = \frac{1}{2\sqrt{\pi \alpha}} G_2(0) e^{\widetilde{F}(0)} y^{-\frac{3}{4} + \frac{\gamma_2 - \gamma_1}{2}} e^{2\alpha\sqrt{y}} \big(1 + O\big(y^{-\frac{1}{2}}\big)\big)$$

as $y \to \infty$. Now it is clear that (5.7) holds.

This completes the proof of Theorem 5.1. □

5.2 Asymptotic Behavior of Fractional Integrals

Let $\alpha \ge 0$ be a real number (the order of fractional integration). Then the fractional integral F_α is a linear operator defined by

$$F_\alpha f(\sigma) = \frac{1}{\Gamma(\alpha)} \int_\sigma^{\infty} (\tau - \sigma)^{\alpha - 1} f(\tau) \, d\tau. \tag{5.20}$$

Let $c > 0$, and suppose M is a function on $[c, \infty)$ given by

$$M(x) = a(x) e^{-b(x)}, \tag{5.21}$$

where a and b are positive functions on $[c, \infty)$, satisfying the following conditions:

1. $x|a'(x)| \le \gamma a(x)$ for some $\gamma > 0$.
2. $b(x) = B(\log x)$, with B a positive increasing function, and $B''(x) \approx 1$ as $x \to \infty$.

Then it is not hard to see that $B'(x) \approx x$, and $B(x) \approx x^2$ as $x \to \infty$.

The next assertion is an Abelian theorem for fractional integrals.

Theorem 5.3 *Let M be defined by* (5.21), *with a and b satisfying conditions* 1 *and* 2 *formulated above. Then for every $\alpha > 0$,*

$$F_\alpha M(\sigma) = \frac{M(\sigma)}{b'(\sigma)^\alpha}\big(1 + O\big((\log\sigma)^{-1}\big)\big) \quad \text{as } \sigma \to \infty. \tag{5.22}$$

Proof of Theorem 5.3 We first observe that

$$a(\rho x) \le \rho^\gamma a(x), \quad \text{if } \rho \ge 1, \quad \text{and} \quad a(\rho x) \le \rho^{-\gamma} a(x), \quad \text{if } \rho < 1. \tag{5.23}$$

In fact with $A(u) = a(e^u)$, we have $|A'| \le \gamma A$, and hence $|A'/A| \le \gamma$. Integrating the previous inequality, we obtain

$$\big|\log A(u+\delta) - \log A(u)\big| \le \gamma|\delta|.$$

If we exponentiate this, we get $a(e^u e^\delta) \le a(u)e^{\gamma|\delta|}$, and we need only take $x = e^u$ and $\rho = e^\delta$.

Next observe that

$$b\big(\sigma(1+y)\big) \ge b(\sigma) + c\log(y+1)\log\sigma, \tag{5.24}$$

for some $c > 0$, with $y \ge -\frac{1}{2}$, and σ large. Moreover,

$$b\big(\sigma(1+y)\big) - b(\sigma) - y\sigma b'(\sigma) = O\big(y^2\log\sigma\big), \tag{5.25}$$

for $0 \le y \le 1$, and σ large.

To prove (5.24), note that

$$b\big(\sigma(1+y)\big) - b(\sigma) = \int_\sigma^{\sigma(1+y)} b'(t)\,dt,$$

while $ub'(u) = B'(\log u)$. In addition $B'(x) \ge cx$ for large x. Similarly, $\sigma^2 b''(\sigma) = O(\log\sigma)$ implies

$$b\big(\sigma(1+y)\big) - b(\sigma) - y\sigma b'(\sigma) = \int_\sigma^{\sigma(1+y)} \big(\sigma(1+y) - t\big)b''(t)\,dt = O\big(y^2\log\sigma\big).$$

Making the change of variables $\tau = (1+y)\sigma$ (with σ fixed) in the integral on the right-hand side of (5.22), we get the following expression:

$$\frac{1}{\Gamma(\alpha)}\sigma^\alpha \int_0^\infty y^{\alpha-1} M\big(\sigma(1+y)\big)\,dy.$$

The main contribution to this integral is for y near zero, and it turns out that it is therefore convenient to write

$$\int_0^\infty y^{\alpha-1} M\big(\sigma(1+y)\big)\,dy = \int_0^{(\log\sigma)^{-\frac{1}{2}}} + \int_{(\log\sigma)^{-\frac{1}{2}}}^\infty = I_1 + I_2.$$

For the error term we use (5.23) and (5.24). It follows directly that the term I_2 is majorized by

$$a(\sigma)e^{-b(\sigma)} \int_{(\log\sigma)^{-\frac{1}{2}}}^\infty y^{\alpha-1}(1+y)^\gamma e^{-c\log(1+y)\log\sigma}\,dy,$$

while the integral is clearly $O(\exp\{-c'(\log\sigma)^{\frac{1}{2}}\})$. Thus

$$I_2 = O\big(M(\sigma)\exp\{-c'(\log\sigma)^{\frac{1}{2}}\}\big). \tag{5.26}$$

To consider I_1, we first replace in it $a(\sigma(1+y))$ by $a(\sigma)$ and $b(\sigma(1+y))$ by $b(\sigma) + y\sigma b'(\sigma)$, giving us the following expression:

$$I_0 = \int_0^{(\log\sigma)^{-\frac{1}{2}}} y^{\alpha-1} a(\sigma)\exp\big\{-\big[b(\sigma) + y\sigma b'(\sigma)\big]\big\}\,dy.$$

It follows that

$$I_0 = a(\sigma)e^{-b(\sigma)} \\ \times \left(\int_0^\infty y^{\alpha-1}\exp\big\{-y\sigma b'(\sigma)\big\}\,dy - \int_{(\log\sigma)^{-\frac{1}{2}}}^\infty y^{\alpha-1}\exp\big\{-y\sigma b'(\sigma)\big\}\,dy\right). \tag{5.27}$$

The first integral in (5.27) is exactly $\Gamma(\alpha)(\sigma b'(\sigma))^{-\alpha}$, while the second integral is clearly

$$O\big(\exp\{-c'(\log\sigma)^{\frac{1}{2}}\}\big)$$

since

$$\sigma b'(\sigma) = B'(\log\sigma) \geq c\log\sigma.$$

Therefore (note $\sigma b'(\sigma) \approx \log\sigma$),

$$I_0 = \Gamma(\alpha)M(\sigma)\big(\sigma b'(\sigma)\big)^{-\alpha}\big(1 + O\big(\exp\{-c'(\log\sigma)^{\frac{1}{2}}\}\big)\big). \tag{5.28}$$

Next, let

$$I_0' = a(\sigma)\int_0^{(\log\sigma)^{-\frac{1}{2}}} y^{\alpha-1}e^{-b(\sigma(1+y))}\,dy.$$

Since $|1-e^{-u}| \leq c|u|$ if $|u| \leq 1$, and (5.25) holds, we see that

$$\left|e^{-b(\sigma(1+y))} - e^{-(b(\sigma)+y\sigma b'(\sigma))}\right| \leq ce^{-(b(\sigma)+y\sigma b'(\sigma))}y^2 \log\sigma,$$

and therefore

$$\left|I_0' - I_0\right| \leq ca(\sigma)\log\sigma \int_0^{(\log\sigma)^{-\frac{1}{2}}} y^{\alpha-1}y^2 e^{-(b(\sigma)+y\sigma b'(\sigma))}\,dy. \tag{5.29}$$

It follows from (5.28) with α replaced by $\alpha+2$ and from (5.29) that

$$I_0' - I_0 = O\big(M(\sigma)(\log\sigma)\big(\sigma b'(\sigma)\big)^{-\alpha-2}\big). \tag{5.30}$$

Using (5.28) and (5.30), we see that

$$I_0' - I_0 = O\big(M(\sigma)\big(\sigma b'(\sigma)\big)^{-\alpha}\big)$$

and also

$$I_0' = O\big(M(\sigma)\big(\sigma b'(\sigma)\big)^{-\alpha}\big). \tag{5.31}$$

Finally, we have

$$I_1 - I_0' = O\left(a(\sigma)\int_0^{(\log\sigma)^{-\frac{1}{2}}} y^{\alpha}e^{-b(\sigma(1+y))}\,dy\right). \tag{5.32}$$

To prove formula (5.32), we observe that

$$a\big(\sigma(1+y)\big) - a(\sigma) = y\sigma a'(\bar{\sigma})$$

for some $\bar{\sigma}$, $\sigma \leq \bar{\sigma} \leq 2\sigma$ (because for large σ, $0 \leq y \leq 1$ in the above integral), and moreover $\sigma a'(\bar{\sigma}) \leq ca(\sigma)$ because $\bar{\sigma}a'(\bar{\sigma}) \leq \gamma a(\bar{\sigma})$, and the estimates in (5.23) hold. Therefore,

$$I_1 - I_0' = O\big(M(\sigma)\big(\sigma b'(\sigma)\big)^{\alpha-1}\big), \tag{5.33}$$

by estimate (5.31) with $\alpha+1$ instead of α. Putting estimates (5.26), (5.28), (5.30), and (5.33) together gives the asserted conclusion, once we note that

$$\exp\big\{-c(\log\sigma)^{\frac{1}{2}}\big\} = O\big((\log\sigma)^{-1}\big)$$

as $\sigma \to \infty$. □

Remark 5.4 Let $\widetilde{M}$ be a function satisfying

$$\widetilde{M}(\sigma) = M(\sigma)\big(1 + O\big(\rho(\sigma)\big)\big)$$

as $\sigma \to \infty$. Here M is a function such as in Theorem 5.3, and ρ is a function for which $\rho(\sigma) \downarrow 0$ as $\sigma \to \infty$. Then for every $\alpha > 0$ the estimate in Theorem 5.3 holds in the following form:

$$F_\alpha \widetilde{M}(\sigma) = \frac{M(\sigma)}{b'(\sigma)^\alpha}\left(1 + O\left(\rho(\sigma) + (\log\sigma)^{-1}\right)\right)$$

as $\sigma \to \infty$.

5.3 Asymptotic Behavior of Integral Operators with Log-Normal Kernels

Let us recall that the stock price distribution density D_t in an uncorrelated stochastic volatility model can be represented by the following formula (see (3.5)):

$$D_t\left(x_0 e^{\mu t} x\right) = \frac{1}{x_0 e^{\mu t}\sqrt{2\pi t}} x^{-\frac{3}{2}} \int_0^\infty y^{-1} m_t(y) \exp\left\{-\left[\frac{\log^2 x}{2ty^2} + \frac{ty^2}{8}\right]\right\} dy, \tag{5.34}$$

where m_t is the mixing distribution density (if this density exists).

We will next prove an Abelian theorem for the special integral operator appearing on the right-hand side of formula (5.34).

Theorem 5.5 *Let A, ζ, and b be positive Borel functions on $[0, \infty)$, and suppose the following conditions hold:*

1. *The functions A and ζ are integrable over any finite sub-interval of $[0, \infty)$.*
2. *The function b is bounded and $\lim_{y\to\infty} b(y) = 0$.*
3. *There exist $y_1 > 0$, $c > 0$, and γ with $0 < \gamma \le 1$ such that ζ and b are differentiable on $[y_1, \infty)$, and in addition,*

$$\left|\zeta'(y)\right| \le c y^{-\gamma} \zeta(y) \quad \text{and} \quad \left|b'(y)\right| \le c y^{-\gamma} b(y)$$

for all $y \ge y_1$.

4. *For every $a > 0$ there exists $y_a > 0$ such that*

$$b(y)\zeta(y) \ge \exp\left\{-a\, y^4\right\}$$

for all $y > y_a$.

5. *There exists a real number l such that*

$$A(y) = e^{ly}\zeta(y)\left(1 + O\left(b(y)\right)\right)$$

as $y \to \infty$.

Then, for every fixed $k > 0$,

$$\int_0^\infty A(y)\exp\left\{-\left(\frac{w^2}{y^2}+k^2y^2\right)\right\}dy$$
$$=\frac{\sqrt{\pi}}{2k}\exp\left\{\frac{l^2}{16k^2}\right\}\zeta\left(k^{-\frac{1}{2}}w^{\frac{1}{2}}\right)\exp\left\{lk^{-\frac{1}{2}}w^{\frac{1}{2}}\right\}e^{-2kw}$$
$$\times\left[1+O\left(w^{-\frac{\gamma}{2}}\right)+O\left(b\left(k^{-\frac{1}{2}}w^{\frac{1}{2}}\right)\right)\right]$$

as $w \to \infty$.

Proof of Theorem 5.5 We first formulate a little stronger result.

Theorem 5.6 *Let ζ be a positive Borel function defined on $[0,\infty)$, and suppose the following conditions hold:*

1. *The function ζ is integrable over any finite sub-interval of $[0,\infty)$.*
2. *There exist $y_1 > 0$, $c > 0$, and γ with $0 < \gamma \le 1$ such that the function ζ is differentiable on $[y_1,\infty)$ and $|\zeta'(y)| \le cy^{-\gamma}\zeta(y)$ for all $y > y_1$.*
3. *For every $a > 0$ there exists $y_a > 0$ such that $\zeta(y) \ge \exp\{-a\,y^4\}$ for all $y > y_a$.*

Let l be a real number, and put $A(y) = e^{ly}\zeta(y)$. Then, for every fixed $k > 0$,

$$\int_0^\infty A(y)\exp\left\{-\left(\frac{w^2}{y^2}+k^2y^2\right)\right\}dy$$
$$=\frac{\sqrt{\pi}}{2k}\exp\left\{\frac{l^2}{16k^2}\right\}\zeta\left(k^{-\frac{1}{2}}w^{\frac{1}{2}}\right)\exp\left\{lk^{-\frac{1}{2}}w^{\frac{1}{2}}\right\}e^{-2kw}$$
$$\times\left(1+O\left(w^{-\frac{\gamma}{2}}\right)\right) \tag{5.35}$$

as $w \to \infty$.

Let us show that Theorem 5.6 implies Theorem 5.5. Suppose Theorem 5.6 holds, and let A, ζ, and b be such as in Theorem 5.5. With no loss of generality, we can assume that

$$A(y) = e^{ly}\zeta(y) + e^{ly}\zeta(y)\tilde{b}(y), \tag{5.36}$$

where $\tilde{b}$ is a real function on $[0,\infty)$ such that $|\tilde{b}(y)| \le \tau b(y)$ for some $\tau > 0$ and all $y > 0$. Since the functions ζ and ζb satisfy the conditions in Theorem 5.6, formula (5.35) holds for the functions $A_1(y) = e^{ly}\zeta(y)$ and $A_2(y) = e^{ly}\zeta(y)b(y)$. This observation and (5.36) imply Theorem 5.5.

We will next prove Theorem 5.6. Let us denote by $I(\omega)$ the integral on the left-hand side of (5.35). If the function A has compact support, then there exists $c > 0$ such that $I(\omega) = O(\exp\{-c\,\omega^2\})$, $\omega \to \infty$. Therefore, for such a function A, the integral $I(\omega)$ can be incorporated into the error term in formula (5.35). This can be seen by taking into account condition 3 in Theorem 5.6. It follows that with no loss of generality we can assume that $y_1 = 0$ in condition 2 in the formulation of Theorem 5.6. We first consider the case when $0 < \gamma < 1$. Since condition 2 holds for the function ζ, it follows that

$$\left|\log\zeta(x+u) - \log\zeta(x)\right| \le c\left|\int_x^{x+u} y^{-\gamma}\,dy\right| \le c'\left|(x+u)^{1-\gamma} - x^{1-\gamma}\right|$$

for $x > 0$ and $x + u > 0$. As a result

$$\zeta(x+u) \le \zeta(x)\exp\{c_1|u|x^{-\gamma}\}, \quad \text{if } |u| \le x, \tag{5.37}$$

and

$$\zeta(x+u) \le \zeta(x)\exp\{c_1|u|^{1-\gamma}\}, \quad \text{if } |u| \ge x. \tag{5.38}$$

Now in considering the integral in (5.35) it is useful to make the following observations. First, by making the change of variables $y \mapsto \frac{y}{k}$, we can reduce to the case $k = 1$. Next

$$\frac{w^2}{y^2} + y^2 = \left(\frac{w}{y} - y\right)^2 + 2w,$$

so making a further change of variables $y \mapsto w^{\frac{1}{2}}y$, we can write the integral in (5.35) as

$$w^{\frac{1}{2}}e^{-2w}\int_0^\infty \exp\left\{-w\left(\frac{1}{y} - y\right)^2\right\}A\left(w^{\frac{1}{2}}y\right)dy.$$

Here the main contribution comes from the neighborhood of $y = 1$, so we make a last change of variables $y = 1 + x$ and write the above integral, except the factor $w^{\frac{1}{2}}e^{-2w}$, as

$$\begin{aligned}\int_{-1}^\infty & e^{-wQ(x)}A\left(w^{\frac{1}{2}}(1+x)\right)dx \\ &= \int_{\{x\ge -1:|x|\le\delta\}} e^{-wQ(x)}A\left(w^{\frac{1}{2}}(1+x)\right)dx \\ &\quad + \int_{\{x\ge -1:|x|>\delta\}} e^{-wQ(x)}A\left(w^{\frac{1}{2}}(1+x)\right)dx = I_1 + I_2,\end{aligned} \tag{5.39}$$

where

$$Q(x) = \left(\frac{1}{1+x} - 1 - x\right)^2$$

and $\delta = w^{\frac{\gamma}{2}-\frac{1}{2}}$.

Let us consider the integral I_2. We will first assume that $x > \delta$. Now easily $Q(x) \ge x^2$, while

$$A\left(w^{\frac{1}{2}}(1+x)\right) = \exp\{lw^{\frac{1}{2}}\}\exp\{lw^{\frac{1}{2}}x\}\zeta\left(w^{\frac{1}{2}}(1+x)\right).$$

Using (5.37) and (5.38), we see that

$$\zeta\left(w^{\frac{1}{2}}(1+x)\right) \le \zeta\left(w^{\frac{1}{2}}\right)\exp\left\{c_1 w^{\frac{1}{2}} x w^{-\frac{\gamma}{2}}\right\}, \quad \text{if } x \le 1,$$

and

$$\zeta\left(w^{\frac{1}{2}}(1+x)\right) \le \zeta\left(w^{\frac{1}{2}}\right)\exp\left\{c_1 \left(w^{\frac{1}{2}} x\right)^{1-\gamma}\right\}, \quad \text{if } x \ge 1.$$

Therefore,

$$\zeta\left(w^{\frac{1}{2}}(1+x)\right) \le \zeta\left(w^{\frac{1}{2}}\right)\exp\left\{c_1 w^{\frac{1-\gamma}{2}} x\right\} \tag{5.40}$$

for all $x > 0$. Then clearly

$$e^{-wQ(x)} A\left(w^{\frac{1}{2}}(1+x)\right) \le \zeta\left(w^{\frac{1}{2}}\right)\exp\left\{l w^{\frac{1}{2}}\right\}\exp\left\{-c_2 w x^2\right\}$$

for all $x > 0$ and large enough w's. Altogether then, the integral taken over $x \ge \delta$ is

$$\begin{aligned} &O\left(\exp\left\{l w^{\frac{1}{2}}\right\}\zeta\left(w^{\frac{1}{2}}\right)\int_{\omega^{\frac{\gamma-1}{2}}}^{\infty} \exp\left\{-c_2 w x^2\right\} dx\right) \\ &\quad = O\left(\exp\left\{l w^{\frac{1}{2}}\right\}\zeta\left(w^{\frac{1}{2}}\right)\exp\left\{-c_3 w^{\gamma}\right\}\right) \end{aligned}$$

as $w \to \infty$, and this expression multiplied by $w^{\frac{1}{2}} e^{-2w}$ is clearly a part of the error term in formula (5.35). The other part of the integral I_2 (where $-1 < x \le -\delta$) is treated similarly.

Let us now consider the main term I_1, and replace in (5.39)

$$A\left(w^{\frac{1}{2}}(1+x)\right) = \exp\left\{l w^{\frac{1}{2}}(1+x)\right\}\zeta\left(w^{\frac{1}{2}}(1+x)\right)$$

by $\exp\{l w^{\frac{1}{2}}\}\exp\{l w^{\frac{1}{2}} x\}\zeta(w^{\frac{1}{2}})$. Then for that integral we get

$$\exp\left\{l w^{\frac{1}{2}}\right\}\zeta\left(w^{\frac{1}{2}}\right)\int_{|x|\le\delta} e^{-wQ(x)}\exp\left\{l w^{\frac{1}{2}} x\right\} dx.$$

Now for small x we have

$$Q(x) = \left(\frac{1}{1+x} - (1+x)\right)^2 = 4x^2 + O\left(x^3\right).$$

Since

$$4wx^2 - w^{\frac{1}{2}} l x = 4w\left(x - \frac{l w^{-\frac{1}{2}}}{8}\right)^2 - \frac{l^2}{16},$$

and the critical point of the above expression, $x = 8^{-1} l w^{-\frac{1}{2}}$, is certainly well inside $|x| \le \delta = w^{\frac{\gamma-1}{2}}$, we can apply the usual arguments of stationary phase (see, e.g. [SSh03], Appendix A) to the integral

$$\int_{|x|\le\delta} e^{-wQ(x)} \exp\{w^{\frac{1}{2}} lx\}\, dx. \tag{5.41}$$

This shows that the previous integral equals

$$\int_{-\infty}^{\infty} \exp\left\{-4w\left(x - \frac{lw^{-\frac{1}{2}}}{8}\right)^2 + \frac{l^2}{16}\right\} dx + O\left(w^{-1}\right)$$
$$+ O\left(\int_{|x|\ge\delta} \exp\{-cwx^2\}\, dx\right). \tag{5.42}$$

Note that the second term in (5.42) estimates the error of replacing $Q(x)$ by $4x^2$ in (5.41). The integral over $|x| \ge \delta$ is $O(\exp\{-c'w^{\gamma}\})$ as before, and hence the result is clearly

$$\frac{1}{2w^{\frac{1}{2}}} \sqrt{\pi} \exp\left\{\frac{l^2}{16}\right\}\left(1 + O\left(w^{-\frac{1}{2}}\right)\right) \quad \text{as } w \to \infty.$$

Finally, we consider the error of replacing $\zeta(w^{\frac{1}{2}}(1+x))$ by $\zeta(w^{\frac{1}{2}})$. By the mean-value theorem, this difference is $w^{\frac{1}{2}} x\zeta'(w^{\frac{1}{2}}(1+\bar{x}))$, where again $|\bar{x}| \le \delta$, and this is $O(w^{\frac{1-\gamma}{2}} x\zeta(w^{\frac{1}{2}}))$, by condition 2 in Theorem 5.6 and (5.40). Hence, the difference contributes to the integral I_1 a quantity

$$O\left(\exp\{lw^{\frac{1}{2}}\}\zeta(w^{\frac{1}{2}})w^{\frac{1-\gamma}{2}} \int_{|x|\le\frac{1}{2}} |x| e^{-wQ(x)} \exp\{lw^{\frac{1}{2}} x\}\, dx\right),$$

which is $O(\exp\{lw^{\frac{1}{2}}\}\zeta(w^{\frac{1}{2}})w^{-\frac{1}{2}}w^{-\frac{\gamma}{2}})$ as $w \to \infty$. This gives us the final result when $0 < \gamma < 1$. The case when $\gamma = 1$ is similar, except that now we take $\delta = \frac{1}{2}$, and use the fact that now $\zeta(x+u) \le c\zeta(x)$ if $|u| \le |x|$, instead of (5.37).

This completes the proof of Theorem 5.6. □

Our primary interest in the next sections is to establish sharp asymptotic formulas with error estimates for mixing distribution densities associated with the classical volatility processes.

5.4 Asymptotic Formulas for Mixing Distribution Densities Associated with Geometric Brownian Motions

Let us first consider a geometric Brownian motion with $\nu = \frac{1}{2}$, $\xi = 1$, and $y_0 = 1$. The next lemma provides an explicit expression for the mixing density m_t in this case.

Lemma 5.7 *The following formula holds*:

$$
\begin{aligned}
& m_t\left(y; \frac{1}{2}, 1, 1\right) \\
&\quad = \frac{1}{\pi t} \exp\left\{\frac{\pi^2}{8t}\right\} y^{-2} \\
&\quad \times \int_{-\infty}^{\infty} \exp\left\{-\frac{\cosh^2 u}{2ty^2}\right\} \cosh u \exp\left\{-\frac{u^2}{2t}\right\} \exp\left\{\frac{i\pi u}{2t}\right\} du. \qquad (5.43)
\end{aligned}
$$

Proof Using the contour shift $y \mapsto y + \frac{\pi i}{2}$ in formula (4.1), we see that

$$
\begin{aligned}
& \int_0^{\infty} e^{-\lambda t y^2} m_t(y)\, dy \\
&\quad = \frac{1}{\sqrt{2\pi t}} e^{\frac{\pi^2}{8t}} \int_{-\infty}^{\infty} \exp\{-\sqrt{2\lambda} \cosh y\} \exp\left\{-\frac{y^2}{2t}\right\} \exp\left\{\frac{i\pi y}{2t}\right\} dy.
\end{aligned}
$$

It follows from the previous equality and (4.77) that

$$
\begin{aligned}
& \int_0^{\infty} e^{-\lambda t y^2} m_t(y)\, dy \\
&\quad = \frac{1}{\sqrt{2\pi t}} e^{\frac{\pi^2}{8t}} \int_{-\infty}^{\infty} \exp\left\{-\frac{y^2}{2t}\right\} \exp\left\{\frac{i\pi y}{2t}\right\} dy \int_0^{\infty} \exp\left\{-\frac{\lambda \cosh^2 y}{2z}\right\} \frac{e^{-z}}{\sqrt{\pi z}}\, dz.
\end{aligned}
$$

Next, replacing λ by $t^{-1}\lambda$ in the formula above and making the substitution

$$
v = (2tz)^{-\frac{1}{2}} \cosh y,
$$

we obtain

$$
\begin{aligned}
& \int_0^{\infty} e^{-\lambda y^2} m_t(y)\, dy \\
&\quad = \frac{1}{t\pi} e^{\frac{\pi^2}{8t}} \int_{-\infty}^{\infty} \exp\left\{-\frac{y^2}{2t}\right\} \exp\left\{\frac{i\pi y}{2t}\right\} \cosh y\, dy \\
&\quad \times \int_0^{\infty} e^{-\lambda v^2} v^{-2} \exp\left\{-\frac{\cosh^2 y}{2tv^2}\right\} dv \\
&\quad = \frac{1}{t\pi} e^{\frac{\pi^2}{8t}} \int_0^{\infty} e^{-\lambda v^2} v^{-2}\, dv \\
&\quad \times \int_{-\infty}^{\infty} \exp\left\{-\frac{\cosh^2 y}{2tv^2}\right\} \exp\left\{-\frac{y^2}{2t}\right\} \exp\left\{\frac{i\pi y}{2t}\right\} \cosh y\, dy. \qquad (5.44)
\end{aligned}
$$

Now, it is not hard to see using (5.44) that formula (5.43) holds. □

The next theorem characterizes the asymptotic behavior of the mixing distribution density m_t associated with the geometric Brownian motion.

Theorem 5.8 *The following formula holds*:

$$m_t(y;\nu,\xi,y_0) = c_1 y^{c_2}(\log y)^{c_3}\exp\left\{-\frac{1}{2t\xi^2}\left(\log\frac{y}{y_0}+\frac{1}{2}\log\log\frac{y}{y_0}\right)^2\right\}$$
$$\times\left(1+O\left((\log y)^{-\frac{1}{2}}\right)\right) \tag{5.45}$$

as $y\to\infty$, *where*

$$c_1 = \frac{1}{\xi\sqrt{\pi t}}2^{-\frac{1}{2t\xi^2}}y_0^{\frac{2\log 2-1}{2t\xi^2}-\alpha}\exp\left\{-\frac{(\log 2)^2}{2t\xi^2}\right\}\exp\left\{-\frac{\alpha^2\xi^2 t}{2}\right\},$$

$$c_2 = \alpha - 1 + \frac{1-2\log 2}{2t\xi^2},\quad c_3 = \frac{\alpha}{2} - \frac{1+2\log 2}{4t\xi^2},\quad \textit{and}\quad \alpha = \frac{2\nu-\xi^2}{2\xi^2}.$$

We will next give an incomplete sketch of the proof of Theorem 5.8. The rigorous proof will be given later. Theorem 5.8 will be first established in the special case where $\nu=\frac{1}{2}$, $\xi=1$, and $y_0=1$. In this case, Lemma 5.7 will play an important role in the proof. After that we deal with the situation where $\xi=1$ and $\nu=\alpha+\frac{1}{2}$. We denote by $m_t^{(\alpha)}$ the mixing distribution density in this case and keep the notation m_t when $\alpha=0$ (i.e. $\nu=\frac{1}{2}$).

In order to understand the behavior of the integral on the right-hand side of (5.43), the following oscillating integral is studied:

$$I(\varepsilon) = \int_{-\infty}^{\infty}\exp\{-\varepsilon(\cosh u)^2\}\exp\left\{-\frac{u^2}{2t}\right\}\cosh u\,\exp\left\{\frac{i\pi u}{2t}\right\}du \tag{5.46}$$

where $\varepsilon>0$. It will be shown that

$$I(\varepsilon) = I_0(\varepsilon)\left(1+O\left(\log\frac{1}{\varepsilon}\right)^{-\frac{1}{2}}\right) \tag{5.47}$$

as $\varepsilon\to 0$, where

$$I_0(\varepsilon) = \left(\frac{\pi}{2}\right)^{\frac{1}{2}}\exp\left\{-\frac{\pi^2}{8t}\right\}\exp\left\{-\frac{N_\varepsilon^2}{2t}+\frac{N_\varepsilon}{2t}\right\}\varepsilon^{-\frac{1}{2}}, \tag{5.48}$$

and N_ε is the solution of the equation

$$\varepsilon\sinh(2N_\varepsilon) = \frac{N_\varepsilon}{t}.$$

The proof requires that we deform the contour of integration for I_ε (the real one) into the complex u-plane, where the principal contributions are then given on the segments $[N_\varepsilon, N_\varepsilon + i\pi]$ and $[-N_\varepsilon, -N_\varepsilon + i\pi]$. With formula (5.47) (and the corresponding asymptotics for $(\frac{d}{d\varepsilon})^k I(\varepsilon))$ one obtains the required result for $\alpha = 0$ and $\alpha = 2k$, respectively, where k is a positive integer. Next, we consider the case where α is different from a non-negative even integer. In this case, we use Dufresne's recurrence formula (see Sect. 5.4.2). Finally, we drop the restriction $\xi = 1$, $y_0 = 1$, using Lemma 5.26 established below.

We will next start proving Theorem 5.8. This theorem will be derived from another assertion, which provides a less explicit asymptotic formula for m_t. Let u_y denote the unique positive solution of the equation

$$\frac{\sinh(2u_y)}{2u_y} = y^2, \quad y > 0, \tag{5.49}$$

and let the function Λ_t be defined by

$$\Lambda_t(y) = \exp\left\{-\frac{u_y^2}{2t} + \frac{u_y}{2t}\right\}$$

where $t, y > 0$.

Theorem 5.9 *The following formula holds*:

$$m_t(y; \nu, \xi, y_0) = c y^{\alpha-1} (\log y)^{\frac{\alpha}{2}} \Lambda_{t\xi^2}\left(\frac{y}{y_0}\right)\left(1 + O\left((\log y)\right)^{-\frac{1}{2}}\right) \tag{5.50}$$

as $y \to \infty$, *where* α *is such as in Theorem* 5.8 *and*

$$c = \frac{1}{y_0^{\alpha} \xi \sqrt{\pi t}} \exp\left\{-\frac{\alpha^2 \xi^2 t}{2}\right\}.$$

The next lemma describes the asymptotic behavior of u_y.

Lemma 5.10 *The following formula holds*:

$$u_y = \log y + \frac{1}{2}\log\log y + \log 2 + O\left(\frac{\log\log y}{\log y}\right) \tag{5.51}$$

as $y \to \infty$.

Remark 5.11 A yet more precise formula is

$$u_y = \log y + \frac{1}{2}\log\log y + \log 2 + \frac{\frac{1}{2}[\frac{1}{2}\log\log y + \log 2]}{\log y} + O\left(\left(\frac{\log\log y}{\log y}\right)^2\right). \tag{5.52}$$

Proof of formula (5.52) Observe that the function $\frac{\sinh(2u)}{2u}$ is strictly increasing for $u > u_0$ and tends to ∞ as $u \to \infty$. Thus, for all sufficiently large positive y, the number u_y exists and we have $u_y \to \infty$ as $y \to \infty$. Since

$$e^{2u_y}\left[1 - e^{-4u_y}\right] = 4y^2 u_y,$$

we have

$$u_y = \log y + \frac{1}{2}\log u_y + \log 2 + O\left(e^{-4u_y}\right) \tag{5.53}$$

as $y \to \infty$. Note that from this we have $u_y \le c \log y$ for sufficiently large y. Next, we plug the expression for u_y in formula (5.53) into the term $\log u_y$ in the same formula, and use the fact that

$$\log(A + B) = \log A + \frac{B}{A} + O\left(\frac{B^2}{A^2}\right),$$

wherever $0 \le B < A$. The result is

$$u_y = \log y + \frac{1}{2}\log\log y + \log 2 + O\left(\frac{\log\log y}{\log y}\right).$$

Inserting this last equation in (5.53) again then gives

$$u_y = \log y + \log 2 + \frac{1}{2}\log\left[\log y + \frac{1}{2}\log\log y + \log 2 + O\left(\frac{\log\log y}{\log y}\right)\right]$$
$$+ O\left(\left(\frac{\log\log y}{\log y}\right)^2\right),$$

which proves formula (5.52). □

The procedure above can of course be iterated indefinitely to give better and better asymptotic descriptions of u_y as $y \to \infty$.

It is not hard to see that Theorem 5.8 follows from Theorem 5.9 and the following lemma.

Lemma 5.12 *For all* $t > 0$,

$$A_t(y) = B_1 y^{B_2} (\log y)^{-B_3} \exp\left\{-\frac{1}{2}\left(\log y + \frac{1}{2}\log\log y\right)^2\right\}$$
$$\times \left(1 + O\left(\frac{(\log\log y)^2}{\log y}\right)\right) \tag{5.54}$$

as $y \to \infty$, *where*

$$B_1 = 2^{-\frac{1}{2t}} \exp\left\{-\frac{(\log 2)^2}{2t}\right\},$$

$$B_2 = \frac{1 - 2\log 2}{2t}, \quad and \quad B_3 = \frac{1 + 2\log 2}{4t}.$$

Proof Using (5.51), we obtain

$$\exp\left\{\frac{u_y}{2t}\right\} = 2^{\frac{1}{2t}} y^{\frac{1}{2t}} (\log y)^{\frac{1}{4t}} \left(1 + O\left(\frac{\log\log y}{\log y}\right)\right). \tag{5.55}$$

Moreover, (5.52) implies

$$\begin{aligned}\exp\left\{-\frac{u_y^2}{2t}\right\} &= 2^{-\frac{1}{2}} \exp\left\{-\frac{(\log 2)^2}{2t}\right\} y^{-\frac{\log 2}{t}} (\log y)^{-\frac{1+\log 2}{2t}} \\ &\quad \times \exp\left\{-\frac{1}{2t}\left(\log y + \frac{1}{2}\log\log y\right)^2\right\} \\ &\quad \times \left(1 + O\left(\frac{(\log\log y)^2}{\log y}\right)\right).\end{aligned} \tag{5.56}$$

Now it is not hard to see that Lemma 5.12 follows from (5.55) and (5.56). □

Oscillating Integral Our next concern will be with the integral in (5.46) We assume that $t > 0$ is fixed, while ε is a variable which ranges over $[0, \infty)$. Our focus will be on the behavior of $I(\varepsilon)$ as $\varepsilon \to 0$. In the applications below we will take $\varepsilon = \frac{1}{2ty^2}$ with $y \to \infty$. We shall see that $I(\varepsilon)$ vanishes of infinite order at $\varepsilon = 0$, and our goal will be to give a precise description of $I(\varepsilon)$ as $\varepsilon \to 0$. In view of the rapid convergence of the integral in (5.46) we see that $I(\varepsilon)$ is a C^∞ function of ε on $[0, \infty)$, and the fact that

$$\left(\frac{d}{d\varepsilon}\right)^k I(\varepsilon)\bigg|_{\varepsilon=0} = 0$$

is equivalent with the assertion

$$\int_{-\infty}^{\infty} (\cosh x)^{2k+1} \exp\left\{-\frac{x^2}{2t}\right\} \exp\left\{\frac{i\pi x}{2t}\right\} dx = 0, \quad k = 0, 1, 2, \ldots. \tag{5.57}$$

To prove this, let

$$F_k(z) = (\cosh z)^{2k+1} \exp\left\{-\frac{z^2}{2t}\right\} \exp\left\{\frac{i\pi z}{2t}\right\}.$$

Note that $F_k(z + i\pi) = -F_k(z)$, since $\cosh(z + i\pi) = -\cosh z$. Now

$$\int_{-\infty}^{\infty} F_k(z)\, dz = \int_{-\infty}^{\infty} F_k(z + i\pi)\, dz,$$

as follows by Cauchy's theorem integrating $F_k(z)$ over the rectangular contour joining the vertices N, $N+i\pi$, $-N+i\pi$, and $-N$, and then letting $N\to\infty$. This proves (5.57).

We are tempted to use the same approach evaluating $I(\varepsilon)$. Set

$$F(z)=\sum_{k=0}^{\infty}\frac{(-\varepsilon)^k}{k!}F_k(z).$$

Once again $F(z+i\pi)=-F(z)$, but the above argument necessarily fails because the contributions of F over the sides $[N,N+i\pi]$ and $[-N+i\pi,-N]$ of the rectangle are in fact rapidly increasing as $N\to\infty$. More precisely, for $z=N+iy$ with N fixed and $0\le y\le\pi$, the factor $\exp\{-\varepsilon(\cosh z)^2\}$ varies from $\exp\{-\varepsilon(\cosh N)^2\}$ (at $y=0$), to essentially $\exp\{\varepsilon(\cosh N)^2\}$ (at $y=\frac{i\pi}{2}$), back to $\exp\{-\varepsilon(\cosh N)^2\}$ (at $y=i\pi$). This leads to the following modification of our argument. We note that by Cauchy's theorem

$$2\int_{-\infty}^{\infty}F(x)\,dx=\int_{\gamma^+}F(z)\,dz+\int_{\gamma^-}F(z)\,dz$$

where γ^+ is the rectangular loop consisting of the segments $[\infty+i\pi,N+i\pi]$, $[N+i\pi,N]$, and $[N,\infty]$, while γ^- is a symmetric contour in the left-half plane. From the above we might guess that the main contribution to the integration over γ^+ and γ^- comes from the integrals over the segments $[N+i\pi,N]$ and $[-N+i\pi,-N]$. Thus, to analyze the integral

$$\int_N^{N+i\pi}F(z)\,dz,$$

we seek $N=N_\varepsilon$ to balance the effects of $\exp\{-\varepsilon(\cosh z)^2\}$ and $\exp\{-\frac{z^2}{2t}\}$ at $z=N+\frac{i\pi}{2}$. Hence we look for the critical point $x=N_\varepsilon$ of $\Phi(x+\frac{i\pi}{2})$ where

$$\Phi(z)=-\varepsilon(\cosh z)^2-\frac{z^2}{2t}.$$

It follows that

$$\Phi'\left(x+\frac{i\pi}{2}\right)=\varepsilon\sinh(2x)-\frac{x}{t},$$

and so

$$\Phi'\left(N_\varepsilon+\frac{i\pi}{2}\right)=0$$

exactly when

$$\varepsilon\sinh(2N_\varepsilon)=\frac{N_\varepsilon}{t}.$$

Note that $N_\varepsilon \to \infty$ as $\varepsilon \to 0$, and as a first approximation, $N_\varepsilon \sim \frac{1}{2}\log\frac{1}{\varepsilon}$; a more precise asymptotic formula for N_ε is given below.

The principal contributions to the asymptotics of $I(\varepsilon)$ will come from integration over the segments $[N_\varepsilon, N_\varepsilon + i\pi]$ and $[-N_\varepsilon, -N_\varepsilon + i\pi]$. The other segments will contribute to the error term.

Our precise result is as follows.

Theorem 5.13 *Let $I(\varepsilon)$ be the integral given by* (5.46). *Then we have*

$$I(\varepsilon) \sim \sqrt{\frac{\pi}{2}} e^{-\frac{\pi^2}{8t}} \exp\left\{-\frac{N_\varepsilon^2}{2t}\right\} \exp\left\{\frac{N_\varepsilon}{2t}\right\} \varepsilon^{-\frac{1}{2}}, \quad \varepsilon \to 0. \tag{5.58}$$

More precisely, if $I_0(\varepsilon)$ is defined by (5.48), *then formula* (5.47) *holds.*

Proof With

$$F(z) = \exp\{-\varepsilon(\cosh z)^2\} \cosh z \exp\left\{-\frac{\pi s^2}{2t}\right\} \exp\left\{\frac{i\pi z}{2t}\right\}$$

we have observed that $F(z + i\pi) = -F(z)$. We also note that

$$\overline{F}(x + iy) = F(-x + iy).$$

Thus, using the contour described above, we have

$$I(\varepsilon) = \Re\left(\int_{\gamma^+} F(z)\,dz\right).$$

Let us consider first the integration over $[N, \infty)$. This and the corresponding integral over $[N + i\pi, \infty]$ will be part of the error term. Now these integrals are majorized by

$$\int_{N_\varepsilon}^{\infty} \exp\{-\varepsilon(\cosh x)^2\} \cosh x \exp\left\{-\frac{x^2}{2t}\right\} dx,$$

which itself is bounded by

$$\exp\left\{-\frac{N_\varepsilon^2}{2t}\right\} \int_0^{\infty} \exp\{-\varepsilon(\cosh x)^2\} \cosh x\, dx.$$

Now

$$\int_0^{\infty} \exp\{-\varepsilon(\cosh x)^2\} \cosh x\, dx = \int_1^{\infty} e^{-\varepsilon u^2} \frac{u}{(u^2-1)^{\frac{1}{2}}}\, du$$

in view of the change of variables $u = \cosh x$. But the size of the last integral is clearly $O(\varepsilon^{-\frac{1}{2}})$ as $\varepsilon \to 0$. Thus the contribution to $I(\varepsilon)$ of the infinite segments $[N, \infty]$ and $[N + i\pi, \infty]$ is

$$O\left(\varepsilon^{-\frac{1}{2}} e^{-\frac{N_\varepsilon^2}{2t}}\right). \tag{5.59}$$

We come to the main term

$$\int_{i\pi+N_\varepsilon}^{N_\varepsilon} \exp\{-\varepsilon(\cosh z)^2\} \exp\left\{-\frac{z^2}{2t}\right\} \cosh z \exp\left\{\frac{i\pi z}{2t}\right\} dz. \tag{5.60}$$

We use the fact that $(\cosh z)^2 = \frac{1}{2}(\cosh 2z + 1)$ and $\cosh z = \frac{e^z+e^{-z}}{2}$. Thus if we replace z by $z + N_\varepsilon$, we see that (5.60) is the sum of the two terms

$$\begin{aligned} &\frac{e^{-\frac{\varepsilon}{2}}}{2} \exp\left\{-\frac{N_\varepsilon^2}{2t}\right\} e^{N_\varepsilon} \int_{i\pi}^{0} e^{P(z)} Q(z)\, dz \\ &\quad + \frac{e^{-\frac{\varepsilon}{2}}}{2} \exp\left\{-\frac{N_\varepsilon^2}{2t}\right\} e^{-N_\varepsilon} \int_{i\pi}^{0} e^{P(z)} Q(z) e^{-2z}\, dz \\ &= I_1 + I_2. \end{aligned}$$

Here

$$P(z) = -\frac{\varepsilon}{2}\cosh(2N_\varepsilon + 2z) - \frac{zN_\varepsilon}{t} + \frac{i\pi N_\varepsilon}{2t} - \frac{\varepsilon}{2}\cosh(2N_\varepsilon)$$

and

$$Q(z) = \exp\left\{-\frac{z^2}{2t}\right\} e^z \exp\left\{\frac{i\pi z}{t}\right\} \exp\left\{\frac{\varepsilon}{2}\cosh(2N_\varepsilon)\right\}.$$

Now set $z = \frac{i\pi}{2} - is$, then $z \in [i\pi, 0]$ when $s \in [-\frac{\pi}{2}, \frac{\pi}{2}]$, and

$$P(z) = p(s) = -\varepsilon\cosh(2N_\varepsilon)(\sin s)^2 + i\varepsilon\sinh(2N_\varepsilon)\left[s - \frac{\sin 2s}{2}\right], \tag{5.61}$$

since

$$\frac{N_\varepsilon}{t} = \varepsilon\sinh(2N_\varepsilon).$$

Also $Q(z) = q(s)$, and

$$Q\left(\frac{i\pi}{2}\right) = q(0) = e^{-\frac{\pi^2}{8t}} i \exp\left\{\frac{\varepsilon}{2}\cosh(N_\varepsilon)\right\}.$$

It follows that

$$\int_{i\pi}^{0} e^{P(z)} Q(z)\, dz = -i \int_{-\frac{\pi}{2}}^{\frac{\pi}{2}} e^{-\lambda(\sin s)^2} \exp\left\{i\mu\left(s - \frac{\sin 2s}{2}\right)\right\} q(s)\, ds \tag{5.62}$$

with $\lambda = \varepsilon\cosh(2N_\varepsilon)$ and $\mu = \varepsilon\sinh(2N_\varepsilon)$.

Our next goal is to estimate I_1 and I_2. We will first prove a lemma.

Lemma 5.14 *The following equality holds*:

$$\int_{i\pi}^{0} e^{P(z)} Q(z)\, dz = -i(\pi)^{\frac{1}{2}}\lambda^{-\frac{1}{2}} q(0)\left(1 + O\left(\lambda^{-\frac{1}{2}}\right)\right) \quad \textit{as } \lambda \to \infty\ (\varepsilon \to 0).$$

Proof of Lemma 5.14 By making the change of variables $u = \sin s$ in the integral on the right-hand side of (5.62), we see that

$$\begin{aligned}
&\int_{i\pi}^{0} e^{P(z)} Q(z)\,dz \\
&\quad = -iq(0)\int_{-1}^{1} e^{-\lambda u^2}\,du \\
&\qquad - i\int_{-1}^{1} e^{-\lambda u^2}\left[\exp\left\{i\mu\left(s-\frac{\sin 2s}{2}\right)\right\}q(\arcsin u)\frac{1}{\sqrt{1-u^2}} - q(0)\right]du \\
&\quad = -iq(0)\int_{-1}^{1} e^{-\lambda u^2}\,du \\
&\qquad - iq(0)\int_{-1}^{1} e^{-\lambda u^2}\left[\exp\left\{i\mu\left(s-\frac{\sin 2s}{2}\right)\right\}\tilde{q}(\arcsin u)\frac{1}{\sqrt{1-u^2}} - 1\right]du
\end{aligned}$$

where $\tilde{q}$ is a smooth function on $[-1, 1]$ with $\tilde{q}(0) = 1$. Next, using the fact that

$$s - \frac{\sin 2s}{2} = O\left(s^3\right)$$

as $s \to 0$, we obtain

$$\begin{aligned}
&\int_{i\pi}^{0} e^{P(z)} Q(z)\,dz \\
&\quad = -iq(0)\left[\int_{-1}^{1} e^{-\lambda u^2}\,du + O\left(\int_{-1}^{1} e^{-\lambda u^2}|u|\,du\right)\right. \\
&\qquad \left. + O\left(\int_{-1}^{1} e^{-\lambda u^2}\mu|u|^3\,du\right)\right].
\end{aligned} \tag{5.63}$$

Now

$$\int_{-1}^{1} e^{-\lambda u^2}\,du = \int_{-\infty}^{\infty} e^{-\lambda u^2}\,du - \int_{|u|\geq 1} e^{-\lambda u^2}\,du = \sqrt{\pi}\lambda^{-\frac{1}{2}} + O\left(e^{-\lambda}\right),$$

while both O terms on the right-hand side of (5.63) are $O(\lambda^{-1})$ (the latter because $\mu \leq \lambda$). This proves the lemma. □

We now gather up the parts making up I_1. For this we recall that

$$q(0) = ie^{-\frac{\pi^2}{8t}}\exp\left\{\frac{\varepsilon}{2}\cosh(2N_\varepsilon)\right\},$$

while

$$\lambda = \varepsilon\cosh(2N_\varepsilon) = \varepsilon\sinh(2N_\varepsilon)\left(1 + O(\varepsilon)\right) = \left(\frac{N_\varepsilon}{t}\right)\left(1 + O(\varepsilon)\right).$$

Also

$$\frac{e^{N_\varepsilon}}{N_\varepsilon^{\frac{1}{2}}} = \left(\frac{2\sin(2N_\varepsilon)}{N_\varepsilon}\right)^{\frac{1}{2}}\left(1+O\left(N_\varepsilon^{-\frac{1}{2}}\right)\right) = \sqrt{2}\varepsilon^{-\frac{1}{2}}t^{-\frac{1}{2}}\left(1+O\left(N_\varepsilon^{-\frac{1}{2}}\right)\right).$$

Then because $\varepsilon = O(N_\varepsilon^{-\frac{1}{2}})$, we have

$$\begin{aligned} I_1 &= \frac{1}{2}\sqrt{\pi}e^{-\frac{\pi^2}{8t}}\exp\left\{-\frac{N_\varepsilon^2}{2t}\right\}e^{N_\varepsilon}\exp\left\{\frac{\varepsilon}{2}\cosh(2N_\varepsilon)\right\}\left(\frac{N_\varepsilon}{t}\right)^{-\frac{1}{2}} \\ &\quad\times\left(1+O\left(N_\varepsilon^{-\frac{1}{2}}\right)\right) \\ &= \frac{\sqrt{2}}{2}\sqrt{\pi}e^{-\frac{\pi^2}{8t}}\varepsilon^{-\frac{1}{2}}\exp\left\{-\frac{N_\varepsilon^2}{2t}\right\}\exp\left\{\frac{N_\varepsilon}{2t}\right\}\left(1+O\left(N_\varepsilon^{-\frac{1}{2}}\right)\right). \end{aligned}$$

A similar evaluation works for I_2 (which gives an extra gain of e^{-2N_ε}). This gain is again $O(N_\varepsilon^{-\frac{1}{2}})$. Next, taking into account the error (5.59), and the fact that $e^{-\frac{N_\varepsilon}{2t}} = O(N_\varepsilon^{-\frac{1}{2}})$, we see that

$$I(\varepsilon) = \frac{\sqrt{2}}{2}\sqrt{\pi}e^{-\frac{\pi^2}{8t}}\exp\left\{-\frac{N_\varepsilon^2}{2t}\right\}\exp\left\{\frac{N_\varepsilon}{2t}\right\}\varepsilon^{-\frac{1}{2}}\left(1+O\left(N_\varepsilon^{-\frac{1}{2}}\right)\right).$$

In addition, we have

$$N_\varepsilon = \frac{1}{2}\log\frac{1}{\varepsilon} + \frac{1}{2}\log\log\frac{1}{\varepsilon} + \frac{1}{2}\log\frac{1}{t} + o(1) \tag{5.64}$$

as $\varepsilon \to 0$. Indeed, take ε to be $\frac{1}{2ty^2}$, (with $y \to \infty$). Then $N_\varepsilon = u_y$, where u_y was defined in (5.49). Now, (5.64) follows from (5.51).

This completes the proof of Theorem 5.13. □

Corollary 5.15 *For each non-negative integer k,*

$$\begin{aligned} \left(\frac{d}{d\varepsilon}\right)^k I(\varepsilon) &= \left(\frac{N_\varepsilon}{2\varepsilon t}\right)^k\sqrt{\frac{\pi}{2}}e^{-\frac{\pi^2}{8t}}\exp\left\{-\frac{N_\varepsilon^2}{2t}\right\}\exp\left\{\frac{N_\varepsilon}{2t}\right\}\varepsilon^{-\frac{1}{2}} \\ &\quad\times\left(1+O\left(\left(\log\frac{1}{\varepsilon}\right)^{-\frac{1}{2}}\right)\right) \end{aligned}$$

as $\varepsilon \to 0$.

This can be proved in the same way as in the case $k = 0$ given above. In fact, using formula (5.46), we see that $(\frac{d}{d\varepsilon})^k$ brings down a factor $(-1)^k(\cosh x)^{2k}$. At the critical point, $z = \frac{i\pi}{2} + N_\varepsilon$, this is

$$\left(\frac{1}{2}\cosh(2N_\varepsilon) - \frac{1}{2}\right)^k \sim 2^{-k}\left(\sinh(2N_\varepsilon)\right)^k = \left(\frac{N_\varepsilon}{2\varepsilon t}\right)^k.$$

The next lemma can be obtained from (5.43) and (5.46).

Lemma 5.16 *For every $y > 0$,*

$$m_t(y) = \frac{1}{\pi t} e^{\frac{\pi^2}{8t}} y^{-2} I\left(\frac{1}{2ty^2}\right).$$

It follows from (5.58), (5.47), the equality

$$N_{\frac{1}{2ty^2}} = u_y,$$

and Lemma 5.16 that formula (5.50) holds. This establishes Theorem 5.9 in the case where $\alpha = 0$.

We now turn our attention to the case $\alpha \neq 0$ in Theorem 5.9. An important tool in this part of the proof is Dufresne's recurrence formula (see Sect. 5.4.5 below). In the next subsections, we gather definitions and facts, which will be used in the proof of Dufresne's recurrence formula.

5.4.1 Hypergeometric Functions

In this subsection, we gather several formulas and facts from the theory of hypergeometric functions. Our presentation is based on Chap. 2 of the book [AAR99] by G.E. Andrews, R. Askey, and B. Roy. This book is a rich source of information about hypergeometric functions.

The shifted factorial (Pochhammer's symbol) $(a)_n$ is defined as follows:

$$(a)_n = a(a+1)\cdots(a+n-1) \quad \text{for } n > 0, \quad (a)_0 = 1.$$

Here a can be any complex number. In the next definition, we assume that a, b, c, and z are complex numbers, and c is not a negative integer or zero.

Definition 5.17 The hypergeometric function ${}_2F_1$ is defined by

$$ {}_2F_1(a, b; c; z) = \sum_{n=0}^{\infty} \frac{(a)_n (b)_n}{(c)_n n!} z^n \tag{5.65}$$

for $|z| < 1$.

Remark 5.18 The power series on the right-hand side of (5.65) converges absolutely, by the ratio test. It is also known that this series converges absolutely if $\Re(c-b-a)>0$ and $|z|=1$. In addition, it converges conditionally if $z=e^{i\theta}\neq 1$ and $0\geq\Re(c-a-b)>-1$, and diverges if $\Re(c-b-a)\leq -1$ (see [AAR99], p. 62).

We will next discuss the analytic continuation of the hypergeometric function ${}_2F_1$. Theorem 5.19 below is due to L. Euler (see [AAR99], p. 65).

Theorem 5.19 *Let $\Re(c)>\Re(b)>0$. Then for $|z|<1$,*

$$ {}_2F_1(a,b;c;z)=\frac{\Gamma(c)}{\Gamma(b)\Gamma(c-b)}\int_0^1 t^{b-1}(1-t)^{c-b-1}(1-zt)^{-a}\,dt. \qquad (5.66)$$

Proof The proof of Euler's theorem uses the Beta function (see Definition 1.28 and formula (1.39)). Expanding $(1-zt)^{-a}$ by the binomial theorem, we see that the right-hand side of (5.66) becomes

$$\frac{\Gamma(c)}{\Gamma(b)\Gamma(c-b)}\sum_{n=0}^{\infty}\frac{(a)_n}{n!}z^n\int_0^1 t^{n+b-1}(1-t)^{c-b-1}\,dt.$$

Next, using (1.39), we rewrite this expression as follows:

$$ {}_2F_1(a,b;c;z)=\frac{\Gamma(c)}{\Gamma(b)}\sum_{n=0}^{\infty}\frac{(a)_n\Gamma(n+b)}{n!\Gamma(n+c)}z^n.$$

This completes the proof of Theorem 5.19. □

Remark 5.20 Using Euler's theorem, we can find the analytic continuation of the function ${}_2F_1$. It follows from the integral representation in (5.66) that for $a\in\mathbb{C}$, $\Re(c)>\Re(b)>0$, there exists the analytic continuation of the function $z\mapsto {}_2F_1(a,b;c;z)$ to the complex plane cut along the real axis from 1 to ∞. Note that the principal branch of the argument is used in the definition of $(1-zt)^{-a}$.

The following equality was established by C.F. Gauss. If a, b, and c are such that $\Re(c-a-b)>0$, then

$$ {}_2F_1(a,b;c;1)=\frac{\Gamma(c)\Gamma(c-a-b)}{\Gamma(c-a)\Gamma(c-b)} \qquad (5.67)$$

(see [AAR99], p. 66).

The function ${}_2F_1$ satisfies Euler's hypergeometric differential equation. This equation is as follows:

$$z(1-z)\frac{d^2f}{dz^2}+\left[c-(a+b+1)z\right]\frac{df}{dz}-abf=0. \qquad (5.68)$$

Euler's equation has three regular singular points at 0, 1, and ∞ (more information on regular singular points of differential equations can be found in [AAR99], p. 639). If c is not an integer, then a basis for series solutions about $z = 0$ for the hypergeometric equation consists of the functions

$$f_1(z) = {}_2F_1(a, b; c; z) \quad \text{and} \quad f_2(z) = z^{1-c}{}_2F_1(a+1-c, b+1-c; 2-c; z)$$

(see [AAR99], p. 77). Moreover, if c is not a negative integer or zero, then analytic solutions about $z = 0$ have the form $f(z) = \alpha\, {}_2F_1(a, b; c; z)$ where $\alpha \in \mathbb{C}$. If c is not a positive integer, then for any solution f of (5.68) such that $f(z) = z^{1-c}g(z)$ with g analytic about zero, we have

$$g(z) = \beta\, {}_2F_1(a+1-c, b+1-c; 2-c; z)$$

where $\beta \in \mathbb{C}$. In addition, a basis for series solutions about $z = 1$ consists of the following functions:

$$f_3(z) = {}_2F_1(a, b; a+b+1-c; 1-x)$$

and

$$f_4(z) = (1-z)^{c-a-b}{}_2F_1(c-a, c-b; c+1-a-b; 1-z).$$

At $z = \infty$, a basis is as follows:

$$f_5(z) = (-z)^{-a}{}_2F_1\left(a, a+1-c; a+1-b; \frac{1}{z}\right)$$

and

$$f_6(z) = (-z)^{-b}{}_2F_1\left(b, b+1-c; b+1-a; \frac{1}{z}\right).$$

We will next formulate several important equalities for hypergeometric functions (see Sect. 2.3 of [AAR99]). The first of these equalities is due to J.F. Pfaff. He proved that

$${}_2F_1(a, b; c; z) = (1-z)^{-a}{}_2F_1\left(a, c-b; c; \frac{z}{z-1}\right). \tag{5.69}$$

Pfaff's formula provides the analytic continuation of the function ${}_2F_1$ from the unit disc $\{z : |z| < 1\}$ into the region $\{z : \Re(z) < \frac{1}{2}\}$. It is assumed in (5.69) that c is not a negative integer or zero. Another useful formula is the following:

$$\begin{aligned} {}_2F_1(a, b; c; z) = {} & \frac{\Gamma(c)\Gamma(c-a-b)}{\Gamma(c-a)\Gamma(c-b)}{}_2F_1(a, b; a+b+1-c; 1-z) \\ & + \frac{\Gamma(c)\Gamma(a+b-c)}{\Gamma(a)\Gamma(c)}(1-z)^{c-a-b} \\ & \times {}_2F_1(c-a, c-b; 1+c-a-b; 1-z). \end{aligned} \tag{5.70}$$

Formulas (5.69) and (5.70) define the analytic continuation of ${}_2F_1$ into the complex plane $\mathbb{C}$ slit along the rays

$$\{z=(x,0)\in\mathbb{C}:1\le x<\infty\},\qquad \{z=(1/2,y)\in\mathbb{C}:\sqrt{3}/2\le\infty<\infty\},$$

and

$$\{z=(1/2,y)\in\mathbb{C}:-\infty<y\le-\sqrt{3}/2\}.$$

It is assumed in (5.70) that c not a negative integer or zero and $c-a-b$ is not an integer. Finally, the formula

$$\begin{aligned}{}_2F_1(a,b;c;z)&=\frac{\Gamma(c)\Gamma(b-a)}{\Gamma(c-a)\Gamma(b)}(-z)^{-a}{}_2F_1\left(a,a-c+1;a-b+1;\frac{1}{z}\right)\\&+\frac{\Gamma(c)\Gamma(a-b)}{\Gamma(a)\Gamma(c-b)}(-z)^{-b}{}_2F_1\left(b,b-c+1;b-a+1;\frac{1}{z}\right)\end{aligned}$$

combined with (5.70) gives the analytic continuation of ${}_2F_1$ into the complex plane $\mathbb{C}$ punctured at $z=\frac{1}{2}+\frac{\sqrt{3}}{2}i$ and $z=\frac{1}{2}+\frac{\sqrt{3}}{2}i$ and slit along the ray

$$\{z=(x,0)\in\mathbb{C}:1\le x<\infty\}.$$

This analytic continuation exists if c is not a negative integer or zero and both $a-b$ and $c-a-b$ are not integers.

5.4.2 Dufresne's Theorems

Let B be a standard Brownian motion and put

$$A_t^{(\beta)}=\int_0^t\exp\{2\beta\tau+2B_\tau\}d\tau$$

where $\beta\in\mathbb{R}$ and $t>0$. In [Duf01], the following family of functions of two variables s and t is considered:

$$h^{\beta,r}(s,t)=e^{\frac{\beta^2t}{2}}\mathbb{E}\left[\left(2A_t^{(\beta)}\right)^{-r}\exp\left\{\frac{s}{2A_t^{(\beta)}}\right\}\right],\quad r\in\mathbb{R},\ s\le0.\tag{5.71}$$

The expression on the right-hand side of (5.71) is finite (see [Duf01]). Set

$$g^{\beta,r}(s,t)=\mathbb{E}\left[\left(2A_t^{(\beta)}\right)^{-r}\exp\left\{\frac{s}{2A_t^{(\beta)}}\right\}\right].$$

Then we have

$$h^{\beta,r}(s,t)=e^{\frac{\beta^2t}{2}}g^{\beta,r}(s,t).\tag{5.72}$$

The Laplace transform of the function $h^{\beta,r}$ with respect to the variable t is defined by

$$\tilde{h}^{\beta,r}(s,\lambda)=\int_0^\infty e^{-\lambda t}h^{\beta,r}(s,t)\,dt. \tag{5.73}$$

The function in (5.73) is analytic for $\Re(s)<1$ (see [Duf01]). Dufresne established several important facts concerning the functions $h^{\beta,r}$. For instance, he described partial differential equations satisfied by $h^{\beta,r}$ and $\tilde{h}^{\beta,r}$ and also found an explicit formula for $\tilde{h}^{\beta,r}$ in terms of the hypergeometric function ${}_2F_1$.

We will next formulate and prove some of the results obtained in [Duf01]. Our proofs mostly follow those in [Duf01], but are sometimes more detailed. In the formulation of Dufresne's results below, the numbers $\beta\in\mathbb{R}$ and $r\in\mathbb{R}$ are fixed, and we put $h=h^{\beta,r}$, $g=g^{\beta,r}$, and $\tilde{h}=\tilde{h}_{\beta,r}$.

Theorem 5.21 *The function h satisfies the following partial differential equation:*

$$\frac{\partial h}{\partial t}+2s(1-s)\frac{\partial^2 h}{\partial s^2}+\big[r+(\beta-2r-1)s\big]\frac{\partial h}{\partial s}-\frac{1}{2}(\beta-2r)^2h=0, \tag{5.74}$$

where $\Re(s)<1$ and $t>0$. In addition,

$$\lim_{t\downarrow 0}h(s,t)=0\quad\text{for } \Re(s)<1. \tag{5.75}$$

Theorem 5.22 *The Laplace transform $\tilde{h}$ converges if*

$$\sqrt{2\lambda}>\max(-\beta,\beta-2r)$$

and if either $\Re(s)<0$, $r\in\mathbb{R}$, or $\Re(s)=0$, $r<1$. It satisfies a hypergeometric equation

$$s(1-s)\frac{\partial^2\tilde{h}}{\partial s^2}+\big[r-(1+2r-\beta)s\big]\frac{\partial\tilde{h}}{\partial s}+\left[\frac{1}{2}\lambda-\frac{1}{4}(\beta-2r)^2\right]\tilde{h}=0. \tag{5.76}$$

Moreover, $\tilde{h}$ is represented by the following formula:

$$\begin{aligned}\tilde{h}(s,\lambda)&=\frac{\Gamma(\alpha_1)\Gamma(\alpha_2+r)}{2(1-s)^{\alpha_2+r}\Gamma(\alpha_1+\alpha_2+1)}\\&\quad\times{}_2F_1\left(\alpha_1,\alpha_2+r,\alpha_1+\alpha_2+1;\frac{1}{1-s}\right),\end{aligned} \tag{5.77}$$

where $\alpha_1=\frac{1}{2}\beta+\sqrt{\frac{1}{2}\lambda}$ and $\alpha_2=\alpha_1-\beta$.

Proof of Theorem 5.21 It will be assumed below that $\Re(s)<0$. Equality (5.75) can be obtained from (5.72) using the monotone convergence theorem.

Put

$$Y_t = \exp\{2\beta t + 2B_t\} \int_0^t \exp\{-(2\beta\tau + 2B_\tau)\}\, d\tau.$$

Then $Y_t = A_t^{(\beta)}$ in law. The previous fact can be established by reversing the time in a Brownian motion. Next, applying the integration by parts formula and Itô's formula, we see that

$$dY_t = \big[(2\beta + 2)Y_t + 1\big]\, dt + 2Y_t\, dB_t$$

and

$$d\left[Y_t^{-r} \exp\left\{\frac{s}{2Y_t}\right\}\right] = a(T_t)\, dt + b(Y_t)\, dB_t,$$

where

$$\begin{aligned} a(y) &= \big[(2\beta + 2)y + 1\big]\frac{\partial}{\partial y}\left(y^{-r} \exp\left\{\frac{s}{2y}\right\}\right) + 2y^2 \frac{\partial^2}{\partial y^2}\left(y^{-r} \exp\left\{\frac{s}{2y}\right\}\right) \\ &= \left[\frac{1}{2}(s^2 - s)y^{-2-r} + \big[-r + (1 + 2r - \beta)s\big]y^{-1-r} + 2r(r - \beta)y^{-r}\right] \\ &\quad \times \exp\left\{\frac{s}{2y}\right\} \end{aligned}$$

and

$$b(y) = 2y\frac{\partial}{\partial y}\left(y^{-r} \exp\left\{\frac{s}{2y}\right\}\right) = -\left(2ry^{-r} + sy^{-r-1}\right) \exp\left\{\frac{s}{2y}\right\}.$$

Since

$$\mathbb{E} \int_0^t b^2(Y_\tau)\, d\tau < \infty,$$

we have

$$\mathbb{E} \int_0^t b(Y_\tau)\, dB_\tau = 0. \tag{5.78}$$

It follows from (5.72) that

$$\frac{\partial h}{\partial t} = \frac{\beta^2}{2} \exp\left\{\frac{\beta^2 t}{2}\right\} g + \exp\left\{\frac{\beta^2 t}{2}\right\} \frac{\partial g}{\partial t}. \tag{5.79}$$

Moreover, taking into account (5.78), we get

$$\frac{\partial g}{\partial t} = \mathbb{E}\left[\frac{1}{2}(s^2 - s)Y_t^{-2-r} + \big[-r + (1 + 2r - \beta)s\big]Y_t^{-1-r} + 2r(r - \beta)Y_t^{-r}\right]$$

$$\times \exp\left\{\frac{s}{2y}\right\} = 2(s^2 - s)\frac{\partial g}{\partial s^2} + 2[-r + (1 + 2r - \beta)s]\frac{\partial g}{\partial s} + 2r(r - \beta)g. \tag{5.80}$$

Now it is clear that (5.79) and (5.80) imply (5.74).

This completes the proof of Theorem 5.21. □

Proof of Theorem 5.22 For the description of the domain of analyticity of the Laplace transform $\tilde{h}$, see [Duf01].

Let us assume that $\Re(s) < 0$ and λ is large enough. Then, multiplying (5.74) by $e^{-\lambda t}$ and integrating, we see that the function $\tilde{h}^{\beta,r}$ satisfies equation (5.76), which is a hypergeometric differential equation with

$$a = \sqrt{\frac{\lambda}{2}} - \frac{\beta - 2r}{2}, \qquad b = -\sqrt{\frac{\lambda}{2}} - \frac{\beta - 2r}{2}, \qquad c = r.$$

Next suppose $r = 0$ and put $\tilde{h} = \tilde{h}^{\beta,0}$. Then the function $\tilde{h}$ satisfies the following equation:

$$s(1 - s)\frac{\partial^2 \tilde{h}}{\partial s^2} + (\beta - 1)s\frac{\partial \tilde{h}}{\partial s} + \left(\frac{1}{2}\lambda - \frac{1}{4}\beta^2\right)\tilde{h} = 0.$$

This is a hypergeometric differential equation with

$$a = -\frac{\beta}{2} + \sqrt{\frac{\lambda}{2}}, \qquad b = -\frac{\beta}{2} - \sqrt{\frac{\lambda}{2}}, \qquad c = 0.$$

After long but straightforward computations, we see that the function

$$g(z) = z^{-a}\tilde{h}\left(\frac{z - 1}{z}\right)$$

satisfies the equation

$$z(1 - z)\frac{\partial^2 g}{\partial z^2} + \left[1 + a - b - (1 + a + c - b)z\right]\frac{\partial g}{\partial z} + a(b - c)g = 0$$

in the disc D of radius $\frac{1}{2}$ centered at the point $z = \frac{1}{2}$. This is a hypergeometric equation with new parameters given by

$$a' = a = -\frac{\beta}{2} + \sqrt{\frac{\lambda}{2}}, \qquad b' = c - b = \frac{\beta}{2} + \sqrt{\frac{\lambda}{2}}, \qquad c' = 1 + a - b = 1 + \sqrt{2\lambda}.$$

It follows from the description of linearly independent solutions of hypergeometric equations in Sect. 5.4.1 that if c' is not an integer, or equivalently, $\lambda \neq \frac{k^2}{2}$, $k \in \mathbb{Z}$, then there exist constants c_1 and c_2 such that

$$z^{-a'}\tilde{h}\left(\frac{z - 1}{z}, \lambda\right) = c_1\, {}_2F_1\left(a', b'; c'; z\right)$$
$$+ c_2\, z^{1-c'}\, {}_2F_1\left(a' + 1 - c', b' + 1 - c'; 2 - c'; z\right)$$

for all $z \in D$. Therefore,

$$\tilde{h}(s,\lambda) = c_1(1-s)^{-a'}{}_2F_1\left(a', b'; c'; \frac{1}{1-s}\right)$$
$$+ c_2(1-s)^{c'-1-a'}{}_2F_1\left(a'+1-c', b'+1-c'; 2-c'; \frac{1}{1-s}\right) \quad (5.81)$$

for all s with $\Re(s) < 0$ and all $c' \notin \mathbb{Z}$. We have $\tilde{h}(s,\lambda) \to 0$ when $s \in \mathbb{R}$ and $s \to -\infty$. On the other hand,

$$-a' = a = -\frac{\beta}{2} + \sqrt{\frac{\lambda}{2}} < 0$$

and

$$c' - 1 - a' = -b = \frac{\beta}{2} + \sqrt{\frac{\lambda}{2}} > 0$$

for large values of λ. Hence

$$(1-s)^{-a'} \to 0 \quad \text{and} \quad (1-s)^{c'-1-a'} \to \infty$$

as $s \to -\infty$. It follows that the constant c_2 in (5.81) is equal to zero. Thus,

$$\tilde{h}(s,\lambda) = c_1(1-s)^{-a'}{}_2F_1\left(a', b'; c'; \frac{1}{1-s}\right) \quad (5.82)$$

for all s with $\Re(s) < 0$ and all $c' \notin \mathbb{Z}$. Formula (5.82) can be extended to all large values of λ by continuity. □

Our next goal is to pass to the limit in (5.82) as $s < 0$ and $s \to 0$. Taking into account (5.67) and the equality $c' - a' - b' = 1$, we see that there exists $\lambda_0 > 0$ such that

$$c_1 = \lim_{s\uparrow 0} \tilde{h}(s,\lambda)\frac{\Gamma(c'-a')\Gamma(c'-b')}{\Gamma(c')\Gamma(c'-a'-b')}$$
$$= \lim_{s\uparrow 0} \tilde{h}(s,\lambda)\Gamma(1+\sqrt{2\lambda})^{-1}\Gamma\left(1+\frac{\beta}{2}+\sqrt{\frac{\lambda}{2}}\right)$$
$$\times \Gamma\left(1-\frac{\beta}{2}+\sqrt{\frac{\lambda}{2}}\right) \quad (5.83)$$

for all $\lambda > \lambda_0$. In addition,

$$\lim_{s\uparrow 0} \tilde{h}(s,\lambda) = \int_0^\infty e^{-\lambda t} \exp\left\{\frac{\beta^2 t}{2}\right\} dt = \frac{2}{2\lambda - \beta^2} \quad (5.84)$$

for $\lambda > \lambda_1$. It follows from (5.82), (5.83), and (5.84) that

$$\tilde{h}(s,\lambda) = \frac{2\Gamma(1+\frac{\beta}{2}+\sqrt{\frac{\lambda}{2}})\Gamma(1-\frac{\beta}{2}+\sqrt{\frac{\lambda}{2}})}{(2\lambda-\beta^2)\Gamma(1+\sqrt{2\lambda})(1-s)^{a'}} {}_2F_1\left(a',b';c';\frac{1}{1-s}\right)$$
$$= \frac{\Gamma(\frac{\beta}{2}+\sqrt{\frac{\lambda}{2}})\Gamma(-\frac{\beta}{2}+\sqrt{\frac{\lambda}{2}})}{2\Gamma(1+\sqrt{2\lambda})(1-s)^{a'}} {}_2F_1\left(a',b';c';\frac{1}{1-s}\right).$$

Therefore,

$$\tilde{h}^{\beta,0}(s,\lambda) = \frac{\Gamma(\frac{\beta}{2}+\sqrt{\frac{\lambda}{2}})\Gamma(-\frac{\beta}{2}+\sqrt{\frac{\lambda}{2}})}{2\Gamma(1+\sqrt{2\lambda})(1-s)^{-\frac{\beta}{2}+\sqrt{\frac{\lambda}{2}}}}$$
$$\times {}_2F_1\left(-\frac{\beta}{2}+\sqrt{\frac{\lambda}{2}}, \frac{\beta}{2}+\sqrt{\frac{\lambda}{2}}; 1+\sqrt{2\lambda}; \frac{1}{1-s}\right)$$
$$= \frac{\Gamma(\frac{\beta}{2}+\sqrt{\frac{\lambda}{2}})\Gamma(-\frac{\beta}{2}+\sqrt{\frac{\lambda}{2}})}{2\Gamma(1+\sqrt{2\lambda})(1-s)^{-\frac{\beta}{2}+\sqrt{\frac{\lambda}{2}}}}$$
$$\times {}_2F_1\left(\frac{\beta}{2}+\sqrt{\frac{\lambda}{2}}, -\frac{\beta}{2}+\sqrt{\frac{\lambda}{2}}; 1+\sqrt{2\lambda}; \frac{1}{1-s}\right). \tag{5.85}$$

Now it is not hard to see that formula (5.85) implies formula (5.77) in the special case where $r = 0$.

We will next establish equality (5.77) for any $r \in \mathbb{R}$. Several standard distributions will appear in the proof.

5.4.3 Exponential, Beta, and Gamma Distributions

In this subsection, we consider three important probability distribution densities.

Exponential Distribution The probability density function of the exponential distribution with the rate parameter $\lambda > 0$ is defined by

$$f_\lambda(x) = \begin{cases} \lambda e^{-\lambda x}, & \text{if } x \geq 0, \\ 0, & \text{if } x < 0. \end{cases}$$

We have

$$\mathbb{E}[X] = \int_0^\infty f_\lambda(x)\,dx = \frac{1}{\lambda}.$$

If a random variable X has the distribution described above, then it is said that X is an exponential variable with mean $\frac{1}{\lambda}$.

Beta Distribution The probability density function of the Beta distribution depends on two shape parameters $\tau > 0$ and $\delta > 0$. This density is supported on the interval $[0, 1]$ and is defined by the following formula:

$$\beta(x; \tau, \delta) = \begin{cases} \frac{1}{B(\alpha,\beta)} x^{\tau-1}(1-x)^{\delta-1}, & \text{if } 0 \le x \le 1, \\ 0, & \text{otherwise} \end{cases}$$

where B is the Beta function (see Definition 1.28). If a random variable X has this distribution density, then we say that X is $\beta(\tau, \delta)$-distributed.

Gamma Distribution The probability density function of the Gamma distribution has already been introduced (see Definition 1.26). Let us recall that this density depends on the shape parameter k and the scale parameter $\theta > 0$ and is given by the following formula:

$$\Gamma(x; k, \theta) = \begin{cases} \frac{1}{\theta^k \Gamma(k)} x^{k-1} e^{-\frac{x}{\theta}}, & \text{if } x \ge 0, \\ 0, & \text{if } x < 0 \end{cases}$$

where Γ is the Gamma function. If a random variable X has this distribution density, then we say that X is $\Gamma(k, \theta)$-distributed.

5.4.4 *Proof of Formula (5.77) for $r \neq 0$*

The following theorem due to Yor (see [Yor92b]) characterizes the law of the time-integral of a geometric Brownian motion subject to an exponentially distributed random change of time. This theorem will be needed in the proof of formula (5.77) for $r \neq 0$.

Theorem 5.23 *Let T_λ be an exponential random variable with mean $\frac{1}{\lambda}$ independent of B. Then*

$$2A_{T_\lambda}^{(\beta)} = \frac{U}{G} \tag{5.86}$$

(in law), where U is a $\beta(1, \alpha_\beta)$-random variable and G is an independent $\Gamma(\tau_\beta, 1)$-random variable with $\alpha_\beta = \frac{1}{2}\beta + \frac{1}{2}(2\lambda + \beta^2)^{\frac{1}{2}}$ and $\tau_\beta = \alpha_\beta - \beta$.

Proof Let U and G be random variables such as in the formulation of Theorem 5.23. Then for $\Re(s) \le 0$, we have

$$\mathbb{E}\left[\left(\frac{G}{U}\right)^r \exp\left\{s\frac{G}{U}\right\}\right]$$
$$= \frac{\alpha_\beta}{\Gamma(\tau_\beta)} \int_0^1 u^{-r}(1-u)^{\alpha_\beta - 1} du \int_0^\infty x^{\tau_\beta + r - 1} \exp\left\{-x + \frac{sx}{u}\right\} dx$$

$$= \frac{\alpha_\beta \Gamma(\tau_\beta + r)}{\Gamma(\tau_\beta)} \int_0^1 u^{-r}(1-u)^{\alpha_\beta - 1}\left(1 - \frac{s}{u}\right)^{-\tau_\beta - r} du$$

$$= \frac{\alpha_\beta \Gamma(\tau_\beta + r)}{\Gamma(\tau_\beta)} \int_0^1 \frac{u^{\tau_\beta}(1-u)^{\alpha_\beta - 1}}{(u-s)^{\tau_\beta + r}} du$$

$$= \frac{\alpha_\beta \Gamma(\tau_\beta + r)}{\Gamma(\tau_\beta)} \int_0^1 \frac{(1-v)^{\tau_\beta} v^{\alpha_\beta - 1}}{((1-s)-v)^{\tau_\beta + r}} dv$$

$$= \frac{\alpha_\beta \Gamma(\tau_\beta + r)}{\Gamma(\tau_\beta)(1-s)^{\tau_\beta + r}} \int_0^1 v^{\alpha_\beta - 1}(1-v)^{\tau_\beta}\left(1 - \frac{v}{1-s}\right)^{-\tau_\beta - r} dv. \quad (5.87)$$

Next, using Euler's Theorem (see (5.66)) with

$$a = \tau_\beta + r, \quad b = \alpha_\beta, \quad c = \alpha_\beta + \tau_\beta + 1, \quad \text{and} \quad z = \frac{1}{1-s},$$

we see that the definition of the hypergeometric series and formula (5.87) imply

$$\mathbb{E}\left[\left(\frac{G}{U}\right)^r \exp\left\{s\frac{G}{U}\right\}\right]$$

$$= \frac{\alpha_\beta \Gamma(\tau_\beta + r)}{\Gamma(\tau_\beta)(1-s)^{\tau_\beta + r}} \frac{\Gamma(\alpha_\beta)\Gamma(\tau_\beta + 1)}{\Gamma(\alpha_\beta + \tau_\beta + 1)} {}_2F_1\left(\alpha_\beta, \tau_\beta + r, \alpha_\beta + \tau_\beta + 1; \frac{1}{1-s}\right)$$

$$= \frac{\alpha_\beta \tau_\beta \Gamma(\alpha_\beta)\Gamma(\tau_\beta + r)}{(1-s)^{\tau_\beta + r}\Gamma(\alpha_\beta + \tau_\beta + 1)} {}_2F_1\left(\alpha_\beta, \tau_\beta + r, \alpha_\beta + \tau_\beta + 1; \frac{1}{1-s}\right).$$

Since $\lambda = 2\alpha_\beta \tau_\beta$, we have

$$\mathbb{E}\left[\left(\frac{G}{U}\right)^r \exp\left\{s\frac{G}{U}\right\}\right]$$

$$= \frac{\lambda \Gamma(\alpha_\beta)\Gamma(\tau_\beta + r)}{2(1-s)^{\tau_\beta + r}\Gamma(\alpha_\beta + \tau_\beta + 1)}$$

$$\times {}_2F_1\left(\alpha_\beta, \tau_\beta + r, \alpha_\beta + \tau_\beta + 1; \frac{1}{1-s}\right). \quad (5.88)$$

A special case of (5.88) where $r = 0$ and formula (5.85) with $\lambda + \frac{\beta^2}{2}$ instead of λ imply the following:

$$\lambda \tilde{h}^{\beta,0}\left(s, \lambda + \frac{1}{2}\beta^2\right) = \mathbb{E}\left[\exp\left\{s\frac{G}{U}\right\}\right]$$

for all s with $\Re(s) < 0$ and $\lambda > \lambda_0$. Therefore,

$$\int_0^\infty \lambda e^{-\lambda t}\mathbb{E}\left[\exp\left\{\frac{s}{2A_t^{(\beta)}}\right\}\right] dt = \mathbb{E}\left[\exp\left\{s\frac{G}{U}\right\}\right]. \quad (5.89)$$

Now it is not hard to see, using (5.89) and the inversion formula for the Laplace transform, that Theorem 5.23 holds. □

We will next prove formula (5.77) for $r \neq 0$. It follows from Theorem 5.23 that

$$\int_0^\infty \lambda e^{-\lambda t}\mathbb{E}\left[\left(2A_t^{(\beta)}\right)^{-r}\exp\left\{\frac{s}{2A_t^{(\beta)}}\right\}\right]dt = \mathbb{E}\left[\left(\frac{G}{U}\right)^r\exp\left\{s\frac{G}{U}\right\}\right] \tag{5.90}$$

for all $\Re(s) \leq 0$. Equality (5.90) is equivalent to the following equality:

$$\lambda\tilde{h}^{\beta,r}\left(s,\lambda+\frac{1}{2}\beta^2\right) = \mathbb{E}\left[\left(\frac{G}{U}\right)^r\exp\left\{s\frac{G}{U}\right\}\right]. \tag{5.91}$$

Taking into account (5.88) and (5.91), we see that formula (5.77) holds for $r \neq 0$.

The proof of Theorem 5.22 is thus completed.

5.4.5 Dufresne's Recurrence Formula

The next statement follows from Dufresne's theorem. Formula (5.92) in Theorem 5.24 is called Dufresne's recurrence formula.

Theorem 5.24 *For all $\mu \in \mathbb{R}$, $r \in \mathbb{R}$, $t > 0$, and $\Re(s) < 1$,*

$$h^{\mu,r}(s,t) = (1-s)^{\mu-r}h^{2r-\mu,r}(s,t). \tag{5.92}$$

Proof By formula (5.77),

$$(1-s)^{r-\mu}\tilde{h}^{\mu,r}(s,\lambda) = \frac{\Gamma(\alpha)\Gamma(\beta+r)}{2(1-s)^\alpha\Gamma(\alpha+\beta+1)}{}_2F_1\left(\alpha,\beta+r,\alpha+\beta+1;\frac{1}{1-s}\right), \tag{5.93}$$

where $\alpha = \frac{1}{2}\mu + \sqrt{\frac{\lambda}{2}}$, $\beta = \alpha - \mu$, and $\tilde{h}^{\mu,r}(s,\lambda)$ is the Laplace transform defined by (5.73). Formula (5.77) also implies that

$$\tilde{h}^{2r-\mu,r}(s,\lambda) = \frac{\Gamma(\tilde{\alpha})\Gamma(\tilde{\beta}+r)}{2(1-s)^{\tilde{\beta}+r}\Gamma(\tilde{\alpha}+\tilde{\beta}+1)}{}_2F_1\left(\tilde{\alpha},\tilde{\beta}+r,\tilde{\alpha}+\tilde{\beta}+1;\frac{1}{1-s}\right).$$

Here

$$\tilde{\alpha} = \frac{1}{2}(2r-\mu)+\sqrt{\frac{\lambda}{2}} = \alpha+r-\mu = \beta+r$$

and $\tilde{\beta} = \tilde{\alpha} - 2r + \mu = \alpha - r$. Since $\tilde{\beta} + r = \alpha$, $\tilde{\alpha} = \beta + r$, and $\tilde{\alpha} + \tilde{\beta} = \alpha + \beta$, we have

$$\tilde{h}^{2r-\mu,r}(s,\lambda) = \frac{\Gamma(\beta+r)\Gamma(\alpha)}{2(1-s)^{\alpha}\Gamma(\alpha+\beta+1)} \times {}_2F_1\left(\beta+r, \alpha, \alpha+\beta+1; \frac{1}{1-s}\right). \tag{5.94}$$

Now it is clear that (5.93), (5.94), and the known formula

$${}_2F_1(a,b,c;z) = {}_2F_1(b,a,c;z)$$

imply the equality

$$(1-s)^{r-\mu}\tilde{h}^{\mu,r}(s,\lambda) = \tilde{h}^{2r-\mu,r}(s,\lambda)$$

for all admissible values of the variables. Finally, the uniqueness theorem for the Laplace transform implies (5.92).

This completes the proof of Theorem 5.24. □

5.4.6 *Equivalent Formulation of Duresne's Recurrence Formula*

Dufresne's formula is a statement about fractional integrals in disguise. Theorem 5.25 formulated below shows that mixing distribution densities associated with geometric Brownian motions with different values of the parameters are linked by fractional integral transforms.

Theorem 5.25 *Dufresne's recurrence formula can be rewritten in the following form*:

$$m_t^{(2r-\beta)}(\sqrt{y}) = c_{r,\beta,t}\, y^{2r-\beta-\frac{1}{2}} \exp\left\{-\frac{1}{2ty}\right\} \times \int_y^{\infty} (\tau - y)^{\beta-r-1} \tau^{-\beta+\frac{1}{2}} \exp\left\{\frac{1}{2t\tau}\right\} m_t^{(\beta)}(\sqrt{\tau})\, d\tau \tag{5.95}$$

where

$$c_{r,\beta,t} = \frac{(2t)^{r-\beta}\exp\{(-2r^2 + 2r\beta)t\}}{\Gamma(\beta - r)},$$

and $r < \beta$.

Proof It is clear that the function $m_t^{(\beta)}$ is the distribution density of the random variable $\{\frac{1}{t}A_t^{(\beta)}\}^{\frac{1}{2}}$. Therefore, (5.71) gives

$$h^{\beta,r}(s,t) = e^{\frac{\beta^2 t}{2}} \int_0^\infty (2ty^2)^{-r} \exp\left\{\frac{s}{2ty^2}\right\} m_t^{(\beta)}(y)\,dy$$

and

$$h^{2r-\beta,r}(s,t) = e^{\frac{(2r-\beta)^2 t}{2}} \int_0^\infty (2ty^2)^{-r} \exp\left\{\frac{s}{2ty^2}\right\} m_t^{(2r-\beta)}(y)\,dy.$$

Now, it is not hard to see that formula (5.92) with $\lambda = 1 - s$ is equivalent to the following formula:

$$\begin{aligned} &e^{\frac{\beta^2 t}{2}} \int_0^\infty y^{-2r} \exp\left\{\frac{1}{2ty^2}\right\} m_t^{(\beta)}(y) \exp\left\{-\frac{\lambda}{2ty^2}\right\} dy \\ &\quad = e^{\frac{(2r-\beta)^2 t}{2}} \lambda^{\beta-r} \int_0^\infty y^{-2r} \exp\left\{\frac{1}{2ty^2}\right\} m_t^{(2r-\beta)}(y) \exp\left\{-\frac{\lambda}{2ty^2}\right\} dy \end{aligned} \tag{5.96}$$

for all $\lambda > 0$.

We will next show that formula (5.96) is equivalent to formula (5.95). Indeed, (5.96) can be rewritten as follows:

$$\begin{aligned} &(2t)^{r-\beta} e^{(-2r\beta+2r^2)t} \lambda^{r-\beta} \int_0^\infty z^{r-\frac{3}{2}} \exp\left\{\frac{z}{2t}\right\} m_t^{(\beta)}(z^{-\frac{1}{2}}) e^{-\lambda z}\,dz \\ &\quad = \int_0^\infty z^{r-\frac{3}{2}} \exp\left\{\frac{z}{2t}\right\} m_t^{(2r-\beta)}(z^{-\frac{1}{2}}) e^{-\lambda z}\,dz \end{aligned} \tag{5.97}$$

for all $\lambda > 0$ (replace λ by $2\lambda t$ and make a substitution $z = y^{-2}$). The function of λ, appearing on the left-hand side of formula (5.97), is the Laplace transform of a certain fractional integral. By the uniqueness theorem for the Laplace transform, we have

$$\begin{aligned} &\frac{(2t)^{r-\beta} e^{(-2r\beta+2r^2)t}}{\Gamma(\beta-r)} \int_0^w (w-z)^{\beta-r-1} z^{r-\frac{3}{2}} \exp\left\{\frac{z}{2t}\right\} m_t^{(\beta)}(z^{-\frac{1}{2}})\,dz \\ &\quad = w^{r-\frac{3}{2}} \exp\left\{\frac{w}{2t}\right\} m_t^{(2r-\beta)}(w^{-\frac{1}{2}}) \end{aligned} \tag{5.98}$$

for all $w > 0$. Finally, we replace w by $\frac{1}{y}$ in (5.98) and make the substitution $z = \frac{1}{\tau}$ in the resulting integral. This establishes the equivalence of (5.98) and (5.95). □

5.4.7 Completion of the Proof of Theorem 5.9

Formula (5.95) implies that for every integer $k \geq 1$,

$$m_t^{(2k)}(\sqrt{y}) = (-2t)^k e^{-2k^2 t} y^{2k-\frac{1}{2}} e^{-\frac{1}{2ty}} \frac{\partial^k}{\partial y^k}\left[y^{\frac{1}{2}} e^{\frac{1}{2ty}} m_t(\sqrt{y})\right]. \tag{5.99}$$

This can be shown by putting $r = k$, $\beta = 2k$, and differentiating k times. Using (5.99) and Lemma 5.16, we obtain

$$m_t^{(2k)}(\sqrt{y}) = \frac{(-2t)^k}{\pi t} e^{-2k^2 t} e^{\frac{\pi^2}{8t}} y^{2k-\frac{1}{2}} e^{-\frac{1}{2ty}} \frac{\partial^k}{\partial y^k}\left[y^{-\frac{1}{2}} e^{\frac{1}{2ty}} I\left(\frac{1}{2ty}\right)\right]. \tag{5.100}$$

It follows from (5.100) and Corollary 5.15 that

$$m_t^{(2k)}(\sqrt{y}) = \frac{1}{\pi t} e^{-2k^2 t} e^{\frac{\pi^2}{8t}} y^{-1} \frac{\partial^k I}{\partial y^k}\left(\frac{1}{2ty}\right)\left(1 + O\left((\log y)^{-\frac{1}{2}}\right)\right).$$

Therefore,

$$m_t^{(2k)}(y) = \frac{1}{\pi t} e^{-2k^2 t} e^{\frac{\pi^2}{8t}} y^{-2} \frac{\partial^k I}{\partial y^k}\left(\frac{1}{2ty^2}\right)\left(1 + O\left((\log y)^{-\frac{1}{2}}\right)\right).$$

Now, using Corollary 5.15, we see that

$$m_t^{(2k)}(y) = \frac{1}{\sqrt{\pi t}} e^{-2k^2 t} y^{2k-1} (\log y)^k \exp\left\{-\frac{u_y^2}{2t}\right\} \exp\left\{\frac{u_y}{2t}\right\}$$
$$\times \left(1 + O\left((\log y)^{-\frac{1}{2}}\right)\right). \tag{5.101}$$

This establishes Theorem 5.9 with $\nu = 2k + \frac{1}{2}$, $\xi = 1$, and $y_0 = 1$.

We will next consider the special case in Theorem 5.9 where $\xi = 1$, $y_0 = 1$, and α is any real number. Let $k \geq 1$ be an integer such that $\alpha < 2k$, and put $\beta = 2k$ and $r = k + \frac{\alpha}{2}$. Using (5.95), we see that

$$m_t^{(\alpha)}(\sqrt{y}) = (2t)^{-k+\frac{\alpha}{2}} \exp\left\{-2t\left(\left(k + \frac{\alpha}{2}\right)^2 - 2\left(k + \frac{\alpha}{2}\right)k\right)\right\} y^{\alpha-\frac{1}{2}} e^{-\frac{1}{2ty}}$$
$$\times \frac{1}{\Gamma(k - \frac{\alpha}{2})} \int_y^\infty (\tau - y)^{k-\frac{\alpha}{2}-1} \tau^{-2k+\frac{1}{2}} e^{\frac{1}{2t\tau}} m_t^{(2k)}(\sqrt{\tau})\, d\tau. \tag{5.102}$$

It follows from (5.101) and (5.102) that

$$m_t^{(\alpha)}(\sqrt{y}) = \frac{1}{\sqrt{\pi t}} e^{-2k^2 t} (2t)^{-k+\frac{\alpha}{2}}$$
$$\times \exp\left\{-2t\left(\left(k + \frac{\alpha}{2}\right)^2 - 2\left(k + \frac{\alpha}{2}\right)k\right)\right\} y^{\alpha-\frac{1}{2}}$$

$$\times \frac{1}{\Gamma(k-\frac{\alpha}{2})}\int_y^\infty (\tau-y)^{k-\frac{\alpha}{2}-1}\tau^{-2k+\frac{1}{2}}m_t^{(2k)}\tau^{k-\frac{1}{2}}(\log\sqrt{\tau})^k$$

$$\times \exp\left\{-\frac{u^2_{\sqrt{\tau}}}{2t}\right\}\exp\left\{\frac{u_{\sqrt{\tau}}}{2t}\right\}d\tau\big(1+O\big((\log y)^{-\frac{1}{2}}\big)\big)$$

and

$$m_t^{(\alpha)}(\sqrt{y}) = \frac{1}{\sqrt{\pi t}}(2t)^{-k+\frac{\alpha}{2}}e^{-\frac{t\alpha^2}{2}}y^{\alpha-\frac{1}{2}}$$

$$\times \frac{1}{\Gamma(k-\frac{\alpha}{2})}\int_y^\infty (\tau-y)^{k-\frac{\alpha}{2}-1}\tau^{-k}(\log\sqrt{\tau})^k$$

$$\times \exp\left\{-\frac{u^2_{\sqrt{\tau}}-u_{\sqrt{\tau}}}{2t}\right\}d\tau\big(1+O\big((\log y)^{-\frac{1}{2}}\big)\big). \tag{5.103}$$

We will next show that Theorem 5.3 can be applied to the fractional integral in formula (5.103). First, we observe that the function $u_{\sqrt{\tau}}$ in this formula can be replaced by the function $\eta(\sqrt{\tau})$ where

$$\eta(y) = \log y + \frac{1}{2}\log\log y + \log 2 + \frac{\frac{1}{2}[\frac{1}{2}\log\log y + \log 2]}{\log y}.$$

This can be done since the functions $u_{\sqrt{\tau}}$ and $\eta(y)$ are asymptotically equivalent (see formula (5.52)). Next, we put

$$a(\sigma) = \sigma^{-k}(\log\sqrt{\sigma})^k$$

and

$$b(\sigma) = \frac{\eta(\sqrt{\sigma})^2 - \eta(\sqrt{\sigma})}{2t}.$$

Then the functions a and b satisfy the conditions in Theorem 5.3. Applying this theorem to (5.103), we obtain

$$\begin{aligned} m_t^{(\alpha)}(\sqrt{y}) &= \frac{1}{\sqrt{\pi t}}(2k)^{-k+\frac{\alpha}{2}}e^{-\frac{t\alpha^2}{2}}y^{\alpha-\frac{1}{2}}y^{-k}(\log\sqrt{y})^k \\ &\quad \times \exp\left\{-\frac{\eta(\sqrt{y})^2-\eta(\sqrt{y})}{2t}\right\}\left(\frac{4ty}{\log y}\right)^{k-\frac{\alpha}{2}}\big(1+O\big((\log y)^{-\frac{1}{2}}\big)\big) \\ &= \frac{1}{\sqrt{\pi t}}e^{-\frac{t\alpha^2}{2}}y^{\frac{\alpha-1}{2}}(\log\sqrt{y})^{\frac{\alpha}{2}}\exp\left\{-\frac{\eta(\sqrt{y})^2-\eta(\sqrt{y})}{2t}\right\} \\ &\quad \times \big(1+O\big((\log y)^{-\frac{1}{2}}\big)\big) \end{aligned}$$

as $y \to \infty$. Hence,

$$m_t^{(\alpha)}(y) = \frac{1}{\sqrt{\pi t}} e^{-\frac{t\alpha^2}{2}} y^{\alpha-1} (\log y)^{\frac{\alpha}{2}} \exp\left\{-\frac{u_y^2}{2t}\right\} \exp\left\{\frac{u_y}{2t}\right\}$$
$$\times \left(1 + O\left((\log y)^{-\frac{1}{2}}\right)\right) \tag{5.104}$$

as $y \to \infty$. This proves Theorem 5.9 in the case where $\xi = 1$ and $y_0 = 1$.

It remains to get rid of the restriction $\xi = 1$, $y_0 = 1$. This can be done using the following simple lemma:

Lemma 5.26 *Let $-\infty < \nu < \infty$, $\xi > 0$, $y_0 > 0$, and $t > 0$. Then for every $y > 0$,*

$$m_t(y; \nu, \xi, y_0) = \frac{1}{y_0} m_{t\xi^2}\left(\frac{y}{y_0}; \frac{\nu}{\xi^2}, 1, 1\right).$$

The proof of Lemma 5.26 is based on the scaling property of Brownian motion. It is left as an exercise for the reader. It is clear that Theorem 5.9 follows from (5.104) and Lemma 5.26.

This completes the proof of Theorem 5.8.

5.5 Asymptotic Behavior of Mixing Distribution Densities Near Zero

The mixing distribution density associated with a geometric Brownian motion decays very rapidly as $y \to 0$.

Theorem 5.27 *Let $-\infty < \nu < \infty$, $\xi > 0$, $y_0 > 0$, and $t > 0$. Then there exists a positive constant $b = b(\nu, \xi, y_0, t)$ such that*

$$m_t(y) = b y^{\beta} \exp\left\{-\frac{y_0^2}{2t\xi^2 y^2}\right\}\left(1 + O\left(y^2\right)\right) \tag{5.105}$$

as $y \to 0$, where $\beta = \frac{2\nu - 2\xi^2}{\xi^2}$.

We will only sketch the proof of Theorem 5.27 in a special case where $\nu = \frac{1}{2}$, $\xi = 1$, $y_0 = 1$, and without the error estimate in formula (5.105). This case can be extended to the case where $-\infty < \nu < \infty$, $\xi = 1$, and $y_0 = 1$, using Dufresne's recurrence formula. Finally, the general case can be obtained from Lemma 5.26. In addition, the following relation is true for the constants appearing in Theorem 5.27:

$$b(\nu, \xi, y_0, t) = y_0^{-2\alpha} b\left(\frac{\nu}{\xi^2}, 1, 1, t\xi^2\right),$$

where $\alpha = \frac{2\nu-\xi^2}{2\xi^2}$. We leave filling in the details as an exercise for the interested reader.

Let us assume that $\nu = \frac{1}{2}$, $\xi = 1$, and $y_0 = 1$. Then $\alpha = \nu - \frac{1}{2}$. The following auxiliary function will be used in the proof:

$$\Phi_t(u) = \exp\left\{-\frac{(\sinh^{-1} u)^2}{2t}\right\} \cos\left(\frac{\pi \sinh^{-1} u}{2t}\right). \tag{5.106}$$

This function is bounded, even, rapidly decreasing, and can be represented by its Taylor series, that is,

$$\Phi_t(u) = \sum_{n=0}^{\infty} \frac{c_{n,t}}{(2n)!} u^{2n}, \quad c_{n,t} = \Phi_t^{(2n)}(0).$$

It follows from formula (5.43) that

$$\begin{aligned} y^{\frac{1}{2}} e^{\frac{1}{2ty}} m_t(\sqrt{y}) &= \frac{1}{\pi t} e^{\frac{\pi^2}{8t}} y^{-\frac{1}{2}} \int_{-\infty}^{\infty} \exp\left\{-\frac{(\sinh u)^2}{2ty}\right\} \cosh u e^{-\frac{u^2}{2t}} e^{\frac{i\pi u}{2t}} \, du \\ &= \frac{1}{\pi t} e^{\frac{\pi^2}{8t}} \int_{-\infty}^{\infty} \exp\left\{-\frac{x^2}{2t}\right\} \exp\left\{-(2t)^{-1}\left(\sinh^{-1}\left(y^{\frac{1}{2}} x\right)\right)^2\right\} \\ &\quad \times \exp\left\{i\pi (2t)^{-1} \sinh^{-1}\left(y^{\frac{1}{2}} x\right)\right\} dx. \end{aligned} \tag{5.107}$$

Using (5.106) and (5.107), we obtain

$$y^{\frac{1}{2}} e^{\frac{1}{2ty}} m_t(\sqrt{y}) = \frac{1}{\pi t} e^{\frac{\pi^2}{8t}} \int_{-\infty}^{\infty} \exp\left\{-\frac{x^2}{2t}\right\} \Phi_t\left(y^{\frac{1}{2}} x\right) dx. \tag{5.108}$$

Since the function Φ is bounded and $\lim_{u\to 0} \Phi(u) = 1$, formula (5.108) implies that

$$m_t(y) \sim b y^{-1} \exp\left\{\frac{\pi^2}{8t}\right\}$$

as $y \to 0$, where

$$b = \frac{\sqrt{2}}{\sqrt{\pi t}} \exp\left\{\frac{\pi^2}{8t}\right\}.$$

This completes the proof of formula (5.105) without the error estimate in a special case where $\nu = \frac{1}{2}$, $\xi = 1$, and $y_0 = 1$.

5.6 Asymptotic Formulas for Mixing Distribution Densities Associated with CIR Processes

In this section, we use formula (4.47) and Theorem 5.1 to analyze the asymptotic behavior of the mixing distribution density m_t associated with a CIR process. It is easy to see that the following equality is equivalent to equality (4.47):

$$\int_0^\infty \exp\{-\lambda y\} y^{-\frac{1}{2}} m_t(2^{\frac{1}{2}} c t^{-\frac{1}{2}} y^{\frac{1}{2}})\, dy$$

$$= \frac{\sqrt{2t}}{c} \exp\left\{\frac{abt}{c^2}\right\}\left(\frac{\sqrt{\lambda+b^2}}{\sqrt{\lambda+b^2}\cosh(\frac{1}{2}t\sqrt{\lambda+b^2}) + b\sinh(\frac{1}{2}t\sqrt{\lambda+b^2})}\right)^{\frac{2a}{c^2}}$$

$$\times \exp\left\{-\frac{y_0 c^{-2}\lambda \sinh(\frac{1}{2}t\sqrt{\lambda+b^2})}{\sqrt{\lambda+b^2}\cosh(\frac{1}{2}t\sqrt{\lambda+b^2}) + b\sinh(\frac{1}{2}t\sqrt{\lambda+b^2})}\right\}. \tag{5.109}$$

Our next goal is to apply Theorem 5.1 to equality (5.109). Before doing this, however, we must give (5.109) an appropriate form and check the validity of the conditions, under which Theorem 5.1 holds.

The following entire function will play an important role in the sequel:

$$\Phi_s(z) = z\cos z + s\sin z, \quad z \in \mathbb{C},$$

where $s \geq -1$. It is clear that the function Φ_s is odd and satisfies $\Phi_s(0) = 0$. We are especially interested in the location of the smallest positive zero of the function Φ_s. It will be shown next that all the zeros of the function Φ_s are real.

Lemma 5.28 *For all $s \geq -1$, the function Φ_s has only real zeros. Moreover, the set of zeros is infinite.*

Proof For every $n \geq 1$ put

$$P_n(z) = z\prod_{k=1}^{n}\left(1 - \frac{z^2}{k^2}\right).$$

Then for all $z \in \mathbb{C}$,

$$P_n(z) \to \frac{\sin \pi z}{\pi} \tag{5.110}$$

as $n \to \infty$ by the product formula. The function P_n is a polynomial of degree $2n+1$, all of whose roots ($z = k$, $k \in \mathbb{Z}$, $|k| \leq n$) are real. Put

$$Q_n(z) = z^{-s+1}\big(z^s P_n(z)\big)', \quad n \geq 1.$$

Then

$$Q_n(z) = sP_n(z) + zP_n'(z), \tag{5.111}$$

and hence Q_n is a polynomial of degree $2n+1$ which vanishes at $z = 0$. It follows from (5.110) and (5.111) that for every $z \in \mathbb{C}$,

$$Q_n(z) \to \frac{s}{\pi}\sin \pi z + z\cos \pi z = \frac{1}{\pi}\Phi_s(\pi z)$$

as $n \to \infty$ uniformly on every compact set in $\mathbb{C}$. By Rolle's theorem, the function $z \mapsto (z^s P_n(z))'$ vanishes at points strictly between k and $k+1$, $-n \le k < n$, since the function $z \mapsto z^s P_n(z)$ vanishes at those points. It follows from the previous considerations that Q_n has all its $2n+1$ roots that are real. Now the desired conclusion that all the roots of Φ_s are real can be obtained using (5.111) and Hurwitz's theorem (see [Con78], Theorem 2.5 in Chapter VII). Here we also need to take into account that the function $\Phi_s(\pi n) \ne 0$ for any nonzero integer n.

The proof above implicitly used the condition $s > -1$ (for otherwise $z^s P_n(z)$ does not vanish at the origin). The result for $s = -1$ can be derived from that for $s > -1$ by a limiting argument.

This completes the proof of Lemma 5.28. □

Definition 5.29 For $s \ge 0$, the smallest positive zero of the function

$$\Phi_s(z) = z\cos z + s\sin z$$

will be denoted by r_s.

It is not hard to see that $r_0 = \frac{\pi}{2}$, and $r_s \uparrow \pi$ as $s \to \infty$. Moreover, the function $s \mapsto r_s$ is differentiable and increasing on $(0, \infty)$. Indeed, the value of r_s for $0 < s < \infty$ is equal to the first coordinate of the point in $\mathbb{R}^2$ where the segment described by

$$y = -\frac{x}{s}, \quad \frac{\pi}{2} < x < \pi,$$

intersects the curve $y = \tan x$. In addition, we have

$$r_s = \phi^{-1}(s), \quad 0 < s < \infty,$$

where

$$\phi(u) = -\frac{u}{\tan u}, \quad \frac{\pi}{2} < u < \pi.$$

It is also clear that

$$\sin(r_0) = 1, \quad \cos(r_0) = 0, \quad \text{and} \quad \Phi_0'(r_0) = -\frac{\pi}{2}.$$

Moreover, if $s > 0$, then

$$\sin(r_s) > 0, \quad \cos(r_s) < 0, \quad \text{and} \quad \Phi_s'(r_s) < 0. \tag{5.112}$$

Remark 5.30 It follows from the last inequality in (5.112) and the equality $\Phi_s(r_s) = 0$ that there exists $\rho > 0$ such that $\Phi_s(v) > 0$ for all v with $r_s - \rho < v < r_s$.

For every $s \ge 0$ the function ρ_s defined by

$$\rho_s(z) = z\cosh z + s\sinh z = -i\Phi_s(iz)$$

has only imaginary zeros. The following notation will be used throughout the remaining part of the present section:

$$\Phi = \Phi_{\frac{1}{2}bt}, \quad r = r_{\frac{tb}{2}}, \; u = u_{b,t} = -4t^{-2} r^2_{\frac{1}{2}tb}.$$

Remark 5.31 It is not hard to see that the function $\lambda \mapsto \Phi(i\frac{1}{2}t\sqrt{\lambda})$ does not have any roots in $(0, \infty)$. Moreover, the number u is the largest nonzero root of this function in $(-\infty, 0]$.

Let us return to formula (5.109). Set $\alpha_1 = 0$, $\alpha_2 = \infty$, $\gamma_1 = 0$, $\gamma_2 = \frac{2a}{c^2}$, $\tau = b^2 - u$, $M(y) = y^{-\frac{1}{2}} m_t(2^{\frac{1}{2}} c t^{-\frac{1}{2}} y^{\frac{1}{2}}) \chi_{\{y>0\}}$,

$$I(z) = \frac{\sqrt{2t}}{c} \exp\left\{\frac{abt}{c^2}\right\} \left(\frac{\sqrt{z+u}}{\sqrt{z+u}\cosh(\frac{1}{2}t\sqrt{z+u}) + b\sinh(\frac{1}{2}t\sqrt{z+u})} \right)^{\frac{2a}{c^2}}$$

$$\times \exp\left\{ -\frac{y_0 c^{-2}(z+u-b^2)\sinh(\frac{1}{2}t\sqrt{z+u})}{\sqrt{z+u}\cosh(\frac{1}{2}t\sqrt{z+u}) + b\sinh(\frac{1}{2}t\sqrt{z+u})} \right\},$$

$$G_1(z) = \frac{\sqrt{z+u}}{\sqrt{z+u}\cosh(\frac{1}{2}t\sqrt{z+u}) + b\sinh(\frac{1}{2}t\sqrt{z+u})}, \tag{5.113}$$

$$G_2(z) = \frac{\sqrt{2t}}{c} \exp\left\{\frac{abt}{c^2}\right\},$$

and

$$F(z) = -\frac{y_0 c^{-2}(z+u-b^2)\sinh(\frac{1}{2}t\sqrt{z+u})}{\sqrt{z+u}\cosh(\frac{1}{2}t\sqrt{z+u}) + b\sinh(\frac{1}{2}t\sqrt{z+u})}. \tag{5.114}$$

We use the principal branch of the argument in the previous formulas. It is clear that (5.2) and (5.4) hold.

It remains to check whether the Tauberian conditions 2–5 in Theorem 5.1 are valid. We restrict ourselves to the functions G_1 and F, since G_2 is a positive constant function. Recall that $u = -4t^{-2} r_{\frac{1}{2}bt} < 0$. The validity of condition 2 will be established first. Note that if λ is a small positive number, then

$$G_1(\lambda) = \frac{\frac{1}{2}t\sqrt{|u| - \lambda}}{\Phi_{\frac{1}{2}bt}(\frac{1}{2}t\sqrt{|u| - \lambda})}$$

and

$$F(\lambda) = \frac{\frac{1}{2}t y_0 c^{-2}(|u| + b^2 - \lambda)\sin(\frac{1}{2}t\sqrt{|u| - \lambda})}{\Phi_{\frac{1}{2}bt}(\frac{1}{2}t\sqrt{|u| - \lambda})}.$$

Next, using Remark 5.30, the inequalities $\frac{\pi}{2} < r_{\frac{1}{2}bt} < \pi$, and the equality $\Phi_{\frac{1}{2}bt}(r_{\frac{1}{2}bt}) = 0$, we see that condition 2 holds. In addition, it is not hard to see that conditions 3 and 4 follow from the definitions of the functions G_1 and F and the properties of the function $\Phi_{\frac{1}{2}bt}$. On the other hand, more work is needed to check condition 6. Let $\delta > 0$ be a small number and suppose $z = z_1 + iz_2$ with $z_1 \in [0, s + \tau - \delta]$ and z_2 positive and large. In order to prove that condition 6 holds, we will first estimate the functions $z_2 \mapsto G_1(iz_2)$ and $z_2 \mapsto \exp\{F(iz_2)\}$ as $z_2 \to \infty$. It is not hard to see that there exist positive constants l_1 and l_2 such that

$$\left|G_1(iz_2)\right| \le l_1 e^{-l_2\sqrt{z_2}} \quad \text{as } z_2 \to \infty. \tag{5.115}$$

It is also true that

$$\begin{aligned}
F(iz_2) &= -\frac{y_0 c^{-2}(iz_2 + u - b^2)\sinh(\frac{1}{2}t\sqrt{iz_2+u})}{\sqrt{iz_2+u}\cosh(\frac{1}{2}t\sqrt{iz_2+u}) + b\sinh(\frac{1}{2}t\sqrt{iz_2+u})} \\
&= -\frac{y_0 c^{-2}\sqrt{iz_2+u}\sinh(\frac{1}{2}t\sqrt{iz_2+u})}{\cosh(\frac{1}{2}t\sqrt{iz_2+u})} + O(1) \\
&= -y_0 c^{-2}\sqrt{iz_2+u} - y_0 c^{-2} \\
&\quad \times \frac{\sqrt{iz_2+u}[\sinh(\frac{1}{2}t\sqrt{iz_2+u}) - \cosh(\frac{1}{2}t\sqrt{iz_2+u})]}{\cosh(\frac{1}{2}t\sqrt{iz_2+u})} + O(1) \\
&= -y_0 c^{-2}\sqrt{iz_2+u} + y_0 c^{-2}\frac{\sqrt{iz_2+u}\exp\{-\frac{1}{2}t\sqrt{iz_2+u}\}}{\cosh(\frac{1}{2}t\sqrt{iz_2+u})} + O(1)
\end{aligned}$$

as $z_2 \to \infty$. Since $\Re(\sqrt{iz_2+u}) > 0$, we have

$$F(iz_2) = -y_0 c^{-2}\sqrt{iz_2+u} + O(1)$$

as $z_2 \to \infty$. Moreover, there exist positive constants l_3 and l_4 such that

$$\left|\exp\{F(iz_2)\}\right| \le l_3 e^{-l_4\sqrt{z_2}} \quad \text{as } z_2 \to \infty. \tag{5.116}$$

Next, reasoning as in the proof of (5.115) and (5.116), we can show that

$$\left|G_1(z_1 + iz_2)\right| \le l_5 e^{-l_6\sqrt{z_2}} \tag{5.117}$$

and

$$\left|\exp\{F(z_1 + iz_2)\}\right| \le l_7 e^{-l_8\sqrt{z_2}} \tag{5.118}$$

as $z_2 \to \infty$, uniformly with respect to $z_1 \in [0, s + \tau - \delta]$. Now, it is not hard to see that the previous two estimates imply the validity of condition 6 in Theorem 5.1. This shows that all the conditions in Theorem 5.1 hold.

In the next two lemmas, we compute the numbers α, β, $\widetilde{G}(0)$ and $\widetilde{F}(0)$.

Lemma 5.32 *The following formulas hold for the residue of the function G_1 at the point $z=0$ and for the number $\widetilde{G}(0)$:*

$$\beta=\frac{8r^3}{t^2\sin r[r^2+\frac{tb}{2}(1+\frac{tb}{2})]} \tag{5.119}$$

and

$$\widetilde{G}(0)=-\frac{r[r^2+\frac{3tb}{2}(1+\frac{tb}{2})]}{2\sin r[r^2+\frac{tb}{2}(1+\frac{tb}{2})]^2}. \tag{5.120}$$

Proof We have

$$\lambda G_1(\lambda)=\frac{it\sqrt{z+u}}{2\Phi(i\frac{1}{2}t\sqrt{z+u})}. \tag{5.121}$$

Recall that

$$\Phi(r)=r\cos r+\frac{tb}{2}\sin r=0.$$

Therefore,

$$\Phi'(r)=\left(1+\frac{tb}{2}\right)\cos r-r\sin r=-\frac{\sin r}{r}\left[r^2+\frac{tb}{b}\left(1+\frac{tb}{2}\right)\right]. \tag{5.122}$$

It is not hard to see that

$$\sqrt{\lambda+u}=\frac{2ir}{t}-\frac{it}{4r}\lambda+\cdots \tag{5.123}$$

and

$$\Phi\left(i\frac{1}{2}t\sqrt{\lambda+u}\right)=\rho_1\lambda+\rho_2\lambda^2+\cdots, \tag{5.124}$$

where

$$\rho_1=-\frac{t^2\sin r}{8r^2}\left[r^2+\frac{tb}{2}\left(1+\frac{tb}{2}\right)\right], \tag{5.125}$$

and

$$\rho_2=\frac{t^4\sin r}{128r^4}\left[r^2-\frac{tb}{2}\left(1+\frac{tb}{2}\right)\right]. \tag{5.126}$$

In the proof of (5.124), we use (5.122) and the equality

$$\Phi''(r)=-\Phi(r)-2\sin r=-2\sin r.$$

Next, we see that (5.121), (5.123), and (5.124) imply the equality

$$\beta+\widetilde{G}(0)\lambda+\cdots=\frac{-r+\frac{t^2}{8r}\lambda+\cdots}{\rho_1+\rho_2\lambda+\cdots}. \tag{5.127}$$

Now, using (5.127) and (5.126), we obtain

$$\beta = -\frac{r}{\rho_1} = \frac{8r^3}{t^2 \sin r[r^2 + \frac{tb}{2}(1 + \frac{tb}{2})]}.$$

This establishes formula (5.119).

To prove (5.120), we first derive the following formula from (5.127):

$$\frac{t^2}{8r} = \rho_1 \widetilde{G}(0) + \beta\rho_2.$$

Solving for $\widetilde{G}(0)$, we get

$$\widetilde{G}(0) = \left(\frac{t^2}{8r} - \beta\rho_2\right)\frac{1}{\rho_1}.$$

Finally, taking into account (5.119), (5.125), and (5.126), and making simplifications, we see that formula (5.120) holds.

This completes the proof of Lemma 5.32. □

Lemma 5.33 *The following formulas hold for the residue of the function F at the point $z = 0$ and for the number $\widetilde{F}(0)$:*

$$\alpha = \frac{4y_0 r^2(4r^2 + t^2b^2)}{t^3c^2[r^2 + \frac{tb}{2}(1 + \frac{tb}{2})]} > 0 \tag{5.128}$$

and

$$\widetilde{F}(0) = -\frac{y_0 K(r, t, b)}{4tc^2[r^2 + \frac{tb}{2}(1 + \frac{tb}{2})]^2} \tag{5.129}$$

where

$$K(r, t, b) = \left[16r^2 - tb\left(4r^2 + t^2b^2\right)\right]\left[r^2 + \frac{tb}{2}\left(1 + \frac{tb}{2}\right)\right] - \left(4r^2 + t^2b^2\right)\left[r^2 - \frac{tb}{2}\left(1 + \frac{tb}{2}\right)\right].$$

Proof It is not hard to see that

$$\sinh\left(\frac{1}{2}t\sqrt{\lambda + u}\right) = i\eta_1 + i\eta_2\lambda + \cdots \tag{5.130}$$

where

$$\eta_1 = \sin r, \quad \text{and} \quad \eta_2 = -\frac{t^2 \cos r}{8r} = \frac{t^3 b \sin r}{16r^2}.$$

It follows from (5.124), (5.130), and from the definition of the function F that

$$\lambda F(\lambda) = \frac{ty_0}{2c^2} \frac{(\lambda + u - b^2)(\eta_1 + \eta_2\lambda + \cdots)}{\rho_1 + \rho_2\lambda + \cdots}$$

$$= \frac{ty_0}{2c^2} \frac{(u - b^2)\eta_1 + [\eta_1 + (u - b^2)\eta_2]\lambda + \cdots}{\rho_1 + \rho_2\lambda + \cdots}.$$

Therefore,

$$\alpha\rho_1 = \frac{ty_0}{2c^2}(u - b^2)\eta_1 \tag{5.131}$$

and

$$\alpha\rho_2 + \widetilde{F}(0)\rho_1 = \frac{ty_0}{2c^2}\left(\eta_1 + (u - b^2)\eta_2\right). \tag{5.132}$$

It is not hard to see that (5.131) implies (5.128). Next, using (5.131) and the expressions for η_1 and η_2 given above, and making simplifications in (5.132), we obtain formula (5.129).

This establishes Lemma 5.33. □

The next assertion provides an asymptotic formula for the density m_t associated with a CIR process. We can derive this formula by applying Theorem 5.1 to (5.109) and taking into account what has already been established above.

Theorem 5.34 *Suppose that $a \geq 0$, $b \geq 0$, and $c > 0$. Then there exist positive constants A, B, and C such that*

$$m_t(y) = Ae^{-Cy^2}e^{By}y^{-\frac{1}{2}+\frac{2a}{c^2}}\left(1 + O\left(y^{-1}\right)\right), \quad y \to \infty. \tag{5.133}$$

Explicit formulas for the constants in formula (5.133) are provided in the following lemma.

Lemma 5.35 *The constants A, B, and C in Theorem 5.34 are as follows:*

$$A = \frac{1}{\sqrt{\pi}}\left(\frac{t}{2c^2}\right)^{\frac{1}{4}+\frac{a}{c^2}} \alpha^{\frac{1}{4}-\frac{a}{c^2}} \beta^{\frac{2a}{c^2}} \exp\left\{\frac{abt}{c^2}\right\} e^{\widetilde{F}(0)},$$

$$B = \frac{\sqrt{2\alpha t}}{c}, \quad \text{and} \quad C = \frac{t^2b^2 + 4r^2_{\frac{tb}{2}}}{2tc^2},$$

where the numbers $\beta > 0$, $\alpha > 0$, and $\widetilde{F}(0)$ are given by (5.119), (5.128), and (5.129), respectively.

Corollary 5.36 *The following formula holds:*

$$m_t(y, a, 0, c, y_0) = Ae^{-Cy^2}e^{By}y^{-\frac{1}{2}+\frac{2a}{c^2}}\left(1 + O\left(y^{-1}\right)\right)$$

as $y \to \infty$, *where*

$$A = \frac{2^{\frac{1}{4}+\frac{a}{c^2}} y_0^{\frac{1}{4}-\frac{a}{c^2}}}{c\sqrt{t}} \exp\left\{-\frac{3y_0}{c^2 t}\right\}, \quad B = \frac{2\sqrt{2y_0}\pi}{c^2 t}, \quad \textit{and} \quad C = \frac{\pi^2}{2c^2 t}.$$

Corollary 5.36 can be derived from Theorem 5.34 and Lemma 5.35 by using the equality $r_0 = \frac{\pi}{2}$, and the fact that for $b = 0$, we have

$$\beta = \frac{4\pi}{t^2}, \quad \alpha = \frac{4y_0\pi^2}{c^2 t^3}, \quad \text{and} \quad \widetilde{F}(0) = -\frac{3y_0}{c^2 t}.$$

5.7 Asymptotic Formulas for Mixing Distribution Densities Associated with Ornstein–Uhlenbeck Processes

Our goal in this section is to find an asymptotic formula, describing the behavior of the mixing distribution density m_t associated with the Ornstein–Uhlenbeck process $Y(q, m, \sigma, y_0)$.

The following statement holds.

Theorem 5.37 *Let Y be an Ornstein–Uhlenbeck process with $q \geq 0$, $m \geq 0$, $y_0 > 0$, and $\sigma > 0$. Then there exist positive constants E, F, and G such that*

$$m_t(y) = E e^{-Gy^2} e^{Fy} \left(1 + O\left(y^{-1}\right)\right) \tag{5.134}$$

as $y \to \infty$.

Remark 5.38 Explicit expressions for the constants E, F, and G in Theorem 5.37 can be found in Lemma 5.42.

Proof The proof of formula (5.134) is much simpler in the case where $m = 0$ than in the general case. For $m = 0$, we can use the following fact established in Sect. 1.16. A squared Ornstein–Uhlenbeck process $Y^2(q, 0, \sigma, y_0)$ has the same law as the CIR process $\widetilde{Y}(\sigma^2, 2q, 2\sigma, y_0^2)$. Therefore if $m = 0$, then we can derive asymptotic formulas for m_t from the corresponding formulas for the CIR process. For instance, formula (5.133), which is valid for the CIR processes, implies formula (5.134) with $m = 0$.

Our next goal is to prove Theorem 5.37 in the general case. The proof is based on formula (4.51) and Theorem 5.1. Recall that for $s \geq 0$, we denoted by r_s the smallest strictly positive zero of the function

$$\Phi_s(z) = z \cos z + s \sin z.$$

It is clear that

$$-i\Phi_s(iz) = z \cosh z + s \sinh z.$$

For $q > 0$ and $t > 0$, put $\Phi = \Phi_{qt}$, $r = r_{qt}$, and $v = v_{q,t} = -r_{qt}^2 t^{-2}$. We will next use the notation in the formulation of Theorem 5.1. It is not hard to see that (4.51) is equivalent to the following formula:

$$\int_{-\infty}^{\infty} e^{-\lambda y} M(y)\, dy = I(\lambda + \tau), \tag{5.135}$$

where $\alpha_1 = 0$, $\alpha_2 = \infty$, $\tau = q^2 - v$, $M(y) = y^{-\frac{1}{2}} m_t(2^{\frac{1}{2}} \sigma t^{-\frac{1}{2}} y^{\frac{1}{2}}) \chi_{\{y>0\}}$, and

$$I(z) = \frac{\sqrt{2t}}{\sigma} e^{\frac{qt}{2}} \left(F_1(z)\right)^{\frac{1}{2}} \exp\left\{F_2(z) + F_3(z) + F_4(z) + F_5(z)\right\}. \tag{5.136}$$

The functions F_k, $1 \le k \le 5$, in (5.136) are given by

$$F_1(z) = \frac{\sqrt{z+v}}{\sqrt{z+v}\cosh(t\sqrt{z+v}) + q\sinh(t\sqrt{z+v})}, \tag{5.137}$$

$$F_2(z) = -\frac{y_0^2(z+v-q^2)\sinh(t\sqrt{z+v})}{2\sigma^2[\sqrt{z+v}\cosh(t\sqrt{z+v}) + q\sinh(t\sqrt{z+v})]}, \tag{5.138}$$

$$F_3(z) = \frac{mqy_0(z+v-q^2)(1-\cosh(t\sqrt{z+v}))}{\sigma^2\sqrt{z+v}[\sqrt{z+v}\cosh(t\sqrt{z+v}) + q\sinh(t\sqrt{z+v})]}, \tag{5.139}$$

$$F_4(z) = \frac{m^2q^2(z+v-q^2)(\sinh(t\sqrt{z+v}) - t\sqrt{z+v}\cosh(t\sqrt{z+v}))}{2\sigma^2(z+v)[\sqrt{z+v}\cosh(t\sqrt{z+v}) + q\sinh(t\sqrt{z+v})]}, \tag{5.140}$$

and

$$F_5(z) = \frac{m^2q^3(z+v-q^2)[2(\cosh(t\sqrt{z+v}) - 1) - t\sqrt{z+v}\sinh(t\sqrt{z+v})]}{2\sigma^2(z+v)^{\frac{3}{2}}[\sqrt{z+v}\cosh(t\sqrt{z+v}) + q\sinh(t\sqrt{z+v})]}. \tag{5.141}$$

Our next goal is to apply Theorem 5.1 to formula (5.135). Set $\gamma_1 = 0$, $\gamma_2 = \frac{1}{2}$, $G_1(z) = F_1(z)$, $G_2(z) = 2^{\frac{1}{2}} t^{\frac{1}{2}} \sigma^{-1} e^{\frac{qt}{2}}$, and $F(z) = F_2(z) + F_3(z) + F_4(z) + F_5(z)$. It is clear that we only need to check the validity of the Tauberian conditions, under which Theorem 5.1 holds. Note that

$$F_1(z) = \frac{it\sqrt{z+v}}{\Phi(it\sqrt{z+v})}, \tag{5.142}$$

$$F_2(z) = -\frac{iy_0^2 t(z+v-q^2)\sinh(t\sqrt{z+v})}{2\sigma^2\Phi(it\sqrt{z+v})}, \tag{5.143}$$

$$F_3(z) = -\frac{imqy_0t(z+v-q^2)[\cosh(t\sqrt{z+v})-1]}{\sigma^2\sqrt{z+v}\Phi(it\sqrt{z+v})}, \tag{5.144}$$

$$F_4(z) = \frac{im^2q^2t(z+v-q^2)[\sinh(t\sqrt{z+v})-t\sqrt{z+v}\cosh(t\sqrt{z+v})]}{2\sigma^2(z+v)\Phi(it\sqrt{z+v})} \tag{5.145}$$

and

$$F_5(z) = \frac{im^2q^3t(z+v-q^2)[4\sinh^2\frac{t\sqrt{z+v}}{2}-t\sqrt{z+v}\sinh(t\sqrt{z+v})]}{2\sigma^2(z+v)^{\frac{3}{2}}\Phi(it\sqrt{z+v})}. \tag{5.146}$$

It is not hard to see that the functions F_j, $1 \le j \le 5$, have removable singularities at the point $z = -q^2(2\sigma^2)^{-1}$.

We will need the following power series representation:

$$\Phi(it\sqrt{z+v}) = \zeta_1 z + \zeta_2 z^2 + \cdots, \tag{5.147}$$

where

$$\zeta_1 = -\frac{t^2\sin r}{2r^2}\left[r^2 + qt(1+qt)\right] \tag{5.148}$$

and

$$\zeta_2 = \frac{t^4\sin r}{8r^4}\left[r^2 - qt(1+qt)\right]. \tag{5.149}$$

Here we use the fact that

$$\Phi(r) = r\cos r + qt\sin r = 0.$$

Let us denote by β the residue of the function $G_1 = F_1$ at $z = 0$. It follows from (5.147) and (5.148) that

$$\beta = \lim_{z\to 0} zG_1(z) = -r\lim_{z\to 0}\frac{z}{\Phi(it\sqrt{z+v})} = -\frac{r}{\zeta_1}$$
$$= \frac{2r^3}{t^2\sin r[r^2+qt(1+qt)]}. \tag{5.150}$$

For every $2 \le j \le 5$ we have

$$F_j(z) = \frac{\alpha_j}{z} + \widetilde{F}_j(z), \quad z \in \mathbb{C}_{-\varepsilon,s+\tau},$$

where $\widetilde{F}_j$ is an analytic function in the strip $\mathbb{C}_{-\varepsilon,s+\tau}$. We will next compute the numbers α_j and $\widetilde{F}_j(0)$. In the computations, we take into account (5.142)–(5.146) and the following formula:

$$\frac{(v-q^2+z)z}{\Phi(it\sqrt{z+v})} = \frac{(v-q^2)+z}{\zeta_1+\zeta_2 z+\cdots} = \tau_1 + \tau_2 z + \cdots. \tag{5.151}$$

In (5.151), ζ_1 and ζ_2 are given by (5.148) and (5.149), respectively, while the constants τ_1 and τ_2 are defined by

$$\tau_1 = \frac{2r^2(r^2+q^2t^2)}{t^4 \sin r[r^2+qt(1+qt)]} \tag{5.152}$$

and

$$\tau_2 = \frac{(r^2+q^2t^2)[r^2-qt(1+qt)]-4r^2[r^2+qt(1+qt)]}{2t^2 \sin r[r^2+qt(1+qt)]^2}. \tag{5.153}$$

Let us recall that

$$v = -\frac{r^2}{t^2} < 0, \qquad \cos(r) < 0, \qquad \sin(r) > 0,$$

$$\Phi(r) = 0, \qquad \Phi'(r) = (1+qt)\cos(r) - r\sin(r),$$

and

$$\Phi''(r) = -2\sin(r) - \Phi(r) = -2\sin(r).$$

The next statements provide explicit formulas for the constants α_j and $\widetilde{F}_j(0)$, $2 \le j \le 5$.

Lemma 5.39 *The following formulas hold for α_j with $2 \le j \le 5$:*

$$\alpha_2 = \tau_1 \frac{y_0^2 t \sin r}{2\sigma^2} > 0, \tag{5.154}$$

$$\alpha_3 = \tau_1 \frac{mqy_0t^2(1-\cos r)}{\sigma^2 r} > 0, \tag{5.155}$$

$$\alpha_4 = \tau_1 \frac{m^2q^2t^3(1+qt)\sin r}{2\sigma^2 r^2} > 0, \tag{5.156}$$

and

$$\alpha_5 = \tau_1 \frac{m^2q^3t^4}{2\sigma^2 r^3}\left(4\sin^2\frac{r}{2} - r\sin r\right) > 0. \tag{5.157}$$

In the previous formulas, τ_1 is given by (5.152).

Lemma 5.40 *The following formulas hold for $\widetilde{F}_j(0)$ with $2 \le j \le 5$:*

$$\widetilde{F}_2(0) = \frac{y_0^2 t}{2\sigma^2}\left[-\tau_1 \frac{t^2\cos r}{2r} + \tau_2 \sin r\right], \tag{5.158}$$

$$\widetilde{F}_3(0) = \frac{mqy_0t^2}{\sigma^2 r}\left[-\tau_1\frac{t^2}{2r^2}(r\sin r + \cos r - 1) + \tau_2(1-\cos r)\right], \tag{5.159}$$

$$\widetilde{F}_4(0) = \frac{m^2q^2t^3}{2\sigma^2 r^2}\left[\tau_1\frac{t^2\sin r}{r^2}\left(1+qt-\frac{r^2}{2}\right) + \tau_2(1+qt)\sin r\right], \tag{5.160}$$

and

$$\widetilde{F}_5(0) = \frac{m^2q^3t^4}{2\sigma^2 r^3}\left[\tau_1\left(\frac{3t^2}{2r^2}\left(4\sin^2\frac{r}{2} - r\sin r\right) - \frac{t^2}{2r}(1+qt)\sin r\right)\right.$$
$$\left. + \tau_2\left(4\sin^2\frac{r}{2} - r\sin r\right)\right]. \tag{5.161}$$

In the previous formulas, τ_1 *and* τ_2 *are given by* (5.152) *and* (5.153), *respectively.*

Proof of Lemmas 5.39 and 5.40 It follows from (5.143), (5.151), and from the formula

$$\sinh(t\sqrt{z+v}) = i\sin r - i\frac{t^2}{2r}(\cos r)z + \cdots \tag{5.162}$$

that

$$zF_2(z) = -\frac{iy_0^2t}{2\sigma^2}(\tau_1 + \tau_2 z + \cdots)\left(i\sin r - i\frac{t^2}{2r}(\cos r)z + \cdots\right)$$
$$= \tau_1\frac{y_0^2t}{2\sigma^2}\sin r + \frac{y_0^2t}{2\sigma^2}\left(\tau_2\sin r - \tau_1\frac{t^2}{2r}\cos r\right)z + \cdots.$$

It follows that α_2 and $\widetilde{F}_2(0)$ are given by formulas (5.154) and (5.158), respectively.

The proof of the remaining formulas in Lemmas 5.39 and 5.40 is similar. Here we use the following power series representations instead of the representation in (5.162):

$$\frac{\cosh(t\sqrt{z+v}) - 1}{\sqrt{z+v}} = it\frac{1-\cos r}{r} - i\frac{t^3(r\sin r + \cos r - 1)}{2r^3}z + \cdots,$$

for (5.155) and (5.159);

$$\frac{\sinh(t\sqrt{z+v}) - t\sqrt{z+v}\cosh(t\sqrt{z+v})}{z+v}$$
$$= -i\frac{t^2(1+qt)\sin r}{r^2} - i\frac{t^4\sin r}{r^4}\left(1+qt-\frac{r^2}{2}\right)z + \cdots,$$

for (5.156) and (5.160); and finally

$$\frac{4\sinh^2\frac{t\sqrt{z+v}}{2} - t\sqrt{z+v}\sinh(t\sqrt{z+v})}{(z+v)^{\frac{3}{2}}}$$

$$= i\frac{t^3}{r^3}\left(r\sin r - 4\sin^2\frac{r}{2}\right)$$

$$+ \left[\frac{i3t^5}{2r^5}\left(r\sin r - 4\sin^2\frac{r}{2}\right) + \frac{it^5}{2r^4}(1+qt)\sin r\right]z + \cdots,$$

for (5.157) and (5.161).

For every $2 \le j \le 4$ the positivity of α_j follows from the formulas in Lemma 5.39. It remains to prove that $\alpha_5 > 0$. We have

$$4\sin^2\frac{r}{2} - r\sin r = 2\sin\frac{r}{2}\left(2\sin\frac{r}{2} - r\cos\frac{r}{2}\right). \tag{5.163}$$

Next, using the inequalities $\frac{\pi}{4} < \frac{r}{2} < \frac{\pi}{2}$, and $x < \tan x$ for $\frac{\pi}{4} < x < \frac{\pi}{2}$, we see that inequality $\alpha_5 > 0$ follows from (5.157) and (5.163).

This completes the proof of Lemmas 5.39 and 5.40. □

Remark 5.41 We have

$$\alpha = \alpha_2 + \alpha_3 + \alpha_4 + \alpha_5, \tag{5.164}$$

where the numbers α_j, $2 \le j \le 5$, are such as in Lemma 5.39. Similarly,

$$\widetilde{F}(0) = \widetilde{F}_2(0) + \widetilde{F}_3(0) + \widetilde{F}_4(0) + \widetilde{F}_5(0). \tag{5.165}$$

Explicit formulas for the numbers $\widetilde{F}_j$, $2 \le j \le 5$ can be found in Lemma 5.40.

It follows from (5.150) and Lemma 5.39 that condition 2 in Theorem 5.1 holds. The validity of condition 1 is straightforward. It is also clear that the function $G_1 = F_1$ is similar to the function G_1 defined by (5.113). Therefore, condition 3 holds. Condition 4 can be checked, using the definitions $G_1 = F_1$ and $F = F_2 + F_3 + F_4 + F_5$, and the fact that the function Φ has a simple zero at $z = r$ (see the last inequality in (5.112)). Since G_2 is a positive constant function, condition 5 is trivial.

It remains to prove condition 6. Note that the functions F_1 and F_2 in (5.137) and (5.138) are similar to the functions G_1 and F in (5.113) and (5.114), respectively. Therefore, estimate (5.117) holds for F_1, while estimate (5.118) holds for F_2. As for the functions F_3, F_4, and F_5 in (5.139), (5.140), and (5.141), one can establish the following estimate for them:

$$\left|F_3(z_1 + iz_2)\right| + \left|F_4(z_1 + iz_2)\right| + \left|F_5(z_1 + iz_2)\right| \le M,$$

for some $M > 0$, where $z_1 \in [0, s + \tau - \delta]$ and z_2 is large enough. Using the estimates mentioned above, we can prove that condition 6 in Theorem 5.1 holds.

It follows from the previous reasoning that Theorem 5.1 can be applied to invert the two-sided Laplace transform in (5.135) approximately. It is not hard to see that this implies Theorem 5.37. □

Moreover, the following lemma holds.

Lemma 5.42 *The constants E, F, and G in Theorem 5.37 are given by*

$$E = \frac{\sqrt{t}}{\sqrt{2\pi}\sigma} \exp\left\{\frac{qt}{2}\right\} \beta^{\frac{1}{2}} \exp\{\widetilde{F}(0)\}, \qquad F = \frac{\sqrt{2\alpha t}}{\sigma}, \qquad G = \frac{r_{qt}^2 + q^2 t^2}{2\sigma^2 t},$$

where the numbers $\beta > 0$, $\alpha > 0$, and $\widetilde{F}(0)$ are defined in (5.150), (5.164), *and* (5.165), *respectively.*

5.8 Constants in Asymptotic Formulas. Simplifications

The constants E, F, and G, appearing in Lemma 5.42, depend on the parameters t, m, q, σ, and y_0. The formulas for E and F in Lemma 5.42 look rather complicated. We will next simplify the formula for the constant α by combining (5.164) and Lemma 5.39. We will also show that the constant F is a linear function of the initial condition y_0 and the long-run mean m.

Lemma 5.43 *The following formula is valid for the constant α:*

$$\alpha = \frac{\tau_1 t}{2\sigma^2 r_{qt}^2 \sin r_{qt}} \left[y_0 r_{qt} \sin r_{qt} + mqt(1 - \cos r_{qt})\right]^2.$$

Moreover

$$F = F_1(t, q, \sigma) y_0 + F_2(t, q, \sigma) m$$

where

$$F_1(t, q, \sigma) = \frac{t\tau_1^{\frac{1}{2}} (\sin r_{qt})^{\frac{1}{2}}}{\sigma^2}$$

and

$$F_2(t, q, \sigma) = \frac{qt^2 \tau_1^{\frac{1}{2}} (1 - \cos r_{qt})}{\sigma^2 r_{qt} (\sin r_{qt})^{\frac{1}{2}}}.$$

The constant τ_1 in the previous formulas is given by (5.152).

Proof Using Lemma 5.39 and (1.35), and making simplifications, we see that

$$\begin{aligned} \alpha = \frac{\tau_1}{2\sigma^2 r^3} \Big\{ & \left[y_0 t^{\frac{1}{2}} r^{\frac{3}{2}} (\sin r)^{\frac{1}{2}}\right]^2 \\ & + 2\left[y_0 t^{\frac{1}{2}} r^{\frac{3}{2}} (\sin r)^{\frac{1}{2}}\right]\left[mqt^{\frac{3}{2}} r^{\frac{1}{2}} (1 - \cos r)(\sin r)^{-\frac{1}{2}}\right] \\ & + m^2 q^2 t^3 \left(r \sin r + 4qt \sin^2 \frac{r}{2}\right) \Big\}. \end{aligned} \tag{5.166}$$

In addition,

$$m^2q^2t^3\left(r\sin r + 4qt\sin^2\frac{r}{2}\right) = \left[mqt^{\frac{3}{2}}r^{\frac{1}{2}}(1-\cos r)(\sin r)^{-\frac{1}{2}}\right]^2. \tag{5.167}$$

Formula (5.167) can be easily checked using the equality $r\cos r = -qt\sin r$. Next using (5.166) and (5.167), we obtain

$$\alpha = \frac{\tau_1}{2\sigma^2 r^3}\left[y_0 t^{\frac{1}{2}}r^{\frac{3}{2}}(\sin r)^{\frac{1}{2}} + mqt^{\frac{3}{2}}r^{\frac{1}{2}}(1-\cos r)(\sin r)^{-\frac{1}{2}}\right]^2. \tag{5.168}$$

Now it is clear that Lemma 5.43 follows from (5.168) and from the formula for the constant F in Lemma 5.42. □

5.9 Notes and References

- Theorem 5.1 in Sect. 5.1 is a Tauberian theorem for the two-sided Laplace transform. It generalizes a similar result for the one-sided transform obtained in [GS10b]. The Abelian theorems in Sects. 5.2 and 5.3 can be found in [GS10a] and [GS10b], respectively.
- The material on the Hull–White model included in Sect. 5.4 is adapted from [GS10a]. Most of the results on the Heston and Stein–Stein models discussed in Sects. 5.6 and 5.7 can be found in [GS10b]. However, our presentation of these results in the present book differs in details from that in [GS10b]. We have also corrected a couple of errors in the expressions for the constants $\widetilde{F}_i(0)$ in Lemma 5.40.
- In [SS91], E.M. Stein and J. Stein introduced and studied the mixing distribution density m_t associated with the Ornstein–Uhlenbeck process. The following asymptotic formula was established in [SS91]:

$$\lim_{y\to\infty}\frac{\log(m_t(y))}{-Gy^2} = 1, \tag{5.169}$$

where the constant G is the same as in Lemma 5.42. Formula (5.169) states that the mixing density $y\mapsto m_t(y)$ is equivalent to the function $y\mapsto e^{-Gy^2}$ in log scale. It is easy to see that formula (5.169) is a consequence of a stronger asymptotic formula obtained in Sect. 5.7 (formula (5.134)). In addition, we have

$$\lim_{\lambda\to\infty}\frac{\log(\mathbb{P}^*(\alpha_t > \lambda))}{-G\lambda^2} = 1, \tag{5.170}$$

where α_t is the realized volatility in the Stein–Stein model given by

$$\alpha_t = \left\{t^{-1}\int_0^t Y_s^2\,ds\right\}^{\frac{1}{2}}, \quad t > 0.$$

J.-D. Deuschel has informed us of the large deviation principle for the family $\{\varepsilon^{\frac{1}{2}}\alpha_1\}_{\varepsilon>0}$ established in the book [DS89] for special Ornstein–Uhlenbeck processes (see Exercise 2.1.13(ii) in [DS89]). It is not hard to see that this large deviation result is equivalent to formula (5.170) in the case considered in [DS89]. In addition, it is stated on p. 286 of [DS89] that the above-mentioned large deviation principle stemmed from a problem posed by J. Stein, and that the problem was solved independently by E.M. Stein and by J.-D. Deuschel and D.W. Stroock. We express our gratitude to J.-D. Deuschel for providing this information.

- Lemma 5.28 was proved in [GS10b]. Similar statements were used in [SS91] and [DY02] without proof.
- It follows from Lemma 5.43 that the parameter F is a linear function of the parameters y_0 and m. This fact was first observed by P. Friz and S. Violante. They have also corrected an error in [GS10b], where an extra factor of t was missing in the expression for the constant α_3. We are indebted to P. Friz and S. Violante for sharing this information with us.

Chapter 6
Asymptotic Analysis of Stock Price Distributions

The stock price distribution is a fundamental object in mathematical finance. A more profound understanding of its properties may shed light on mysteries of stochastic volatility. This chapter is focused on sharp asymptotic formulas for stock price densities in classical stochastic volatility models. Such formulas have many consequences. For example, they can be used to study the tail behavior of stock price distributions, the asymptotics of call and put pricing functions, and the smile asymptotics in stochastic volatility models.

One of the key results in the present chapter is formula (6.2). This formula characterizes stock price asymptotics in the Heston model. It was originally obtained in [GS10b] for the uncorrelated Heston model, and later generalized in [FGGS11] to the case where the correlation coefficient ρ is negative. The proofs of formula (6.2) in [FGGS11] (for ρ with $-1 < \rho \leq 0$) and in [GS10b] (for $\rho = 0$) are qualitatively different. We include these proofs in Sects. 6.1.1–6.1.7 and Sect. 6.2, respectively. In the case where $\rho = 0$, the explicit expressions obtained for the constants in formula (6.2) in two alternative proofs of the asymptotic formula for the Heston density are different. We reconcile these expressions in Sect. 6.3.

Sharp asymptotic formulas can also be found for the stock price density in the Stein–Stein model (see Sect. 6.4). The main references here are the paper [GS10b] where the uncorrelated Stein–Stein model is considered, and the paper [DFJV11] concerning correlated Stein–Stein models. For the Hull–White model, only the case where $\rho = 0$ has been studied (see [GS10a, GS06]). The results on the asymptotic behavior of the stock price density in the uncorrelated Hull–White model are presented in Sect. 6.5.

6.1 Asymptotic Formulas for Stock Price Densities in Heston Models

Let us consider the Heston model given by

$$\begin{cases} dX_t = \sqrt{Y_t} X_t \, dW_t, \\ dY_t = (a - bY_t) \, dt + c\sqrt{Y_t} \, dZ_t, \end{cases} \tag{6.1}$$

A. Gulisashvili, *Analytically Tractable Stochastic Stock Price Models*,
Springer Finance, DOI 10.1007/978-3-642-31214-4_6,
© Springer-Verlag Berlin Heidelberg 2012

where W and Y are standard Browninan motions under a risk-neutral measure $\mathbb{P}$. For the sake of simplicity, we assume that $x_0 = 1$ and $-1 < \rho \leq 0$, where x_0 is the initial condition for the stock price process, and ρ is the correlation coefficient for W and Z. The previous restriction on ρ covers essentially all practical applications of the Heston model.

The next assertion characterizes the asymptotic behavior of the stock price density in the negatively correlated Heston model (see [FGGS11]).

Theorem 6.1 *For every $T > 0$ there exist positive constants A_1, A_2, and A_3 such that the following asymptotic formula holds for the distribution density D_T of the stock price X_T in the Heston model described by* (6.1):

$$D_T(x) = A_1 x^{-A_3} e^{A_2 \sqrt{\log x}} (\log x)^{-\frac{3}{4}+\frac{a}{c^2}} \left(1 + O\left((\log x)^{-\frac{1}{2}}\right)\right) \tag{6.2}$$

as $x \to \infty$.

Remark 6.2 The case where $\rho = -1$ is excluded form consideration for the following reason. The proof of formula (6.2) given below relies heavily on the existence of exploding moments of the stock price in the Heston model. However, if $\rho = -1$, then the stock price moments in the Heston model do not explode (see [K-R11]).

We will prove Theorem 6.1 in the next subsections. The methods used in the proof are based almost entirely on affine principles; at practically no point do we need knowledge of the (explicit, but cumbersome) closed form expression of the moment generating function of $\log X_T$ or, equivalently, the Mellin transform of X_T (see formula (4.92)). Instead, all the necessary information on the Mellin transform will be extracted by analyzing the corresponding Riccati equations near criticality, using higher-order Euler estimates. We then take into account the Tauberian principle that the precise behavior of the transformed function near the singularity (the leading order of which is exactly described by the critical slope!) contains entire asymptotic information about the original function, and combine this principle with a classical saddle point computation to establish Theorem 6.1. Explicit formulas for the constants A_1, A_2, and A_3 are given in (6.41) and (6.46).

6.1.1 Heston Models as Affine Models and Moment Explosions

We will next briefly discuss the behavior of the stock price moments in the Heston model. Given $s \geq 1$, define the explosion time for the moment of order s by

$$T^*(s) = \sup\left\{t \geq 0 : \mathbb{E}\left[X_t^s\right] < \infty\right\},$$

and for any $T > 0$, define the upper critical moment $s_+(T)$ by

$$s_+ = s_+(T) = \sup\left\{s \geq 1 : \mathbb{E}\left[X_T^s\right] < \infty\right\}. \tag{6.3}$$

For the Heston model, and many other stochastic volatility models, the explosion time T^* is explicitly known. The critical moment, for fixed T, is then found numerically from $T^*(s_+(T)) = T$. The previous equality shows that $s_+(t) \geq 1$ is the (generalized) inverse of the (decreasing) function $T^*(\cdot)$. We can also introduce the lower critical moment as follows:

$$s_- = s_-(T) = \inf\{s \leq 0 : \mathbb{E}[S_T^s] < \infty\}. \tag{6.4}$$

Definition 6.3 Let $T > 0$ be fixed. The quantities

$$\sigma = -\frac{\partial T^*(s)}{\partial s}\bigg|_{s=s_+} \quad \text{and} \quad \kappa = \frac{\partial^2 T^*(s)}{\partial s^2}\bigg|_{s=s_+} \tag{6.5}$$

are called the upper critical slope and the upper critical curvature, respectively. Similarly, the lower critical slope and curvature are defined by

$$\sigma_- = -\frac{\partial T^*(s)}{\partial s}\bigg|_{s=s_-} \quad \text{and} \quad \kappa_- = \frac{\partial^2 T^*(s)}{\partial s^2}\bigg|_{s=s_-},$$

respectively.

Note that the critical slopes and curvatures depend on T.

Our next goal is to identify the smallest positive singularity of the function on the right-hand side of (4.87), and to analyze the asymptotic behavior of this function near the singularity. An elementary computation shows that

$$2c^2 \min_{\eta \in [0,\infty]} R(s,\eta) = -\left[(s\rho c - b)^2 - c^2\left(s^2 - s\right)\right].$$

Set

$$\Delta(s) = (s\rho c - b)^2 - c^2\left(s^2 - s\right) \quad \text{and} \quad \chi(s) = s\rho c - b. \tag{6.6}$$

A typical situation in applications ($\rho \leq 0$, and $b > 0$) implies that χ is negative for $s \geq 0$. We thus assume in the sequel that $\chi(s) < 0$ for all $s \geq 0$. This assumption allows us to use the following formula from [K-R11], Theorem 4.2:

$$T^*(s) = \begin{cases} +\infty & \text{if } \Delta(s) \geq 0, \\ \int_0^\infty 1/R(s,\eta)\, d\eta & \text{if } \Delta(s) < 0. \end{cases} \tag{6.7}$$

The integral in (6.7) can be represented as follows. For $\Delta(s) < 0$, we have

$$T^*(s) = \frac{2}{\sqrt{-\Delta(s)}}\left(\arctan\frac{\sqrt{-\Delta(s)}}{\chi(s)} + \pi\right). \tag{6.8}$$

Moreover, the derivative

$$\frac{\partial T^*}{\partial s} = -\int_0^\infty \frac{\partial_s R}{R^2}(s,\eta)\, d\eta$$

can be computed explicitly. Indeed, from (6.8) we get

$$\frac{\partial T^*}{\partial s} = -T^*(s)\frac{2\rho c(s\rho c - b) - c^2(2s-1)}{2\Delta(s)} - \frac{[c^2(2s-1) - 2\rho c(s\rho c - b)](s\rho c - b) + 2\rho c\Delta(s)}{\Delta(s)[(s\rho c - b)^2 - \Delta(s)]}. \tag{6.9}$$

Since $T^*(s_+) = T$, formula (6.9) implies that

$$\sigma = -\frac{\partial T^*}{\partial s}(s_+) = \frac{R_1}{R_2}, \tag{6.10}$$

where

$$\begin{aligned} R_1 = {} & Tc^2 s_+(s_+ - 1)\big[c^2(2s_+ - 1) - 2\rho c(s_+\rho c - b)\big] \\ & - 2(s_+\rho c - b)\big[c^2(2s_+ - 1) - 2\rho c(s_+\rho c - b)\big] \\ & + 4\rho c\big[c^2 s_+(s_+ - 1) - (s_+\rho c - b)^2\big] \end{aligned}$$

and

$$R_2 = 2c^2 s_+(s_+ - 1)\big[c^2 s_+(s_+ - 1) - (s_+\rho c - b)^2\big].$$

Let $s \geq 1$, and recall that $T^*(s)$ is the explosion time of ψ. On the other hand, using the Riccati ODE for ψ, we see that

$$(1/\psi)^{\cdot} = -\frac{\dot{\psi}}{\psi^2} = -\frac{R(s,\psi)}{\psi^2}.$$

Since

$$\frac{R(s,u)}{u^2} \to c^2/2$$

as $u \to \infty$, we obtain

$$\psi(s,t) \sim \frac{1}{\frac{c^2}{2}(T^*(s) - t)} \tag{6.11}$$

as $t \uparrow T^*(s)$, uniformly on bounded subintervals of $[1,\infty)$. Next, fix $T > 0$. Then $T = T^*(s_+)$. Using the fact that the function T^* is continuously differentiable (and even C^2) with respect to s, we get

$$T^*(s) - T = T^*(s) - T^*(s_+) = (s_+ - s)\big(\sigma + O(s_+ - s)\big) \sim \sigma(s_+ - s) \tag{6.12}$$

as $s \uparrow s_+$, where σ is the critical slope. Hence

$$\psi(s,T) \sim \frac{2}{(s_+ - s)c^2\sigma} \tag{6.13}$$

as $s \uparrow s_+$. It follows from (6.11) and (6.13) that

$$\phi(s,t) = \int_0^t a\psi(s,\vartheta)\,d\vartheta$$

has a logarithmic blowup:

$$\phi(s,t) \sim -\frac{2a}{c^2}\log\bigl(T^*(s)-t\bigr)$$

as $t \uparrow T^*(s)$, or

$$\phi(s,T) \sim -\frac{2a}{c^2}\log\bigl((s^*-s)\sigma\bigr)$$

as $s \uparrow s_+$.

The next lemma refines these asymptotic results.

Lemma 6.4 *Let $T > 0$, and suppose $s \uparrow s_+$. Then the following formulas hold*:

$$\begin{aligned}\psi(s,T) &= \frac{2}{(s_+-s)c^2\sigma} - \frac{s_+\rho c - b}{c^2} - \frac{\kappa}{c^2\sigma^2} + O(s_+-s),\\ \phi(s,T) &= \frac{2a}{c^2}\log\frac{1}{s_+-s} + \frac{2a}{c^2}\log\frac{T}{\sigma}\\ &\quad + a\int_0^T\left(\psi(s_+,\vartheta) - \frac{2}{c^2(T-\vartheta)}\right)d\vartheta + O(s_+-s).\end{aligned} \tag{6.14}$$

Proof The main idea is to use second-order Euler estimates for the Riccati ODEs near criticality to describe the limiting behavior of $\psi(s,t)$ and $\phi(s,t)$ as $t \uparrow T^*(s)$. Then the proof can be completed by taking into account (6.12). More precisely, let us introduce the time-to-criticality

$$\tau = T^*(s) - t,$$

and set

$$\hat{\psi}(s,\tau) = \psi\bigl(s, T^*(s)-\tau\bigr).$$

Observe that $1/\hat{\psi}(s,0) = 0$ and

$$\begin{aligned}(1/\hat{\psi})^{\cdot} &= -\frac{(\hat{\psi})^{\cdot}}{\hat{\psi}^2} = \frac{1}{\hat{\psi}^2}R(s,\hat{\psi})\\ &= \frac{c^2}{2} + \frac{s\rho c - b}{\hat{\psi}} + \frac{s^2-s}{2\hat{\psi}^2} = W(s,1/\hat{\psi}),\end{aligned}$$

where

$$W(s,u) = \frac{c^2}{2} + (s\rho c - b)u + \frac{s^2-s}{2}u^2.$$

A higher-order Euler scheme for this ODE yields

$$(1/\hat{\psi})(s,\tau) = (1/\hat{\psi})(s,0) + W(s,0)\tau + W(s,0)W'(s,0)\tau^2/2 + o(\tau^2)$$

as $\tau \to 0$ and s stays in a bounded interval. Since $W(s,0) = \frac{c^2}{2}$ and $W'(s,0) = b + s\rho c$, we obtain

$$\begin{aligned} 1/\hat{\psi}(s,\tau) &= \frac{c^2}{2}\tau\left(1 + \frac{s\rho c - b}{2}\tau + o(\tau^2)\right) \\ &= \frac{c^2}{2}\tau\left(1 - \frac{s\rho c - b}{2}\tau + o(\tau^2)\right)^{-1}. \end{aligned}$$

It follows that

$$\begin{aligned} \hat{\psi}(s,\tau) &= \frac{2}{c^2\tau}\left(1 - \frac{s\rho c - b}{2}\tau + o(\tau^2)\right) \\ &= \frac{2}{c^2\tau} - \frac{s\rho c - b}{c^2} + O(\tau) \end{aligned} \tag{6.15}$$

as $\tau = T^*(s) - t \downarrow 0$. Note that

$$\begin{aligned} \frac{1}{\tau} &= \left(\sigma(s_+ - s) + \frac{1}{2}\kappa(s_+ - s)^2 + O\left((s_+ - s)^3\right)\right)^{-1} \\ &= \frac{1}{\sigma(s_+ - s)} - \frac{\kappa}{2\sigma^2} + O(s_+ - s). \end{aligned}$$

Therefore

$$\psi(s,T) = \frac{2}{c^2\sigma(s_+ - s)} - \frac{s_+\rho c - b}{c^2} - \frac{\kappa}{c^2\sigma^2} + O(s_+ - s)$$

as $s \uparrow s_+$. For the expansion of

$$\phi(s,t) = \int_0^t a\psi(s,\vartheta)\,d\vartheta,$$

we obtain

$$\begin{aligned} \phi(s,t) &= a\int_0^t \left(\psi(s,\vartheta) - \frac{2}{c^2(T^*(s) - \vartheta)}\right) d\vartheta + \frac{2a}{c^2}\int_0^t \frac{1}{T^*(s) - \vartheta}\,d\vartheta \\ &= \frac{2a}{c^2}\log\frac{1}{T^*(s) - t} + \frac{2a}{c^2}\log T^*(s) \\ &\quad + a\int_0^t \left(\psi(s,\vartheta) - \frac{2}{c^2(T^*(s) - \vartheta)}\right) d\vartheta \end{aligned}$$

$$
\begin{aligned}
&= \frac{2a}{c^2} \log \frac{1}{T^*(s) - t} + \frac{2a}{c^2} \log T^*(s) \\
&\quad + a \int_0^{T^*(s)} \left(\psi(s, \vartheta) - \frac{2}{c^2 (T^*(s) - \vartheta)} \right) d\vartheta \\
&\quad + O\big(T^*(s) - t\big).
\end{aligned} \tag{6.16}
$$

To establish the last equality in (6.16), we note that the integrand in the integral

$$
\int_t^{T^*(s)} \left(\psi(s, \vartheta) - \frac{2}{c^2 (T^*(s) - \vartheta)} \right) d\vartheta
$$

has an expansion resulting from (6.15). This expansion may be integrated termwise (see [dBru81]), which gives the $O(T^*(s) - t)$ estimate. Next, using (6.12) and (6.16), we see that formula (6.14) holds. □

Remark 6.5 Lemma 6.4 also holds as s tends to s_+ in the complex plane under the condition $\Re(s) < s_+$.

6.1.2 Saddle Point Method and Mellin Inversion

As noted in Sect. 6.1.1, we can represent the Mellin transform MD_T as follows:

$$
\log MD_T(u) = \phi(u - 1, T) + y_0 \psi(u - 1, T)
$$

(see formula (4.87)). Therefore, the density D_T can be recovered using the Mellin inversion formula (4.86). This gives

$$
D_T(x) = \frac{1}{2\pi i} \int_{s - i\infty}^{s + i\infty} e^{-uL + \phi(u-1,T) + y_0 \psi(u-1,T)} \, du. \tag{6.17}
$$

In (6.17), we use the notation $L = \log x$, and assume $\Re(s) \in (s_-(T), s_+(T))$, where $s_+(T)$ and $s_-(T)$ are the upper critical moment and the lower critical moment defined by (6.3) and (6.4), respectively.

We will next justify the applicability of formula (6.17) to the Heston model. It is known (see [Luc07]) that all the singularities of the Mellin transform of the stock price density in the Heston model are located on the real line. Therefore, the function

$$
u \mapsto \exp\big\{\phi(u - 1, T) + y_0 \psi(u - 1, T)\big\}
$$

is analytic everywhere in the complex plane except the points of singularity on the real line. The next lemma implies that the integral on the right-hand side of formula (6.17) exists (see Remark 6.7 below). This lemma will also be used in the tail estimates for the saddle point method. By symmetry, it clearly suffices to consider the upper tail ($\Im(u) > 0$).

Lemma 6.6 *Let $T > 0$ and $1 \le s_1 \le \Re(s) \le s_2$. Then the following estimate holds as $\Im(s) \to \infty$:*

$$\left|e^{\phi(s,T)+y_0\psi(s,T)}\right| = O\left(e^{-C\Im(s)}\right),$$

where the constant $C > 0$ depends on T, s_1, s_2, and y_0.

Proof Let $s = \xi + iy$ and suppose $y > 0$. We will first estimate the function ψ. Recall that

$$\dot{\psi} = \frac{1}{2}\left(s^2 - s\right) + \frac{c^2}{2}\psi^2 - b\psi + s\psi\rho c \quad \text{with } \psi(\xi, 0) = 0.$$

Set $\psi = f + ig$ and $\gamma = b - \xi\rho c$. Then $\gamma \ge 0$, and we have

$$\dot{f} = \frac{1}{2}\left(\xi^2 - y^2 - \xi\right) + \frac{c^2}{2}\left(f^2 - g^2\right) - \gamma f, \quad f(s, 0) = 0,$$
$$\dot{g} = \frac{1}{2}(2\xi y - y) + c^2 fg - \gamma g, \quad g(s, 0) = 0.$$

Our goal is to show that there exists a positive continuously differentiable function $t \mapsto C(t)$ on $[0, T]$ such that

$$f(s, t) \le -C(t)y, \tag{6.18}$$

where $s = \xi + iy$, $1 \le s_1 \le \xi \le s_2$, and y is large enough. We first observe that f satisfies the differential inequality

$$\begin{aligned} \dot{f} &\le \frac{1}{2}\left(\xi^2 - y^2 - \xi\right) + \frac{c^2}{2}f^2 - \gamma f \\ &\le -\frac{1}{3}y^2 + \frac{c^2}{2}f^2 - \gamma f \end{aligned} \tag{6.19}$$

for $y > \tilde{y}$, where $\tilde{y}$ depends only on s_1 and s_2. Set

$$V(y, r) = -\frac{1}{3}y^2 + \frac{c^2}{2}r^2 - \gamma r.$$

Then (6.19) can be rewritten as follows:

$$\dot{f}(s, t) \le V\left(y, f(s, t)\right) \tag{6.20}$$

where $s = \xi + iy$.

Our next goal is to find a function $C(t)$, $t \in [0, T]$, such that $C(0) = 0$, C is strictly positive for $t \in (0, T]$, and the function F defined by $F(y, t) = -C(t)y$ satisfies the differential inequality

$$V(y, F) \le \dot{F}. \tag{6.21}$$

Let us first suppose that such a function C exists. Then it is clear that given $s = \xi + iy$, the initial data $F(y,0) = f(s,0) = 0$ match. Now we can use the ODE comparison results and derive from (6.20) and (6.21) that (6.18) holds, which implies the following estimate:

$$\left|e^{y_0 \psi(s,T)}\right| = e^{y_0 f(s,T)} \le e^{-y_0 C(T) \Im(s)} \tag{6.22}$$

for all $s = \xi + iy$ with y large enough and $s_1 \le \xi \le s_2$.

We now look for the function C satisfying the equation

$$\dot{C}(t) = -\gamma C(t) + \theta,$$

where θ is a positive constant, and $C(0) = 0$. The solution to this equation is given by

$$C(t) = \begin{cases} \theta \gamma^{-1}(1 - e^{-\gamma t}) & \text{if } \gamma > 0, \\ \theta t & \text{if } \gamma = 0. \end{cases}$$

It follows that for $t \in (0, T]$, we have $0 < C(t) \le T\theta$. Next, choosing $\theta > 0$ so that

$$-\frac{1}{3} + \frac{c^2}{2} T^2 \theta^2 = -\frac{1}{4},$$

we obtain

$$\begin{aligned} V\big(y, F(y,t)\big) &\le -\frac{1}{3} y^2 + \frac{c^2}{2} T^2 \theta^2 y^2 + \gamma C(t) y \\ &= -\frac{1}{4} y^2 + \big(\theta - \dot{C}(t)\big) y \\ &\le -\dot{C}(t) y = \dot{F}(y,t). \end{aligned} \tag{6.23}$$

In (6.23), y is large enough and depends only on θ, and hence on the model parameter c and on T. This shows that the function F satisfies the differential inequality in (6.21), and it follows that estimates (6.18) and (6.22) hold.

Finally, we note that

$$\Re\big(\phi(s,T)\big) = a \int_0^T f(s,t)\,dt \le ay\left(-\int_0^T C(t)\,dt\right) = -ay\widetilde{C}(T).$$

Therefore, for $\Im(s)$ large enough,

$$\left|e^{\phi(s,T) + y_0 \psi(s,T)}\right| \le \exp\big\{-\big(a\widetilde{C}(T) + y_0 C(T)\big)\Im(s)\big\}.$$

The proof of Lemma 6.6 is thus completed. □

Remark 6.7 Using Lemma 6.6, we see that the integral in (6.17) exists. Indeed, its integrand decays exponentially at $\pm i\infty$. Moreover, if $u - 1$ is imaginary, then the characteristic function of the random variable $\widetilde{X}_T = \log(X_T)$ decays exponentially. It follows that $\widetilde{X}_T$ (and therefore X_T) admits a smooth density.

6.1.3 Finding the Saddle Point

We will study the asymptotic behavior of the expression in (6.17) by the saddle point (or steepest descent) method. The main idea is to deform the contour of integration into a path of steepest descent from a saddle point of the integrand. In cases where the method can be applied successfully, the saddle becomes steeper and more pronounced as the parameter (x in our case) increases. We then replace the integrand with a local expansion around the saddle point. The resulting integral, taken over a small part of the contour containing the saddle point, is easy to evaluate asymptotically. Finally, in order to characterize the asymptotics of the original integral, it suffices to show that its tails are negligible.

A (real) saddle point of the integrand in formula (6.17) can be found by making its derivative equal to zero. It usually suffices to calculate an approximate saddle point. This is what we are going to do next. Note that Lemma 6.4 and Remark 6.5 imply the following formula:

$$\phi(u-1,T)+y_0\psi(u-1,T)=\frac{\beta^2}{u^*-u}+\frac{2a}{c^2}\log\frac{1}{u^*-u}+\Gamma+O\left(u^*-u\right) \qquad (6.24)$$

as $u \to u^* = s_+ + 1 = A_3$ with $\Re(u) < u^*$. Here we put

$$\beta^2=\frac{2y_0}{c^2\sigma} \qquad (6.25)$$

and

$$\begin{aligned}\Gamma &= -y_0\left(\frac{s_+\rho c-b}{c^2}+\frac{\kappa}{c^2\sigma^2}\right)+\frac{2a}{c^2}\log\frac{T}{\sigma}\\ &\quad +a\int_0^T\left(\psi(s_+,\vartheta)-\frac{2}{c^2(T-\vartheta)}\right)d\vartheta. \end{aligned} \qquad (6.26)$$

Keeping only the dominant term in (6.24), we get the approximate saddle point equation:

$$\left[x^{-u}\exp\left\{\frac{\beta^2}{u^*-u}\right\}\right]'=0,$$

or, equivalently,

$$-L+\frac{\beta^2}{(u^*-u)^2}=0. \qquad (6.27)$$

The solution to (6.27) is given by

$$\hat{u}=\hat{u}(x)=u^*-\beta L^{-1/2}. \qquad (6.28)$$

The expression in (6.28) is the approximate saddle point of the integrand.

6.1.4 Local Expansion Around the Saddle Point

Our objective in the present subsection is to expand the function

$$u \mapsto \phi(u-1,T) + y_0\psi(u-1,T)$$

at the point $u = \hat{u}$. Put $u = \hat{u} + iy$, and recall that we use the following notation: $\sigma = -\partial_s T^*|_{s_+}$ and $L = \log x$. Since the approximate saddle point $\hat{u}$ approaches u^* as $L \to \infty$, we may expand the function above using (6.24). To make the expansion valid uniformly with respect to the new integration parameter y, we confine y to the following small interval:

$$|y| < L^{-\alpha}, \quad \frac{2}{3} < \alpha < \frac{3}{4}. \tag{6.29}$$

The choice of the upper bound on α in (6.29) is influenced by the tail estimate in Lemma 6.9 below. Since $u^* - u = \beta L^{-1/2} - iy$, we have

$$\begin{aligned}\frac{1}{u^*-u} &= \beta^{-1}L^{1/2}\big(1 - i\beta^{-1}L^{1/2}y\big)^{-1} \\ &= \beta^{-1}L^{1/2}\big(1 + i\beta^{-1}L^{1/2}y - \beta^{-2}Ly^2 + O\big(L^{3/2-3\alpha}\big)\big) \\ &= \beta^{-1}L^{1/2} + i\beta^{-2}Ly - \beta^{-3}L^{3/2}y^2 + O\big(L^{2-3\alpha}\big).\end{aligned} \tag{6.30}$$

It follows that

$$\begin{aligned}\log\frac{1}{u^*-u} &= \log\big[\beta^{-1}L^{1/2}\big(1 + O\big(L^{1/2-\alpha}\big)\big)\big] \\ &= \frac{1}{2}\log L - \log\beta + O\big(L^{1/2-\alpha}\big).\end{aligned}$$

Next, plugging the previous expansions with $u = \hat{u} + iy$ into (6.24), we obtain the following asymptotic formula:

$$\begin{aligned}&\phi(\hat{u}-1+iy,T) + y_0\psi(\hat{u}-1+iy,T) \\ &\quad = \beta L^{1/2} + \frac{a}{c^2}\log L + iLy - \beta^{-1}L^{3/2}y^2 - \frac{2a}{c^2}\log\beta \\ &\qquad + \Gamma + O\big(L^{2-3\alpha}\big).\end{aligned} \tag{6.31}$$

6.1.5 Saddle Point Approximation of the Density

For the sake of simplicity, we will first obtain formula (6.2) with a weaker error estimate $O((\log x)^{-1/4+\varepsilon})$, where $\varepsilon > 0$ is arbitrary. Then it will be explained how to get the stronger estimate $O((\log x)^{-1/2})$.

Our first goal is to shift the contour in the Mellin inversion formula (6.17) through the saddle point $\hat{u}$. This gives

$$D_T(x) = \frac{1}{2\pi i} \int_{\hat{u}-i\infty}^{\hat{u}+i\infty} e^{-uL+\phi(u-1,T)+y_0\psi(u-1,T)}\, du. \tag{6.32}$$

Therefore,

$$D_T(x) = x^{-\hat{u}} \frac{1}{2\pi} \int_{-\infty}^{\infty} e^{-iyL+\phi(\hat{u}+iy-1,T)+y_0\psi(\hat{u}+iy-1,T)}\, dy. \tag{6.33}$$

The factor $x^{-\hat{u}} \approx x^{-u^*} = x^{-A_3}$ yields the leading-order decay in (6.2). Its exponent corresponds to the location of the dominating singularity of the Mellin transform. The lower-order factors are dictated by the type of the singularity at $u = u^*$ as will be shown later.

6.1.6 Tail Estimates

This subsection contains two technical lemmas which will be used in the estimates of the tail integral.

Lemma 6.8 *Let $B > 0$ and set $L = \log x$. Then the part of the integral in* (6.32) *where $\Im(u) > B$ is $O(x^{-A_3} \exp\{\beta L^{1/2}\})$.*

Proof If $\tilde{B} > B$ is a sufficiently large positive constant, then it easily follows from Lemma 6.6 that

$$\begin{aligned}\left| \int_{\hat{u}+i\tilde{B}}^{\hat{u}+i\infty} e^{-uL+\phi(u-1)+y_0\psi(u-1)}\, du \right| &\le C x^{-A_3} \exp\{\beta L^{1/2}\} \int_{\tilde{B}}^{\infty} e^{-Cy}\, dy \\ &= O\big(x^{-A_3} \exp\{\beta L^{1/2}\}\big).\end{aligned}$$

Moreover, since the Mellin transform of D_T does not have singularities outside the real line (see [Luc07]), we have

$$\left| \int_{\hat{u}+iB}^{\hat{u}+i\tilde{B}} e^{-uL+\phi+y_0\psi}\, du \right| = O\big(e^{-\hat{u}L}\big) = O\big(x^{-A_3} \exp\{\beta L^{1/2}\}\big).$$

This completes the proof of Lemma 6.8. □

Lemma 6.8 shows that the part of the tail integral where $\Im(u) > B$ is asymptotically much smaller than the central part. We will next estimate the entire tail integral.

Lemma 6.9 *The following estimate holds for the tail integral*:

$$\left|\int_{\hat{u}+iL^{-\alpha}}^{\hat{u}+i\infty} e^{-uL+\phi+y_0\psi}\,du\right| = x^{-A_3}\exp\left\{2\beta L^{1/2} - \frac{1}{2}\beta^{-1}L^{3/2-2\alpha} + O(\log L)\right\}.$$

Proof Our goal is to prove that there exists a constant $B > 0$ such that the absolute value of the part of the tail integral where $L^{-\alpha} < \Im(u) < B$ is given by

$$x^{-A_3}\exp\left\{2\beta L^{1/2} - \frac{1}{2}\beta^{-1}L^{3/2-2\alpha} + O(\log L)\right\}. \tag{6.34}$$

It suffices to establish the previous statement, since Lemma 6.8 shows that the absolute value of the integral from $\hat{u} + iB$ to $\hat{u} + i\infty$ is asymptotically smaller than the expression in (6.34). Indeed, dividing (6.34) by $x^{-A_3}\exp\{\beta L^{1/2}\}$, we get the term $\exp\{\beta L^{1/2} + O(L^{3/2-2\alpha})\}$, and it is not hard to see that this term tends to infinity. Note that $3/2 - 2\alpha < 1/2$ by (6.29). Next, using Lemma 6.4 and Remark 6.5, we see that for some constant $\gamma > 0$,

$$e^{\phi(u-1,T)+y_0\psi(u-1,T)} = O\left(\exp\left\{\frac{\beta^2}{A_3 - u} - \gamma\log(A_3 - u)\right\}\right)$$

as u tends to $u^* = s_+ + 1 = A_3$ inside the analyticity strip. More precisely, there exists a constant $C > 0$ such that for a sufficiently small number $B > 0$ and for all u in the analyticity strip with $|\Im(u)| < B$ and $\Re(u) > u^* - B$, we have

$$\left|e^{\phi(u-1)+y_0\psi(u-1)}\right| \le C|A_3 - u|^{-\gamma}\exp\left\{\Re\left(\frac{\beta^2}{A_3 - u}\right)\right\}.$$

It follows that

$$\begin{aligned}
&\left|\int_{\hat{u}+iL^{-\alpha}}^{\hat{u}+iB} e^{-uL+\phi+y_0\psi}\,du\right| \\
&\quad\le Cx^{-A_3}\exp\{\beta L^{1/2}\}\int_{L^{-\alpha}}^{B}\left|A_3 - (\hat{u}+iy)\right|^{-\gamma}\exp\left\{\Re\left(\frac{\beta^2}{A_3 - (\hat{u}+iy)}\right)\right\}dy \\
&\quad\le Cx^{-A_3}\exp\{\beta L^{1/2}\}L^{\gamma/2}\exp\left\{\frac{\beta^2(A_3-\hat{u})}{(A_3-\hat{u})^2 + L^{-2\alpha}}\right\} \\
&\quad= Cx^{-A_3}\exp\{2\beta L^{1/2} - \beta^{-1}L^{3/2-2\alpha} + O(\log L)\}.
\end{aligned}$$

Here we use the fact that the factor $|A_3 - (\hat{u}+iy)|^{-\gamma}$ grows like a power of L. This can be seen from

$$\beta L^{-\frac{1}{2}} = A_3 - \hat{u} \le \left|A_3 - (\hat{u}+iy)\right|.$$

Furthermore, the quantity

$$\Re\left(\frac{\beta^2}{A_3 - (\hat{u}+iy)}\right) = \frac{\beta^2(A_3-\hat{u})}{(A_3-\hat{u})^2 + y^2} \tag{6.35}$$

decreases with respect to $|y|$. Therefore, we can estimate the integral of the function in (6.35) over the interval $[L^{-\alpha}, B]$ by the value of its integrand at $L^{-\alpha}$ times the length of the integration path. The latter is absorbed into C, and the former is given by

$$\frac{\beta^2(A_3-\hat{u})}{(A_3-\hat{u})^2+L^{-2\alpha}} = \beta L^{1/2} - \frac{\beta L^{1/2}}{\beta^2 L^{2\alpha-1}+1}$$
$$= \beta L^{1/2} - \beta^{-1} L^{3/2-2\alpha} + O\big(L^{5/2-4\alpha}\big).$$

Finally, we write the factor $L^{\gamma/2}$ as $\exp\{O(\log L)\}$.

This completes the proof of Lemma 6.9. □

The "tail" of the integral in (6.33), corresponding to $|y| > L^{-\alpha}$, can be estimated using Lemma 6.9. It follows that

$$D_T(x) = x^{-\hat{u}} \frac{1}{2\pi} \int_{-L^{-\alpha}}^{L^{-\alpha}} e^{-iyL+\phi(\hat{u}+iy-1,T)+y_0\psi(\hat{u}+iy-1,T)}\, dy$$
$$+ x^{-A_3} \exp\big\{2\beta L^{1/2} - \beta^{-1} L^{3/2-2\alpha} + O(\log L)\big\}.$$

Next, using (6.31) and the equality

$$x^{-\hat{u}} \exp\big\{\beta L^{1/2}\big\} = x^{-u^*} \exp\big\{2\beta L^{1/2}\big\},$$

we obtain

$$D_T(x) = \frac{\exp\{\Gamma\}}{2\pi} x^{-u^*} e^{2\beta L^{1/2}} \beta^{-2a/c^2} L^{a/c^2}$$
$$\times \int_{-L^{-\alpha}}^{L^{-\alpha}} \exp\big\{-\beta^{-1} L^{3/2} y^2\big\}\, dy \big(1 + O\big(L^{2-3\alpha}\big)\big)$$
$$+ x^{-A_3} \exp\big\{2\beta L^{1/2} - \beta^{-1} L^{3/2-2\alpha} + O(\log L)\big\}. \tag{6.36}$$

Evaluating the Gaussian integral, we get

$$\int_{-L^{-\alpha}}^{L^{-\alpha}} \exp\big\{-\beta^{-1} L^{3/2} y^2\big\}\, dy = \beta^{1/2} L^{-3/4} \int_{-\beta^{-1/2} L^{3/4-\alpha}}^{\beta^{-1/2} L^{3/4-\alpha}} \exp\big\{-w^2\big\}\, dw$$
$$\sim \beta^{1/2} L^{-3/4} \int_{-\infty}^{\infty} \exp\big\{-w^2\big\}\, dw$$
$$= \sqrt{\pi} \beta^{1/2} L^{-3/4}. \tag{6.37}$$

Here we use the fact that the tails of the Gaussian integral are exponentially small in L. Taking into account (6.36) and (6.37), we can compare the main part of the asymptotic expansion and the two error terms:

$$\text{const} \times x^{-A_3} L^{a/c^2 - 3/4} \exp\big\{2\beta L^{1/2}\big\}$$

(main part),

$$x^{-A_3} L^{a/c^2-3/4} \exp\{2\beta L^{1/2}\} O(L^{2-3\alpha})$$

(error from the local expansion), and

$$x^{-A_3} \exp\{2\beta L^{1/2} - \beta^{-1} L^{3/2-2\alpha} + O(\log L)\}$$

(error from the tail estimate). Since $2 - 3\alpha < 0$, the expression on the second line is asymptotically smaller than the main part. In addition, since $3/2 - 2\alpha > 0$, the quantity $\exp\{-\beta^{-1} L^{3/2-2\alpha}\}$ decays faster than any power of L. This shows that the expression on the third line is negligible in comparison with the error term in the local expansion. Hence, it suffices to keep only the error term resulting from the local expansion. As a result, the error term in the asymptotic formula for D_T is $O(L^{2-3\alpha}) = O(L^{-1/4+\varepsilon})$ (take α close to $\frac{3}{4}$). More precisely, using (6.36) and (6.37), we obtain the following formula:

$$\begin{aligned} D_T(x) = &\left[\frac{\exp\{\Gamma\}}{2\pi}\sqrt{\pi}\beta^{1/2-2a/c^2}\right] x^{-(s_+ +1)} e^{2\beta L^{1/2}} L^{-3/4+a/c^2} \\ &\times \left(1 + O\left(L^{-1/4+\varepsilon}\right)\right) \end{aligned} \tag{6.38}$$

as $L \to \infty$. It follows from (6.38) that formula (6.2) holds with

$$A_1 = \frac{\exp\{\Gamma\}}{2\pi}\sqrt{\pi}\beta^{1/2-2a/c^2} \tag{6.39}$$

and with a weaker error estimate.

We will next show how to obtain the relative error $O((\log x)^{-1/2})$ in formula (6.2). Taking two more terms in the expansion (6.30) of $1/(u^* - u)$, we get

$$\begin{aligned} \frac{1}{u^* - u} &= \beta^{-1} L^{1/2}\left(1 - i\beta^{-1} L^{1/2} y\right)^{-1} \\ &= \beta^{-1} L^{1/2}\left(1 + i\beta^{-1} L^{1/2} y - \beta^{-2} L y^2 - i\beta^{-3} L^{3/2} y^3 \right. \\ &\quad \left. + \beta^{-4} L^2 y^4 + O\left(L^{5/2-5\alpha}\right)\right) \\ &= \beta^{-1} L^{1/2} + i\beta^{-2} L y - \beta^{-3} L^{3/2} y^2 - i\beta^{-4} L^2 y^3 \\ &\quad + \beta^{-5} L^{5/2} y^4 + O\left(L^{3-5\alpha}\right). \end{aligned}$$

Expanding the logarithm, we obtain

$$\begin{aligned} \log\frac{1}{u^* - u} &= \log\left[\beta^{-1} L^{1/2}\left(1 + i\beta^{-1} L^{1/2} y - \beta^{-2} L y^2 + O\left(L^{3/2-3\alpha}\right)\right)\right] \\ &= \frac{1}{2}\log L - \log\beta + i\beta^{-1} L^{1/2} y - \frac{1}{2}\beta^{-2} L y^2 + O\left(L^{3/2-3\alpha}\right). \end{aligned}$$

Next, using the previous two expansions in (6.24), we obtain a refined expansion of the integrand:

$$\begin{aligned} &x^{-\hat{u}-iy} \exp\{\phi(\hat{u}-1+iy,T)+y_0\psi(\hat{u}-1+it,T)\} \\ &\quad = x^{-u^*}\exp\left\{2\beta L^{1/2}+\frac{a}{c^2}\log L-\beta^{-1}L^{3/2}y^2-\frac{2a}{c^2}\log\beta+\Gamma\right\} \\ &\quad\quad \times \left(1+c_1L^2y^3+c_2L^{5/2}y^4+c_3L^{1/2}y+c_4Ly^2+c_5L^{-1/2}+O\left(L^{-3/4+\varepsilon}\right)\right), \end{aligned} \tag{6.40}$$

for some constants $c_1,\dots,c_5$. Note that the terms with c_1 and c_2 come from $(u^*-u)^{-1}$, those involving c_3 and c_4 from $\log(u^*-u)^{-1}$, and the one with c_5 from u^*-u. Here we use the fact that the O-term in (6.24) is of the form $c(u^*-u)+O((u^*-u)^2)$. This can be obtained by applying a third-order Taylor expansion and reasoning as in the previous subsections. Using the same ideas as in the proof of the weaker error estimate, we see that the main term and the error term from the tail estimate remain the same. In addition, the error term from the local expansion can be obtained as follows. First we integrate the functions in formula (6.40), and then take into account that

$$\int_{L^{-\alpha}}^{L^{-\alpha}} y^3\exp\{-\beta^{-1}L^{3/2}y^2\}\,dy=\int_{L^{-\alpha}}^{L^{-\alpha}} y\exp\{-\beta^{-1}L^{3/2}y^2\}\,dy=0.$$

The two integrals resulting from the y^2 and y^4-terms in (6.40) can be easily calculated. They yield a relative contribution of $L^{-1/2}$, which can be combined with the term $c_5L^{-1/2}$. It follows that the absolute error term from the local expansion is

$$x^{-A_3}L^{a/c^2-3/4}\exp\{2\beta L^{1/2}\}O\left(L^{-1/2}\right).$$

Remark 6.10 It is not hard to see by analyzing the previous part of the proof of Theorem 6.1 that the constants A_2 and A_3 are given by

$$A_3=s_++1 \quad\text{and}\quad A_2=2\sqrt{2y_0}c^{-1}(\sigma)^{-\frac{1}{2}}. \tag{6.41}$$

6.1.7 Explicit Formula for the Constant A_1

We will next find an explicit formula for the constant A_1 appearing in formula (6.2). This will complete the proof of Theorem 6.1. Taking into account (6.25) and (6.39), we see that it suffices to compute $\exp\{\Gamma\}$. For any $t\in(0,T)$, put

$$J_t=a\int_0^t\left(\psi(s_+,\vartheta)-\frac{2}{c^2(T-\vartheta)}\right)d\vartheta. \tag{6.42}$$

It follows from (6.16) that

$$J_t = \phi(s_+, t) - \frac{2a}{c^2} \log \frac{T}{T-t}.$$

Now (4.90) and (6.42) give

$$J_t = \frac{2a}{c^2} \log \left\{ \frac{(T-t)\exp\{(b - c\rho s_+)\frac{t}{2}\} P(s_+)}{T[P(s_+)\cosh \frac{tP(s_+)}{2} + (b - c\rho s_+)\sinh \frac{tP(s_+)}{2}]} \right\}. \tag{6.43}$$

Using l'Hôpital's rule, we get

$$\begin{aligned} L &= \lim_{t \uparrow T} \frac{T-t}{P(s_+)\cosh \frac{tP(s_+)}{2} + (b - c\rho s_+)\sinh \frac{tP(s_+)}{2}} \\ &= \lim_{t \uparrow T} \frac{-2}{P(s_+)^2 \sinh \frac{tP(s_+)}{2} + (b - c\rho s_+)P(s_+)\cosh \frac{tP(s_+)}{2}}. \end{aligned} \tag{6.44}$$

Note that the denominator in the expression under the first limit sign in (6.44) tends to zero, since the function φ explodes for $u = s_+(T)$ and $t = T$. We have

$$P(s_+)\cosh \frac{TP(s_+)}{2} = (c\rho s_+ - b)\sinh \frac{TP(s_+)}{2},$$

and therefore

$$L = \frac{2}{(c\rho s_+ - b)^2 \sinh \frac{TP(s_+)}{2} - P(s_+)^2 \sinh \frac{TP(s_+)}{2}}. \tag{6.45}$$

It follows from (6.42)–(6.45) that

$$\begin{aligned} \lim_{t \uparrow T} e^{J_t} &= \exp\left\{ a \int_0^T \left(\psi(s_+, \vartheta) - \frac{2}{c^2(T-\vartheta)} \right) d\vartheta \right\} \\ &= \exp\left\{ \frac{(b - c\rho s_+)aT}{c^2} \right\} \\ &\quad \times \left\{ \frac{2\sqrt{(b - c\rho s_+)^2 + c^2(s_+ - s_+^2)}}{Tc^2 s_+(s_+ - 1)\sinh[\frac{T}{2}\sqrt{(b - c\rho s_+)^2 + c^2(s_+ - s_+^2)}]} \right\}^{\frac{2a}{c^2}}. \end{aligned}$$

Next, using (6.26), we see that

$$\begin{aligned} e^{\Gamma} = \left(\frac{T}{\sigma} \right)^{\frac{2a}{c^2}} &\exp\left\{ -y_0 \left(\frac{c\rho s_+ - b}{c^2} + \frac{\kappa}{c^2 \sigma^2} \right) \right\} \exp\left\{ \frac{(b - c\rho s_+)aT}{c^2} \right\} \\ &\times \left\{ \frac{2\sqrt{(b - c\rho s_+)^2 + c^2(s_+ - s_+^2)}}{Tc^2 s_+(s_+ - 1)\sinh[\frac{T}{2}\sqrt{(b - c\rho s_+)^2 + c^2(s_+ - s_+^2)}]} \right\}^{\frac{2a}{c^2}}. \end{aligned}$$

Therefore, (6.39) implies that

$$
\begin{aligned}
A_1 = {} & \frac{1}{\sqrt{\pi}} 2^{-\frac{3}{4}-\frac{a}{c^2}} y_0^{\frac{1}{4}-\frac{a}{c^2}} c^{\frac{2a}{c^2}-\frac{1}{2}} \sigma^{-\frac{a}{c^2}-\frac{1}{4}} \\
& \times \exp\left\{ -y_0 \left(\frac{c\rho s_+ - b}{c^2} + \frac{\kappa}{c^2 \sigma^2} \right) - \frac{aT}{c^2}(c\rho s_+ - b) \right\} \\
& \times \left\{ \frac{2\sqrt{(b - c\rho s_+)^2 + c^2(s_+ - s_+^2)}}{c^2 s_+ (s_+ - 1) \sinh[\frac{T}{2}\sqrt{(b - c\rho s_+)^2 + c^2(s_+ - s_+^2)}]} \right\}^{\frac{2a}{c^2}}. \qquad (6.46)
\end{aligned}
$$

The proof of Theorem 6.1 is thus completed. □

Remark 6.11 It was shown above that the constant σ can be expressed in terms of the constant s_+ and the Heston model parameters (see formula (6.10)). The same statement is valid for the constant κ. Indeed, it suffices to differentiate the functions, appearing in formula (6.9), with respect to the variable s and plug $s = s_+$ into the resulting formula. Consequently, (6.46) yields an explicit formula, representing the constant A_1 in terms of s_+ and the Heston model parameters.

Our next goal is to briefly discuss the behavior of the Heston density $D_T(x)$ near zero. As $x \downarrow 0$, the integrand in (6.17) has a saddle point that approaches the singularity $s_- + 1$ at a speed of $(-\log x)^{-1/2}$. Here the symbol s_- stands for the lower critical moment defined by (6.4). Next, reasoning exactly as in the case where $x \to \infty$, we obtain the following asymptotic formula:

$$
\begin{aligned}
D_T(x) = {} & \widetilde{A}_1 x^{\widetilde{A}_3} \exp\{\widetilde{A}_2 \sqrt{-\log x}\} (-\log x)^{a/c^2 - 3/4} \\
& \times \left(1 + O\left((-\log x)^{-1/2}\right)\right) \qquad (6.47)
\end{aligned}
$$

as $x \downarrow 0$, where

$$
\widetilde{A}_3 = -(s_- + 1), \qquad \widetilde{A}_2 = 2\frac{\sqrt{2y_0}}{c\sqrt{\sigma_-}},
$$

$$
\begin{aligned}
\widetilde{A}_1 = {} & \frac{1}{2\sqrt{\pi}} (2y_0)^{1/4 - a/c^2} c^{2a/c^2 - 1/2} \sigma_-^{-a/c^2 - 1/4} \\
& \times \exp\left\{ -y_0 \left(\frac{s_- \rho c - b}{c^2} + \frac{\kappa_-}{c^2 \sigma_-^2} \right) - \frac{aT}{c^2}(c\rho s_- - b) \right\} \\
& \times \left(\frac{2\sqrt{b^2 - 2bc\rho s_- + c^2 s_-(1 - (1 - \rho^2)s_-)}}{c^2 s_-(s_- - 1) \sinh \frac{T}{2}\sqrt{b^2 - 2bc\rho s_- + c^2 s_-(1 - (1 - \rho^2)s_-)}} \right)^{2a/c^2}.
\end{aligned}
$$

Remark 6.12 The density $\widetilde{D}_T$ of the log-price $\widetilde{X}_T = \log X_T$ is given by $\widetilde{D}_T(x) = e^x D_T(e^x)$. Using (6.2) and (6.47), we get

$$\widetilde{D}_T(x) = A_1 \exp\{-(A_3 - 1)x\} \exp\{A_2\sqrt{x}\} x^{a/c^2-3/4}\big(1 + O(x^{-1/2})\big)$$

as $x \to \infty$, and

$$\widetilde{D}_T(x) = \widetilde{A}_1 \exp\{-(\widetilde{A}_3 + 1)|x|\} \exp\{\widetilde{A}_2\sqrt{|x|}\}|x|^{a/c^2-3/4}\big(1 + O(|x|^{-1/2})\big)$$

as $x \to -\infty$.

6.2 Asymptotic Formulas for Stock Price Densities in Uncorrelated Heston Models

In the present section, we provide an alternative proof of Theorem 6.1 for the Heston model with $\rho = 0$. The methods employed in the alternative proof differ significantly from those used in the proof of Theorem 6.1 in the general case. The former proof is based on the integral representation of the stock price density in an uncorrelated stochastic volatility model (see (3.5)) and on the asymptotic formula for the mixing distribution density in the Heston model (see (5.133)).

An alternative proof of Theorem 6.1 for the uncorrelated Heston model. Consider the Heston model with $\mu \in \mathbb{R}^1$, $a \ge 0$, $b \ge 0$, $c > 0$, $x_0 > 0$, $y_0 > 0$, and $\rho = 0$. Put $A(y) = y^{-1} m_T(y) e^{Cy^2}$, $k = \sqrt{C + \frac{T}{8}}$, $l = B$, $\zeta(y) = Ay^{-\frac{3}{2}+\frac{2a}{c^2}}$, and $b(y) = y^{-\frac{1}{2}}$, where the constants A, B, and C are the same as in Lemma 5.35 with $t = T$. It is not hard to see that condition 2 in Theorem 5.5 follows from formula (5.133). In addition, it is clear that condition 3 with $\gamma = 1$ holds. The next lemma shows that condition 1 in Theorem 5.5 also holds.

Lemma 6.13 *For every $T > 0$ and $s > 0$ the following inequality holds*:

$$\int_0^s y^{-1} m_T(y)\, dy < \infty.$$

Proof Suppose the Laplace transform of a positive function h is integrable over the interval $(1, \infty)$. Then it is clear that

$$\int_0^s h(y) y^{-1}\, dy < \infty$$

for all $s > 0$. Put

$$h(y) = y^{-\frac{1}{2}} m_T\big(2^{\frac{1}{2}} c t^{-\frac{1}{2}} y^{\frac{1}{2}}\big), \quad y > 0.$$

The Laplace transform of this function is integrable over $(1, \infty)$ (this follows from (5.109)). Taking into account the previous reasoning, we complete the proof of Lemma 6.13. □

Since all the conditions in Theorem 5.5 hold, we can apply this theorem and get the following formula:

$$\int_0^\infty y^{-1} m_T(y) \exp\left\{-\left(\frac{z^2}{y^2} + \frac{Ty^2}{8}\right)\right\} dy$$
$$= A\frac{\sqrt{\pi}}{2k} \exp\left\{\frac{B^2}{16k^2}\right\} k^{\frac{3}{4}-\frac{a}{c^2}} z^{-\frac{3}{4}+\frac{a}{c^2}} \exp\left\{Bk^{-\frac{1}{2}}\sqrt{z}\right\} e^{-2kz}$$
$$\times \left(1 + O\left(z^{-\frac{1}{2}}\right)\right) \tag{6.48}$$

as $z \to \infty$. Next, replacing z by $\frac{\log x}{\sqrt{2T}}$ in formula (6.48) and taking into account formula (3.5) and the equality $k = \frac{\sqrt{8C+T}}{2\sqrt{2}}$, we obtain

$$D_T\left(x_0 e^{\mu T} x\right) = \frac{A}{x_0 e^{\mu T}} 2^{-\frac{3}{4}+\frac{a}{c^2}} T^{-\frac{1}{8}-\frac{a}{2c^2}} (8C+T)^{-\frac{1}{8}-\frac{a}{2c^2}} \exp\left\{\frac{B^2}{2(8C+T)}\right\}$$
$$\times (\log x)^{-\frac{3}{4}+\frac{a}{c^2}} \exp\left\{\frac{B\sqrt{2}}{T^{\frac{1}{4}}(8C+T)^{\frac{1}{4}}}\sqrt{\log x}\right\}$$
$$\times x^{-\left(\frac{3}{2}+\frac{\sqrt{8C+T}}{2\sqrt{T}}\right)} \left(1 + O\left((\log x)^{-\frac{1}{2}}\right)\right) \tag{6.49}$$

as $x \to \infty$.

It is clear that formula (6.49) with $x_0 = 1$ and $\mu = 0$ implies Theorem 6.1 for the uncorrelated Heston model. Moreover, the following lemma holds:

Lemma 6.14 *The constants A_1, A_2, and A_3, obtained in the alternative proof of Theorem* 6.1 *for the uncorrelated Heston model, are given by*

$$A_1 = A\left(x_0 e^{\mu T}\right)^{A_3 - 1} 2^{-\frac{3}{4}+\frac{a}{c^2}} T^{-\frac{1}{8}-\frac{a}{2c^2}} (8C+T)^{-\frac{1}{8}-\frac{a}{2c^2}}$$
$$\times \exp\left\{\frac{B^2}{2(8C+T)}\right\}, \tag{6.50}$$

$$A_2 = \frac{B\sqrt{2}}{T^{\frac{1}{4}}(8C+T)^{\frac{1}{4}}}, \quad \text{and} \quad A_3 = \frac{3}{2} + \frac{\sqrt{8C+T}}{2\sqrt{T}}, \tag{6.51}$$

where A, B, and C are defined in Lemma 5.35 *with $t = T$.*

Remark 6.15 Formula (6.49) and the symmetry property of D_T (formula (3.6)) imply that

$$D_T(x) = A_1 \left(x_0 e^{\mu T}\right)^{3 - 2A_3} e^{A_2 \sqrt{\log \frac{1}{x}}} \left(\log \frac{1}{x}\right)^{-\frac{3}{4}+\frac{a}{c^2}} x^{A_3 - 3}$$

$$\times \left(1 + O\left(\left(\log \frac{1}{x}\right)^{-\frac{1}{2}}\right)\right), \tag{6.52}$$

as $x \to 0$. The constants A_1, A_2, and A_3 in (6.52) are the same as in Lemma 6.14.

6.3 The Constants A_1, A_2 and A_3 Obtained by Different Methods Are Equal

The asymptotic behavior of the stock price density in the Heston model is influenced by three constants: A_1, A_2, and A_3 (see Theorem 6.1). Recall that for the uncorrelated Heston model, we obtained two representations for each of those constants (see formulas (6.41), (6.46), and the formulas in Lemma 6.14). At the first glance, they do not look similar at all. Therefore, it would be important to reconcile differently looking formulas, representing the constants A_1, A_2, and A_3.

Lemma 6.16 *The expressions, describing the constants* A_1, A_2, *and* A_3 *in* (6.41) *and* (6.46), *and the corresponding expressions in Lemma* 6.14 *are equal.*

Proof For the constant A_3, Lemma 6.16 can be easily established by taking into account the fact that the number s_+ is the order of the critical moment of the stock price in the Heston model.

We will next reconcile the constants A_2. On the one hand, if $\rho = 0$, then formula (6.10), (6.41) imply that

$$A_2^2 = \frac{16 y_0 s_+ (s_+ - 1)[c^2 s_+ (s_+ - 1) - b^2]}{c^2 (2s_+ - 1)[T c^2 s_+ (s_+ - 1) + 2b]}. \tag{6.53}$$

On the other hand, the constant A_2, appearing in the alternative proof of Theorem 6.1 for the uncorrelated Heston model, is given by

$$A_2 = \frac{\sqrt{2} B}{T^{\frac{1}{4}} (8C + T)^{\frac{1}{4}}} \tag{6.54}$$

where

$$B = \frac{\sqrt{2T\alpha}}{c} \quad \text{and} \quad C = \frac{4r^2 + T^2 b^2}{2Tc^2}$$

(see (6.51)). Therefore

$$8C + T = \frac{4(4r^2 + T^2 b^2) + T^2 c^2}{T c^2}.$$

We will also need the equality

$$\alpha = \frac{4 y_0 r^2 (4r^2 + T^2 b^2)}{T^3 c^2 [r^2 + \frac{Tb}{2}(1 + \frac{Tb}{2})]}$$

(see (5.128)) and the fact that the numbers r and s_+ are related as follows:

$$r = \frac{T}{2}\left[c^2 s_+(s_+ - 1) - b^2\right]^{\frac{1}{2}}. \tag{6.55}$$

Equality (6.55) can be derived from the formula for the constant A_3 given in (6.51) and the equality $A_3 = s_+ + 1$. Next, using (6.54), we obtain

$$A_2^2 = \frac{16 y_0 r^2 (4r^2 + T^2 b^2)}{c^3 T^2 [4(4r^2 + T^2 b^2) + T^2 c^2]^{\frac{1}{2}} [r^2 + \frac{Tb}{2}(1 + \frac{Tb}{2})]}. \tag{6.56}$$

Finally, it is not hard to prove that (6.55) and (6.56) imply the following equality:

$$A_2^2 = \frac{16 y_0 s_+ (s_+ - 1)[c^2 s_+ (s_+ - 1) - b^2]}{c^2 (2s_+ - 1)[T c^2 s_+ (s_+ - 1) + 2b]}. \tag{6.57}$$

Comparing formulas (6.53) and (6.57), we see that the constants A_2 obtained by different methods are equal.

We will finally reconcile the constants A_1, appearing in the different proofs of the asymptotic formula for the stock price density in the Heston model. Let us first see what expressions have to be reconciled. On the one hand, the constant A_1 in Theorem 6.1 is given by formula (6.46). Taking into account the condition $\rho = 0$ and assuming that s_+, σ, and κ have been computed under the constraint $\rho = 0$, we get the following representation formula for A_1:

$$\begin{aligned} A_1 = {} & \frac{1}{\sqrt{\pi}} 2^{-\frac{3}{4} - \frac{a}{c^2}} y_0^{\frac{1}{4} - \frac{a}{c^2}} c^{\frac{2a}{c^2} - \frac{1}{2}} \sigma^{-\frac{a}{c^2} - \frac{1}{4}} \exp\left\{\frac{y_0 b}{c^2} + \frac{\kappa}{c^2 \sigma^2} + \frac{abT}{c^2}\right\} \\ & \times \left\{\frac{2\sqrt{c^2(s_+^2 - s_+) - b^2}}{c^2 s_+ (s_+ - 1) \sin[\frac{T}{2}\sqrt{c^2(s_+^2 - s_+) - b^2}]}\right\}^{\frac{2a}{c^2}}. \end{aligned} \tag{6.58}$$

On the other hand, the constant A_1 in the alternative proof of Theorem 6.1 has the following representation:

$$A_1 = A 2^{-\frac{3}{4} + \frac{a}{c^2}} T^{-\frac{1}{8} - \frac{a}{2c^2}} (8C + T)^{-\frac{1}{8} - \frac{a}{2c^2}} \exp\left\{\frac{B^2}{2(8C + T)}\right\}, \tag{6.59}$$

where the constants A, B, and C are defined in Lemma 5.35 (see formula (6.50) with $t = T$). Therefore, it suffices to prove that the expressions on the right-hand side of (6.58) and (6.59) are equal.

Our first goal is to rewrite formula (6.58) in terms of the parameter $r = r_{\frac{Tb}{2}}$ and the model constants. Let us note that (6.55) implies the following equalities:

$$c^2 s_+ (s_+ - 1) - b^2 = \frac{4r^4}{T^2}, \tag{6.60}$$

$$c^2 s_+(s_+ - 1) = \frac{4r^2 + T^2 b^2}{T^2}, \tag{6.61}$$

and

$$2s_+ - 1 = \frac{\sqrt{16r^2 + 4T^2b^2 + T^2c^2}}{Tc}. \tag{6.62}$$

Moreover, (6.10) with $\rho = 0$, (6.60), (6.61), and (6.62) give

$$\sigma = \frac{cT^2(16r^2 + 4T^2b^2 + T^2c^2)^{\frac{1}{2}}(4r^2 + T^2b^2 + 2Tb)}{br^2(4r^2 + T^2b^2)}. \tag{6.63}$$

We will next compute κ. It is not hard to see that if $\rho = 0$, then (6.9) and (6.6) imply

$$\begin{aligned}\frac{\partial T^*}{\partial s} &= T^*(s)\frac{c^2(2s-1)}{2(b^2 - c^2(s^2 - s))} + \frac{b(2s-1)}{(b^2 - c^2(s^2-s))(s^2-s)} \\ &= \frac{2s-1}{b^2 - c^2(s^2-s)}\left[\frac{c^2T^*(s)}{2} + \frac{b}{s^2-s}\right].\end{aligned} \tag{6.64}$$

Differentiating the functions appearing in (6.64) and plugging $s = s_+$ into the resulting formula, we obtain

$$\kappa = \frac{T^3c^2L(r,T,b,c)}{64r^4(4r^2+T^2b^2)^2} \tag{6.65}$$

where

$$\begin{aligned}L(r,T,b,c) &= 2\big(8r^2 + 4Tb^2 + T^2c^2\big)\big(4r^2 + T^2b^2 + 2Tb\big)\big(4r^2 + T^2b^2\big) \\ &\quad + \big(16r^2 + 4T^2b^2 + T^2c^2\big) \\ &\quad \times \big[\big(4r^2 + T^2b^2 + 2Tb\big)\big(4r^2 + T^2b^2\big) + 16Tbr^2\big].\end{aligned}$$

In the proof of (6.65), we take into account the equalities

$$\kappa = \frac{\partial^2 T^+}{\partial s^2}, \qquad \sigma = -\frac{\partial T^*}{\partial s}, \qquad T^*(s_+) = T,$$

and also (6.60), (6.61), and (6.62). Now, formulas (6.58), (6.63), and (6.65) imply that

$$\begin{aligned}A_1 &= M_1(y_0, r, T, a, b, c)\exp\left\{\frac{abT}{c^2}\right\} \\ &\quad \times \exp\left\{\frac{y_0 M_2(y_0, r, T, b, c)}{c^2T(16r^2 + 4T^2b^2 + T^2c^2)(4r^2 + T^2b^2 + 2Tb)^2}\right\},\end{aligned} \tag{6.66}$$

where

$$
\begin{aligned}
&M_1(y_0, r, T, a, b, c) \\
&= \pi^{-\frac{1}{2}} 2^{-\frac{3}{4}-\frac{a}{c^2}} y_0^{\frac{1}{4}-\frac{a}{c^2}} c^{\frac{2a}{c^2}-\frac{1}{2}} \\
&\quad \times \left[\frac{cT^2(16r^2+4T^2b^2+T^2c^2)^{\frac{1}{2}}(4r^2+T^2b^2+2Tb)}{8r^2(4r^2+T^2b^2)}\right]^{-\frac{a}{c^2}-\frac{1}{4}} \\
&\quad \times \left(\frac{4rT}{\sin r(4r^2+T^2b^2)}\right)^{\frac{2a}{c^2}}
\end{aligned}
\tag{6.67}
$$

and

$$
\begin{aligned}
&M_2(y_0, r, T, b, c) \\
&= Tb\left(16r^2+4T^2b^2+T^2c^2\right)\left(4r^2+T^2b^2+2Tb\right)^2 \\
&\quad - 2\left(8r^2+4T^2b^2+2Tb\right)\left(4r^2+T^2b^2\right) - \left(16r^2+4T^2b^2+T^2c^2\right) \\
&\quad \times \left[\left(4r^2+T^2b^2+2Tb\right)\left(4r^2+T^2b^2\right)+16Tbr^2\right].
\end{aligned}
\tag{6.68}
$$

Simplifying the expression on the right-hand side of (6.67), we obtain

$$
\begin{aligned}
M_1(y_0, r, T, a, b, c) &= \pi^{-\frac{1}{2}} 2^{\frac{6a}{c^2}} y_0^{\frac{1}{4}-\frac{a}{c^2}} c^{\frac{a}{c^2}-\frac{3}{4}} T^{-\frac{1}{2}} \left(16r^2+4T^2b^2+T^2c^2\right)^{-\frac{1}{8}-\frac{a}{2c^2}} \\
&\quad \times \left(4r^2+T^2b^2\right)^{\frac{1}{4}-\frac{a}{c^2}} \left(4r^2+T^2b^2+2Tb\right)^{-\frac{1}{4}-\frac{a}{c^2}}.
\end{aligned}
\tag{6.69}
$$

Next, we turn our attention to the expression on the right-hand side of (6.59). Let us put

$$
\widetilde{N}_2(y_0, r, T, b, c) = \widetilde{F}(0) + \frac{B^2}{2(8C+T)}.
$$

Then (5.128), (5.129), and the formulas for the constants B and C in Lemma 5.35, imply the following equality:

$$
\begin{aligned}
\widetilde{N}_2(y_0, r, T, b, c) &= \frac{y_0}{Tc^2(4r^2+2T^2b^2+2Tb)^2(16r^2+4T^2b^2+T^2c^2)} \\
&\quad \times N_2(y_0, r, T, b, c)
\end{aligned}
\tag{6.70}
$$

where

$$
\begin{aligned}
&N_2(y_0, r, T, b, c) \\
&= 16r^2\left(4r^2+T^2b^2\right)\left(4r^2+T^2b^2+2Tb\right) \\
&\quad + \left(16r^2+4T^2b^2+T^2c^2\right)\left[Tb\left(4r^2+T^2b^2\right)\left(4r^2+T^2b^2+2Tb\right)\right.
\end{aligned}
$$

$$+ \left(4r^2 + T^2b^2\right)\left(4r^2 - T^2b^2 - 2Tb\right) - 16r^2\left(4r^2 + T^2b^2 + 2Tb\right)\Big]. \quad (6.71)$$

It is not hard to show that

$$A_1 = N_1(y_0, r, T, a, b, c) \exp\left\{\frac{abT}{c^2}\right\}$$
$$\times \exp\left\{\frac{y_0 N_2(y_0, r, T, b, c)}{c^2 T(16r^2 + 4T^2b^2 + T^2c^2)(4r^2 + T^2b^2 + 2Tb)^2}\right\} \quad (6.72)$$

with

$$N_1(y_0, r, T, a, b, c) = \pi^{-\frac{1}{2}} 2^{\frac{6a}{c^2}} y_0^{\frac{1}{4} - \frac{a}{c^2}} c^{\frac{a}{c^2} - \frac{3}{4}} T^{-\frac{1}{2}} \left(16r^2 + 4T^2b^2 + T^2c^2\right)^{-\frac{1}{8} - \frac{a}{2c^2}}$$
$$\times \left(4r^2 + T^2b^2\right)^{\frac{1}{4} - \frac{a}{c^2}} \left(4r^2 + T^2b^2 + 2Tb\right)^{-\frac{1}{4} - \frac{a}{c^2}}. \quad (6.73)$$

In the proof of (6.72), we use (6.59), (5.119), (5.128), (6.70), (6.71), and the formula for the constant A in Lemma 5.35. Comparing equalities (6.69) and (6.73), we see that

$$M_1(y_0, r, T, a, b, c) = N_1(y_0, r, T, a, b, c). \quad (6.74)$$

Our next goal is to prove that

$$M_2(y_0, r, T, b, c) = N_2(y_0, r, T, b, c). \quad (6.75)$$

It is not hard to see by analyzing (6.68) and (6.71) that the following conditions hold:

$$M_2(y_0, r, T, b, c) = \eta_1(y_0, T, b, c)r^6 + \eta_2(y_0, T, b, c)r^4$$
$$+ \eta_3(y_0, T, b, c)r^2 + \eta_4(y_0, T, b, c)$$

and

$$N_2(y_0, r, T, a, b, c) = \zeta_1(y_0, T, b, c)r^6 + \zeta_2(y_0, T, b, c)r^4$$
$$+ \zeta_3(y_0, T, b, c)r^2 + \zeta_4(y_0, T, b, c).$$

The coefficients in front of the powers of r in the previous formulas can be computed explicitly. The resulting formulas are as follows:

$$\eta_1(y_0, T, b, c) = \zeta_1(y_0, T, b, c) = 256Tb - 512,$$

$$\eta_2(y_0, T, b, c) = \zeta_2(y_0, T, b, c)$$
$$= 192T^3b^3 - 192T^2b^2 - 512Tb + 16T^3bc^2 - 48T^2c^2,$$

$$\eta_3(y_0,T,b,c) = \zeta_3(y_0,T,b,c)$$
$$= 48T^5b^5 - 160T^3b^3 + 8T^5b^3c^2 - 40T^3bc^2,$$

and

$$\eta_4(y_0,T,b,c) = \zeta_4(y_0,T,b,c)$$
$$= 4T^7b^7 + 4T^6b^6 - 8T^5b^5 + T^7b^5c^2 + T^6b^4c^2 - 2T^5b^3c^2.$$

Now it is clear that (6.75) can be easily derived from the previous formulas. Finally, we observe that (6.66), (6.72), (6.74), and (6.75) imply that the constants A_1, appearing in the different proofs of Theorem 6.1 for the uncorrelated Heston model, are equal.

This completes the proof of Lemma 6.16. □

6.4 Asymptotic Formulas for Stock Price Densities in Stein–Stein Models

Let us consider the Stein–Stein model with $\mu \in \mathbb{R}^1$, $q \geq 0$, $m \geq 0$, $\sigma > 0$, $-1 < \rho \leq 0$, $x_0 > 0$, and $y_0 > 0$. The next theorem characterizes the asymptotic behavior of the stock price distribution density D_t in this model.

Theorem 6.17 *For every $t > 0$ there exist positive constants B_1, B_2, and B_3 such that*

$$D_t(x) = B_1 e^{B_2\sqrt{\log x}}(\log x)^{-\frac{1}{2}}x^{-B_3}\left(1 + O\left((\log x)^{-\frac{1}{2}}\right)\right), \tag{6.76}$$

as $x \to \infty$.

Proof Let us begin with the uncorrelated Stein–Stein model. We will first show that it is possible to apply Theorem 5.5 with

$$A(y) = y^{-1}m_t(y)e^{Gy^2}, \qquad k = \sqrt{G + \frac{t}{8}}, \qquad l = F, \qquad \zeta(y) = Ey^{-1},$$

and $b(y) = y^{-\frac{1}{2}}$, where the constants E, F, and G are such as in Lemma 5.42. It is not hard to see that condition 2 in Theorem 5.5 follows from formula (5.134). In addition, it is clear that condition 3 with $\gamma = 1$ holds. The validity of condition 1 in Theorem 5.5 can be shown by reasoning as in the proof of Lemma 6.13.

It follows from Theorem 5.5 that

$$\int_0^\infty y^{-1}m_t(y)\exp\left\{-\left(\frac{z^2}{y^2} + \frac{ty^2}{8}\right)\right\}dy$$
$$= E\frac{\sqrt{\pi}}{2k}\exp\left\{\frac{F^2}{16k^2}\right\}k^{\frac{1}{2}}z^{-\frac{1}{2}}\exp\left\{Fk^{-\frac{1}{2}}\sqrt{z}\right\}e^{-2kz}\left(1 + O\left(z^{-\frac{1}{2}}\right)\right) \tag{6.77}$$

as $z \to \infty$. Replacing z by $\frac{\log x}{\sqrt{2t}}$ in formula (6.77) and taking into account formula (3.5) and the equality $k = \frac{\sqrt{8G+t}}{2\sqrt{2}}$, we obtain

$$\begin{aligned} D_t\left(x_0 e^{\mu t} x\right) &= \frac{E}{x_0 e^{\mu t}} 2^{-\frac{1}{2}} t^{-\frac{1}{4}} (8G+t)^{-\frac{1}{4}} \exp\left\{\frac{F^2}{2(8G+t)}\right\} \\ &\quad \times (\log x)^{-\frac{1}{2}} \exp\left\{\frac{F\sqrt{2}}{t^{\frac{1}{4}}(8G+t)^{\frac{1}{4}}} \sqrt{\log x}\right\} x^{-\left(\frac{3}{2}+\frac{\sqrt{8G+t}}{2\sqrt{t}}\right)} \\ &\quad \times \left(1 + O\left((\log x)^{-\frac{1}{2}}\right)\right) \end{aligned} \tag{6.78}$$

as $x \to \infty$.

Now, it is clear that formula (6.78) implies Theorem 6.17 for the uncorrelated model. □

In addition, the following lemma holds.

Lemma 6.18 *The constants B_1, B_2, and B_3 are given by*

$$B_1 = E\left(x_0 e^{\mu t}\right)^{B_3 - 1} 2^{-\frac{1}{2}} t^{-\frac{1}{4}} (8G+t)^{-\frac{1}{4}} \exp\left\{\frac{F^2}{2(8G+t)}\right\},$$

$$B_2 = \frac{F\sqrt{2}}{t^{\frac{1}{4}}(8G+t)^{\frac{1}{4}}}, \quad \text{and} \quad B_3 = \frac{3}{2} + \frac{\sqrt{8G+t}}{2\sqrt{t}},$$

where the numbers E, F, and G are defined in Lemma 5.42.

The constant B_2 is a linear function of y_0 and m. This will be shown next.

Lemma 6.19 *The constant B_2 in Theorem* 6.17 *can be represented as follows*:

$$B_2 = H_1(t,q,\sigma) y_0 + H_2(t,q,\sigma) m$$

where

$$H_1(t,q,\sigma) = \frac{2(r_{qt}^2 + q^2 t^2)^{\frac{1}{2}} r_{qt}}{t\sigma^{\frac{3}{2}} (4r_{qt}^2 + 4q^2 t^2 + \sigma^2 t^2)^{\frac{1}{4}} (r_{qt}^2 + qt(1+qt))^{\frac{1}{2}}}$$

and

$$H_2(t,q,\sigma) = \frac{2q(r_{qt}^2 + q^2 t^2)^{\frac{1}{2}} (1 - \cos r_{qt})}{\sigma^{\frac{3}{2}} \sin r_{qt} (4r_{qt}^2 + 4q^2 t^2 + \sigma^2 t^2)^{\frac{1}{4}} (r_{qt}^2 + qt(1+qt))^{\frac{1}{2}}}.$$

Proof Put $r = r_{qt}$, and recall that

$$G = \frac{r^2 + q^2 t^2}{2\sigma^2 t^2}$$

(see Lemma 6.18). Note that G does not depend on y_0 and m. We have

$$8G + t = \frac{4r^2 + 4q^2 t^2 + \sigma^2 t^2}{\sigma^2 t}. \tag{6.79}$$

Next, using (5.152), (6.79), and the expression for the constant B_2 in Lemma 6.18, and making simplifications, we see that Lemma 6.19 holds. □

The behavior of the density D_t near zero in the uncorrelated Stein–Stein model is characterized in the next assertion.

Theorem 6.20 *For every* $t > 0$,

$$D_t(x) = B_1 \left(x_0 e^{\mu t}\right)^{3-2B_3} e^{B_2 \sqrt{\log x}} (\log x)^{-\frac{1}{2}} x^{-B_3} \times \left(1 + O\left((\log x)^{-\frac{1}{2}}\right)\right) \tag{6.80}$$

as $x \to 0$. *The constants* B_1, B_2, *and* B_3 *in* (6.80) *are the same as in Lemma* 6.18.

It is easy to see that Theorem 6.20 follows from Theorem 6.17 and formula (3.6).

We will next turn our attention to the Stein–Stein model with $-1 < \rho < 0$. Let us recall that by Lemma 2.7 the stock price density in the correlated Stein–Stein model with $m = 0$ (the model in (2.17)) coincides with the stock price density in the special Heston model described by (2.18). It follows from the previous statement that Theorem 6.17, established above for the uncorrelated Stein–Stein model, also holds for the correlated Stein–Stein model with $m = 0$. Indeed, it is not hard to see that Lemma 2.7 and formula (6.2) imply formula (6.76) for the correlated Stein–Stein model with $m = 0$. In this formula, the constants B_1, B_2, and B_3 coincide with the constants A_1, A_2, and A_3 appearing in formula (6.2) for the Heston model with

$$a = \sigma^2, \quad b = 2q, \quad \text{and} \quad c = 2\sigma. \tag{6.81}$$

In addition, the initial conditions for the stock price process and the variance process in the Heston model described above should be x_0 and y_0^2, where x_0 and y_0 are the initial conditions for the stock price process and the volatility process in the original correlated Stein–Stein model. It follows from the formulas in (6.81) that $-\frac{3}{4} + \frac{a}{c^2} = -\frac{1}{2}$. The previous equality explains why the power of the logarithmic function in formula (6.76) for the Stein–Stein model with $m = 0$ equals $-\frac{1}{2}$.

For the correlated Stein–Stein model with $m \neq 0$ and $-1 < \rho \leq 0$, formula (6.76) was recently established by J.D. Deuschel, P. Friz, A. Jacquier, and S. Violante (see [DFJV11]). These authors used the Laplace method on Wiener space, going back to G. Ben Arous (see [BA88a, BA88b]), to obtain small-noise asymptotic expansions for marginal distribution densities of projected diffusion processes under certain conditions (see Theorem 9 in [DFJV11]). If special scaling relations hold for the projected process, than the small-noise asymptotics for the density can be converted into the tail asymptotics (see Corollary 14 in [DFJV11]). It was shown in Sect. 4.6 of [DFJV11] that the above-mentioned sufficient conditions and scaling relations hold for the Stein–Stein model. The resulting asymptotic expansion for the Stein–Stein density coincides with that in formula (6.76) (see [DFJV11]) for more details).

Remark 6.21 It is not excluded that Theorem 5.1 could be applied to the explicit formulas for the Mellin transform of stock price distribution densities in the correlated Heston and Stein–Stein models (see Lemmas 4.16 and 4.18). This would provide alternative proofs of formulas (6.2) and (6.76).

6.5 Asymptotic Formulas for Stock Price Densities in Uncorrelated Hull–White Models

Let us consider the uncorrelated Hull–White model with $\mu \in \mathbb{R}^1$, $\nu \in \mathbb{R}^1$, $\xi > 0$, $x_0 > 0$, and $y_0 > 0$. The next theorem describes the asymptotic behavior of the stock price distribution density D_t in the Hull–White model.

Theorem 6.22 *For every* $t > 0$,

$$\begin{aligned} D_t(x) &= Cx^{-2}(\log x)^{\frac{c_2-1}{2}}(\log\log x)^{c_3} \\ &\quad \times \exp\left\{-\frac{1}{2t\xi^2}\left(\log\left[\frac{1}{y_0}\sqrt{\frac{2}{t}\log\frac{x}{x_0e^{\mu t}}}\right]\right.\right. \\ &\quad \left.\left. + \frac{1}{2}\log\log\left[\frac{1}{y_0}\sqrt{\frac{2}{t}\log\frac{x}{x_0e^{\mu t}}}\right]\right)^2\right\} \\ &\quad \times \left(1 + O\left((\log\log x)^{-\frac{1}{2}}\right)\right) \end{aligned} \tag{6.82}$$

as $x \to \infty$. *In formula* (6.82), *the constant* C *is given by*

$$C = x_0 e^{\mu t} c_1 2^{\frac{c_2-1-2c_3}{2}} t^{-\frac{c_2+1}{2}}, \tag{6.83}$$

and the constants c_1, c_2, *and* c_3 *are the same as in Theorem* 5.8.

Proof We will first check that the conditions is Theorem 5.5 hold with $A(y) = y^{-1}m_t(y)$, $l = 0$,

$$\zeta(y) = c_1 y^{c_2-1}(\log y)^{c_3}\exp\left\{-\frac{1}{2t\xi^2}\left(\log\frac{y}{y_0} + \frac{1}{2}\log\log\frac{y}{y_0}\right)^2\right\},$$

$b(y) = (\log(2+y))^{-\frac{1}{2}}$, $0 < \gamma < 1$, $k = \sqrt{\frac{t}{8}}$, and $w = \frac{\log x}{\sqrt{2t}}$. Indeed, it is clear from Theorem 5.27 and the definition of ζ that condition 1 in Theorem 5.5 holds. Conditions 2 and 3 follow from the definitions of b and ζ, and finally, condition 4 in Theorem 5.5 follows from Theorem 5.8. Next, using formula (5.34), applying Theorem 5.5, and simplifying the resulting expression, we get formula (6.82).

This completes the proof of Theorem 6.22. □

Remark 6.23 It follows from (6.82) that

$$D_t(x) \approx x^{-2}(\log x)^{\tau_1}(\log\log x)^{\tau_2} \times \exp\left\{-\frac{1}{8t\xi^2}(\log\log x + \log\log\log x)^2\right\} \tag{6.84}$$

as $x \to \infty$, where

$$\tau_1 = \frac{c_2 - 1}{2} + \frac{1}{2t\xi^2}\log(y_0\sqrt{t}) \quad \text{and} \quad \tau_2 = c_3 + \frac{1}{2t\xi^2}\log(y_0\sqrt{t}).$$

Indeed, using the mean value theorem, we can prove that

$$\begin{aligned}
&\left(\log\left[\frac{1}{y_0}\sqrt{\frac{2}{t}\log\frac{x}{x_0e^{\mu t}}}\right] + \frac{1}{2}\log\log\left[\frac{1}{y_0}\sqrt{\frac{2}{t}\log\frac{x}{x_0e^{\mu t}}}\right]\right)^2 \\
&= \left(\log\left[\frac{1}{y_0}\sqrt{\frac{2}{t}\log\frac{x}{x_0e^{\mu t}}}\right] + \frac{1}{2}\log\left(\frac{1}{2}\log\log\frac{x}{x_0e^{\mu t}}\right)\right)^2 + O(1) \\
&= \left(\log\frac{1}{y_0} + \frac{1}{2}\log\frac{2}{t} + \frac{1}{2}\log\log x + \frac{1}{2}\log\frac{1}{2} + \frac{1}{2}\log\log\log x\right)^2 + O(1) \\
&= \left(\log\frac{1}{y_0\sqrt{t}} + \frac{1}{2}\log\log x + \frac{1}{2}\log\log\log x\right)^2 + O(1)
\end{aligned} \tag{6.85}$$

as $x \to \infty$. It follows from (6.82) and (6.85) that

$$D_t(x) \approx x^{-2}(\log x)^{\frac{c_2-1}{2}}(\log\log x)^{c_3}\exp\left\{-\frac{1}{8t\xi^2}(\log\log x + \log\log\log x)^2\right\} \times \exp\left\{\frac{1}{2t\xi^2}\left[\log(y_0\sqrt{t})\right](\log\log x + \log\log\log x)\right\} \tag{6.86}$$

as $x \to \infty$. Now it is not hard to see that (6.86) implies (6.84).

A similar asymptotic formula for the stock price distribution density D_t in the Hull–White model holds in the case where $x \to 0$. This formula can be obtained using (6.82) and the symmetry property of D_t (see (3.6).

Theorem 6.24 *For every $t > 0$,*

$$D_t(x) = C\left(x_0 e^{\mu t}\right)^{-1} x^{-1}\left(\log\frac{1}{x}\right)^{\frac{c_2-1}{2}}\left(\log\log\frac{1}{x}\right)^{c_3}$$
$$\times \exp\left\{-\frac{1}{2t\xi^2}\left(\log\left[\frac{1}{y_0}\sqrt{\frac{2}{t}\log\frac{x_0 e^{\mu t}}{x}}\right]\right.\right.$$
$$\left.\left.+\frac{1}{2}\log\log\left[\frac{1}{y_0}\sqrt{\frac{2}{t}\log\frac{x_0 e^{\mu t}}{x}}\right]\right)^2\right\}$$
$$\times\left(1+O\left(\left(\log\log\frac{1}{x}\right)^{-\frac{1}{2}}\right)\right) \tag{6.87}$$

as $x \to 0$. In formula (6.87), *the constant C is given by* (6.83) *and the constants c_1, c_2, and c_3 are the same as in Theorem* 5.8.

6.6 Comparison of Stock Price Densities

In this section, we consider only uncorrelated models. It follows from (6.82) that for the Hull–White model, the function $D_t(x)$ behaves near infinity roughly like the function x^{-2}. This is an extremely slow decay. For instance, no uncorrelated stochastic volatility model has the stock price density decaying like $x^{-2+\epsilon}$ with $\epsilon > 0$. This fact can be obtained using the symmetry property of D_t (see (3.6)). Indeed, if the stock price density decays near infinity like $x^{-2+\epsilon}$, then near zero it behaves like $x^{-1-\epsilon}$, which contradicts the integrability of D_t.

For the Heston model, the stock price distribution density $D_t(x)$ behaves near infinity roughly as the function x^{-A_3} and near zero like the function x^{A_3-3} (see Theorem 6.1).

For the Stein–Stein model, Theorem 6.17 implies that $D_t(x)$ behaves at infinity roughly like the function x^{-B_3} and at zero like the function x^{B_3-3}.

Recall that the constants A_3 and B_3, characterizing the power type decay of the stock price distribution density in the Heston model and the Hull–White model, respectively, are given by the following formulas:

$$A_3 = A_3(t,b,c) = \frac{3}{2} + \frac{\sqrt{8C+t}}{2\sqrt{t}} \quad \text{where } C = C(t,b,c) = \frac{t}{2c^2}\left(b^2 + \frac{4}{t^2}r^2_{\frac{tb}{2}}\right)$$

and

$$B_3 = B_3(t,b,c) = \frac{3}{2} + \frac{\sqrt{8G+t}}{2\sqrt{t}} \quad \text{where } G = G(t,q,\sigma) = \frac{t}{2\sigma^2}\left(q^2 + \frac{1}{t^2}r^2_{qt}\right),$$

where r_s denotes the smallest positive root of the entire function

$$z \mapsto z\cos z + s\sin z$$

(see Lemmas 5.35, 6.14, 5.42, and 6.18). Note that A_3 does not depend on a, while B_3 does not depend on m. We have $A_3 > 2$ and $B_3 > 2$. Therefore, the tail of the stock price distribution in the Hull–White model is heavier than the corresponding tail in the Heston and the Stein–Stein model.

6.7 The Constants A_3 and B_3

The following equalities characterize the long-time and the short-time behavior of the constant A_3:

$$\lim_{t\to 0} A_3(t,b,c) = \infty; \qquad \lim_{t\to\infty} A_3(t,b,c) = \infty \quad \text{for } b > 0;$$

$$\lim_{t\to\infty} A_3(t,0,c) = 2.$$

For the constant B_3, we have

$$\lim_{t\to 0} B_3(t,q,\sigma) = \infty; \qquad \lim_{t\to\infty} B_3(t,q,\sigma) = \infty.$$

We can also compute the limits of A_3 and B_3 with respect to the model parameters. This gives

$$\lim_{b\to 0} A_3(t,b,c) = \frac{3}{2} + \frac{\sqrt{4\pi^2 + c^2t^2}}{2ct} > 2, \qquad \lim_{b\to -\infty} A_3(t,b,c) = \infty,$$

and

$$\lim_{c\to 0} A_3(t,b,c) = \infty, \qquad \lim_{c\to\infty} A_3(t,b,c) = 2$$

for the constant A_3. Similar formulas hold for the constant B_3. More precisely,

$$\lim_{q\to 0} B_3(t,q,\sigma) = \frac{3}{2} + \frac{\sqrt{\pi^2 + \sigma^2t^2}}{2\sigma t} > 2, \qquad \lim_{q\to\infty} B_3(t,q,\sigma) = \infty$$

and

$$\lim_{\sigma\to 0} B_3(t,q,\sigma) = \infty, \qquad \lim_{\sigma\to\infty} B_3(t,q,\sigma) = 2.$$

6.8 Notes and References

- The asymptotic behavior of the distribution density of the log-price in the Heston model was characterized in log scale in the paper [DY02] of A.A. Drăgulescu and V.M. Yakovenko.
- Moment explosions in stochastic models were studied in [AP07, LM07, GK10, FK-R10, K-R11].

- The saddle point method is explained in [dBru81, Mur84, BH95, Mil06, FS09].
- The constants A_1, A_2, and A_3 are represented by different expressions in two alternative proofs of formula (6.2) for the uncorrelated Heston model. The reconciliation of the alternative formulas for the constants A_2 and A_3 was performed originally in [FGGS11]. For the constant A_1, this reconciliation is done in Sect. 6.3 of this book for the first time.

Chapter 7
Regularly Varying Functions and Pareto-Type Distributions

The theory of regular variation was developed by J. Karamata. Regularly varying functions have a deceptively simple structure. A function defined on $(0, \infty)$ is called regularly varying if it can be represented as the product of a power function and a function that behaves like the logarithmic function. However, in spite of their apparent simplicity, regularly varying functions possess numerous interesting properties. We refer the reader to the monograph [BGT87] by N.H. Bingham, C.M. Goldy, and J.L. Teugels, which is an encyclopedic source of information on regularly varying functions.

In this chapter, we gather selected definitions and facts from the theory of regular variation (see Sects. 7.1–7.5). They will be used in the rest of the book. Section 7.6 concerns Pareto-type distributions of statistics, which are distributions with regularly varying tails. We also introduce a new notion of weak Pareto-type distributions (see Definition 7.26). Finally, in Sect. 7.7 we show that distributions of the stock price in the Hull–White, Stein–Stein, and Heston models are of Pareto type.

7.1 Regularly Varying Functions

In the present book, we consider only positive regularly varying functions.

Definition 7.1 Let f be a positive measurable function on $[a, \infty)$ with $a > 0$, and let α be a real number. It is said that the function f is regularly varying at infinity with index α, if for any $\lambda > 0$,

$$\lim_{x \to \infty} \frac{f(\lambda x)}{f(x)} = \lambda^{\alpha}.$$

The class of all regularly varying functions with index α is denoted by R_{α}. A function L from the class R_0 is called slowly varying. A slowly varying func-

A. Gulisashvili, *Analytically Tractable Stochastic Stock Price Models*,
Springer Finance, DOI 10.1007/978-3-642-31214-4_7,
© Springer-Verlag Berlin Heidelberg 2012

tion L satisfies the following condition: for all $\lambda > 0$,

$$\lim_{x\to\infty} \frac{L(\lambda x)}{L(x)} = 1. \tag{7.1}$$

It is clear that $f \in R_\alpha$ if and only if $f(x) = x^\alpha L(x)$ where the function L is slowly varying. Simple examples of slowly varying functions are: $(\log x)^\alpha$, $(\log\log x)^\alpha$, $(\log\log\log x)^\alpha$, etc., where $\alpha \in \mathbb{R}$. The function $\exp\{A(\log x)^\alpha\}$ with $A \in \mathbb{R}$ belongs to the class R_0 if and only if $\alpha < 1$.

More examples of functions from the class R_α can be obtained using the so-called proximate orders. The next definition is taken from [Lev64].

Definition 7.2 A differentiable function ρ on (a, ∞) is called a proximate order if there exists $\alpha = \alpha_\rho \in \mathbb{R}$ such that

$$\lim_{x\to\infty} \rho(x) = \alpha \quad \text{and} \quad \lim_{x\to\infty} (x \log x)\rho'(x) = 0. \tag{7.2}$$

We will next prove that for the function $f(x) = x^{\rho(x)}$ where ρ is a proximate order, we have $f \in R_\alpha$. Indeed,

$$\frac{f(\lambda x)}{f(x)} = \exp\{[\rho(\lambda x) - \rho(x)]\log x + \rho(\lambda x)\log\lambda\}. \tag{7.3}$$

It is not hard to see that the second condition in (7.2) gives

$$\log x[\rho(\lambda x) - \rho(x)] = \log x \int_x^{\lambda x} \rho'(u)\,du \to 0 \tag{7.4}$$

as $x \to \infty$. Now, taking into account (7.3), (7.4) and the first condition in (7.2), we obtain $f \in R_\alpha$.

A function from the class R_0 grows slower than any power x^α with $\alpha > 0$. Conversely, a function, growing slower than any power can be majorized by a slowly varying function. The previous statement can be derived from the following theorem.

Theorem 7.3 *Let f be a measurable positive function on $[1, \infty)$, and suppose for every $\alpha > 0$, $f(x) = o(x^\alpha)$ as $x \to \infty$. Then there exists a slowly varying function l such that $f(x) = o(l(x))$ as $x \to \infty$.*

Proof With no loss of generality, we can assume that $f(x) \geq 1$ for all $x > 1$. It follows from the conditions in Theorem 7.3 that there exists a positive sequence $a_k \uparrow \infty$ as $k \to \infty$ such that $a_1 = 1$ and $kf(x) \leq x^{\frac{1}{k}}$ for all $x \geq a_k$. Put

$$l(x) = \exp\left\{\int_1^x \frac{\varepsilon(t)}{t}\,dt\right\} \tag{7.5}$$

where $\varepsilon(x) = \frac{1}{k}$ for $a_k \leq x < a_{k+1}$, $k \geq 1$. Then it is easy to see that $l \in R_0$.

Let $m \geq 1$ and $x \geq a_m$. Denote by n the integer for which $a_n \leq x < a_{n+1}$. Then

$$mf(x) \leq nf(x) \leq x^{\frac{1}{n}} \leq x^{\varepsilon(x)} \leq l(x). \tag{7.6}$$

In the proof of (7.6), we use (7.5), the definition of a_n, and the fact that the function ε is decreasing. It follows that $f(x) \leq \frac{1}{m} l(x)$ for all $x \geq a_m$, and hence $f(x) = o(l(x))$ as $x \to \infty$. This completes the proof of Theorem 7.3. □

We will next prove two fundamental theorems of the theory of regularly varying functions.

Theorem 7.4 *Let L be a slowly varying function on $[a, \infty)$. Then for every b with $a < b < \infty$,*

$$\lim_{x \to \infty} \sup_{\lambda \in [a,b]} \left| \frac{L(\lambda x)}{L(x)} - 1 \right| = 0. \tag{7.7}$$

Theorem 7.5 *Let L be a slowly varying function on $[a, \infty)$. Then there exist a number $b > 0$, a bounded measurable function η on $[b, \infty)$ with*

$$\lim_{x \to \infty} \eta(x) = c \in \mathbb{R},$$

and a continuous function ε on $[b, \infty)$ with $\lim_{x \to \infty} \varepsilon(x) = 0$ such that

$$L(x) = \exp\left\{\eta(x) + \int_b^x \frac{\varepsilon(t)}{t} dt\right\}, \quad x > b.$$

Theorems 7.4 and 7.5 are called the uniform convergence theorem and the representation theorem for slowly varying functions, respectively.

Proof of Theorem 7.4 In the proof, the symbol l will stand for the Lebesgue measure on the real line. Let us put

$$g(x) = \log L(e^x), \quad x \in [\log a, \infty). \tag{7.8}$$

Then condition (7.1) is equivalent to the following:

$$\lim_{x \to \infty} \left| g(x+h) - g(x) \right| = 0, \quad h \in \mathbb{R}. \tag{7.9}$$

Similarly, condition (7.7) can be reformulated as follows. For every bounded interval $I \subset \mathbb{R}$,

$$\lim_{x \to \infty} \sup_{h \in I} \left| g(x+h) - g(x) \right| = 0. \tag{7.10}$$

It is not hard to see that with no loss of generality, we can assume that $I = [0, 1]$. We will prove (7.10) reasoning by contradiction. Suppose (7.10) does not hold. Then there exist $\delta > 0$, a sequence $x_k \uparrow \infty$, and a sequence $h_k \in [0, 1]$ such that

$$\left| g(x_k + h_k) - g(x_k) \right| > \delta \tag{7.11}$$

for all $k \geq 1$. For every integer $m \geq 1$ define measurable sets by

$$U_m = \left\{ h \in [0,2] : \left| g(x_k + h) - g(x_k) \right| < \frac{\delta}{2}, k \geq m \right\}$$

and

$$V_m = \left\{ h \in [0,2] : \left| g(x_k + h_k + h) - g(x_k + h_k) \right| < \frac{\delta}{2}, k \geq m \right\}.$$

Then (7.9) implies that $U_m, V_m \uparrow [0,2]$ as $m \to \infty$, and hence $l(U_p) > \frac{3}{2}, l(V_p) > \frac{3}{2}$ for any sufficiently large integer p. Set $\widetilde{V}_p = V_p + h_p$. Since $U_p \subset [0,3]$, $\widetilde{V}_p \subset [0,3]$, $l(U_p) = \frac{3}{2}$, and $l(\widetilde{V}_p) = \frac{3}{2}$, the sets U_p and $\widetilde{V}_p$ have a nonempty intersection. Hence, there exists $h \in U_p$ such that $h - h_p \in V_p$. This means that

$$\left| g(x_p + h) - g(x_p) \right| < \frac{\delta}{2} \quad \text{and} \quad \left| g(x_p + h) - g(x_p + h_p) \right| < \frac{\delta}{2}. \tag{7.12}$$

It follows from (7.12) that $|g(x_p + h_p) - g(x_p)| < \delta$ which contradicts (7.11).

The proof of Theorem 7.4 is thus completed. □

Proof of Theorem 7.5 We can reformulate the conclusion in Theorem 7.5 in terms of the function g in (7.8) as follows. There exist a number $b_1 > \log a$, a bounded measurable function η_1 on $[b_1, \infty)$ with

$$\lim_{x \to \infty} \eta_1(x) = c_1 \in \mathbb{R},$$

and a continuous function ε_1 on $[b_1, \infty)$ with $\lim_{x \to \infty} \varepsilon_1(x) = 0$ such that

$$g(x) = \eta_1(x) + \int_{b_1}^{x} \varepsilon_1(t)\, dt, \quad x > b_1. \tag{7.13}$$

It will be shown next that the function g admits such a representation. First note that g is locally bounded on the interval $[b_1, \infty)$, if b_1 is a sufficiently large number. Indeed, there exists b_1 such that for all $x > b_1$ and $0 \leq h \leq 1$, we have $|g(x + h) - g(x)| < 1$ (this follows from Theorem 7.4). Hence, for every $k \geq 1$, $|g(x)| \leq |g(b_1)| + k$, $x \in [b_1, b_1 + k]$. This proves the local boundedness of g. It is easy to see that if b_1 is large enough and $x > b_1$, then we have

$$g(x) = \int_{x}^{x+1} \big(g(x) - g(t)\big)\, dt + \int_{b_1}^{x} \big(g(t+1) - g(t)\big)\, dt + \int_{b_1}^{b_1+1} g(t)\, dt. \tag{7.14}$$

Put $\varepsilon_1(t) = g(t+1) - g(t)$ and

$$\eta_1(t) = \int_{x}^{x+1} \big(g(x) - g(t)\big)\, dt + \int_{b_1}^{b_1+1} g(t)\, dt.$$

Then Theorem 7.4 and the local boundedness of g on $[b_1,\infty)$ imply the local boundedness of η_1 on $[b_1,\infty)$. Moreover,

$$\lim_{x\to\infty}\eta_1(x)=\int_{b_1}^{b_1+1}g(t)\,dt.$$

Therefore, η_1 is a bounded measurable function on $[b_1,\infty)$. Similarly, ε_1 is a bounded measurable function satisfying $\lim_{x\to\infty}\varepsilon_1(x)=0$. Since

$$g(x)=\eta_1(x)+\int_{b_1}^{x}\varepsilon_1(t)\,dt, \tag{7.15}$$

we can represent the function g by the formula in (7.13). However, it is not clear whether the function ε_1 is continuous on $[b_1,\infty)$. We will next explain how to modify the proof in order to include the continuity condition. Note that if the function g is continuous on $[b_1,\infty)$, then so is the function ε_1.

Put

$$g^*(x)=\int_{b_1}^{x}\big(g(t+1)-g(t)\big)\,dt=\int_{b_1}^{x}\varepsilon_1(t)\,dt.$$

It is clear that g^* is a continuous function on $[b_1,\infty)$. Moreover, for all h,

$$g^*(x+h)-g^*(x)\to 0 \quad \text{as } x\to\infty. \tag{7.16}$$

Indeed, it is not hard to see that if $h>0$, then

$$g^*(x+h)-g^*(x)=\int_0^h\big(g(x+t+1)-g(x)\big)\,dt-\int_0^h\big(g(x+t)-g(x)\big)\,dt.$$

It follows from Theorem 7.4 that condition (7.16) holds for $h>0$. The case where $h<0$ can be treated similarly. This establishes (7.16).

Since g^* is continuous and condition (7.16) holds, we can apply the special case of Theorem 7.4 that has already been established. We have

$$g^*(x)=\eta_1^*(x)+\int_{b_1}^{x}\varepsilon_1^*(t)\,dt.$$

It follows that there exists a representation (7.15) of the function g with $\eta_1=g-g^*+\eta_1^*$ and $\varepsilon_1=\varepsilon_1^*$. It remains to prove that the function η_1 satisfies the required conditions. This can be seen from the following formula:

$$g(x)-g^*(x)=\int_x^{x+1}\big(g(x)-g(t)\big)\,dt+\int_{b_1}^{b_1+1}g(t)\,dt$$

(see (7.14)). Indeed, Theorem 7.4 shows that the function $g-g^*$ is locally bounded and moreover

$$\lim_{x\to\infty}\big(g(x)-g^*(x)\big)=\int_{b_1}^{b_1+1}g(t)\,dt.$$

This completes the proof of Theorem 7.5. □

We will next formulate Karamata's theorem concerning the asymptotic behavior of integrals of regularly varying functions (see [BGT87], Theorems 1.5.11 and 1.6.1).

Theorem 7.6 *The following are true:*

1. *Let $f \in R_\alpha$ with $\alpha < -1$. Then*

$$\lim_{x\to\infty} \frac{xf(x)}{\int_x^\infty f(u)\,du} = -(\alpha+1).$$

2. *Let f be a positive locally integrable function on $[x_0,\infty)$ such that*

$$\lim_{x\to\infty} \frac{xf(x)}{\int_x^\infty f(u)\,du} = \beta$$

where $\beta > 0$. Then $f \in R_{-\beta-1}$.

7.2 Class R_{-1} and Regularly Varying Majorants of Integrable Monotone Functions

Let f be a nonnegative nonincreasing Lebesgue integrable function on $(0,\infty)$. Then there exists a constant c such that

$$f(x) \le \frac{c}{x} \quad \text{for all } x > 0.$$

Indeed, we have

$$f(x) \le \frac{1}{x}\int_0^x f(u)\,du \le \frac{c}{x}$$

where $c = \int_0^\infty f(u)\,du$. It follows that the function f has a majorant from the class R_{-1}. However, this majorant is not an integrable function on $(0,\infty)$.

An interesting question whether any function f such as above has an integrable majorant from the class R_{-1} was answered affirmatively in [Den06].

Theorem 7.7 *Let f be a nonnegative nonincreasing function on $(0,\infty)$ from the space $L^1(0,\infty)$. Then there exists a nonincreasing continuous function h on $(0,\infty)$ such that*

1. *$f(x) \le h(x)$ for all $x \in (0,\infty)$.*
2. *$h \in L^1(0,\infty)$.*
3. *$h \in R_{-1}$.*

We will first derive a useful corollary from Theorem 7.7 and then prove the theorem.

Corollary 7.8 *Let f be a nonnegative nonincreasing function on $[0,\infty)$ such that $f \in L^1(0,\infty)$. Then there exists a nonincreasing slowly varying function l such that $l(x) \to 0$ as $x \to \infty$ and*

$$\int_x^\infty f(u)\,du \le l(x)$$

for all $x \ge 0$.

Proof It is known (see [BGT87], Proposition 1.5.9b) that for a function from the class R_{-1}, part 1 of Karamata's theorem (Theorem 7.6) is valid in the following form. □

Theorem 7.9 *Let $f \in R_{-1}$ and suppose*

$$\int_{x_0}^\infty f(u)\,du < \infty$$

for some $x_0 \ge 0$. Then the function $x \to \int_x^\infty f(u)\,du$ is slowly varying and

$$\lim_{x\to\infty} \frac{\int_x^\infty f(u)\,du}{xf(x)} = \infty.$$

To prove Corollary 7.8, we first apply Theorem 7.7 to majorize the function f by a function h, satisfying the conditions in Theorem 7.7, then observe that

$$\int_x^\infty f(u)\,du \le \int_x^\infty h(u)\,du, \quad x \ge 0,$$

and finally take into account that the function l defined by

$$l(x) = \int_x^\infty h(u)\,du$$

is slowly varying (see Theorem 7.9).

Proof of Theorem 7.7 We will assume with no loss of generality that the function f is bounded and strictly positive, and construct a majorant h on the interval $[a,\infty)$ for some $a > 0$. Our first goal is to replace the function f by an appropriate piecewise constant function. Let us define a function r on $(0,\infty)$ by $r(x) = y_n$ for $x_{n-1} \le x \le x_n$ where $x_0 = 0$, $x_n = 2^n$, and $y_n = f(x_{n-1})$ for all $n \ge 1$. Then the function r is piecewise constant and is an integrable majorant of f. The integrability of r follows from

$$\int_0^\infty r(x)\,dx = \sum_{n=1}^\infty y_n(x_n - x_{n-1}) = 2f(0) + \sum_{n=2}^\infty f(x_{n-1})2^{n-1}$$

$$\leq 2f(0) + 2\sum_{n=2}^{\infty} \int_{x_{n-2}}^{x_{n-1}} f(u)\,du < \infty.$$

It is also true that

$$\sum_{n=1}^{\infty} y_n x_n \leq y_1 x_1 + 4\sum_{n=2}^{\infty} \int_{x_{n-2}}^{x_{n-1}} f(u)\,du < \infty.$$

Therefore, we can choose a positive sequence $\{\varepsilon_n\}$ such that $\varepsilon_n \downarrow 0$ and

$$\sum_{n=1}^{\infty} \varepsilon_n^{-1} y_n x_n < \infty. \tag{7.17}$$

Next, another auxiliary function will be constructed. This function is a majorant of r and simultaneously a piecewise defined function, where the pieces are constant multiples of powers of the function x^{-1}. Let $m_0 = 0$, $m_1 = 1$, and for $k \geq 1$, set

$$m_{k+1} = \sup\left\{ n > m_k : r(x) \leq y_n \left(\frac{x_n}{x}\right)^{1-\varepsilon_n} \text{ for all } x_{m_k} \leq x < x_n \right\}. \tag{7.18}$$

Then it is not hard to see that

$$m_{k+1} = \sup\left\{ n > m_k : y_i \leq y_n \left(\frac{x_n}{x_i}\right)^{1-\varepsilon_n} \text{ for all } m_k < i \leq n \right\}.$$

The integer m_{k+1} is finite. Indeed, suppose $m_{k+1} = \infty$. Then there exists a sequence $n_j \to \infty$ such that

$$y_i \leq y_{n_j} \left(\frac{x_{n_j}}{x_i}\right)^{1-\varepsilon_{n_j}} \quad \text{for all } m_k < i \leq n_j.$$

Therefore,

$$y_{m_k} x_{m_k}^{1-\varepsilon_{n_j}} \leq y_{n_j} x_{n_j}^{1-\varepsilon_{n_j}} \leq y_{n_j} x_{n_j} \to 0$$

as $j \to \infty$, which leads to a contradiction since $y_{m_k} > 0$, $x_{m_k} > 0$, and $\varepsilon_{n_j} \to 0$ as $j \to \infty$.

The auxiliary function g is defined as follows:

$$g(x) = \begin{cases} y_{m_{k+1}} (\frac{x_{m_{k+1}}}{x})^{1-\varepsilon_{m_{k+1}}} & \text{if } x_{m_k} \leq x < x_{m_{k+1}}, \\ \max(y_1, 1 + y_2 (\frac{x_{m_2}}{x_{m_1}})^{1-\varepsilon_{m_2}}) & \text{if } 0 \leq x < x_{m_1}. \end{cases} \tag{7.19}$$

Using the definitions of r and g and (7.18), we see that $r(x) \leq g(x)$ for all $x \geq 0$. Moreover g is a nonincreasing function. Indeed, the definitions imply that g does not increase on the interval $[0, x_{m_2})$. To prove that g does not increase on $[0, \infty)$, it suffices to show that

$$y_{m_k} > y_{m_{k+1}} \left(\frac{x_{m_{k+1}}}{x_{m_k}} \right)^{1-\varepsilon_{m_{k+1}}} \tag{7.20}$$

for all $k \geq 1$.

In order to establish (7.20), we will reason by contradiction. Let us suppose that the inequality

$$y_{m_k} \leq y_{m_{k+1}} \left(\frac{x_{m_{k+1}}}{x_{m_k}} \right)^{1-\varepsilon_{m_{k+1}}} \tag{7.21}$$

holds for some $k \geq 2$. Then using (7.21) and taking i with $m_{k-1} < i \leq m_k$, we get the following estimates:

$$\begin{aligned} y_i &\leq y_{m_k} \left(\frac{x_{m_k}}{x_i} \right)^{1-\varepsilon_{m_k}} \leq y_{m_{k+1}} \left(\frac{x_{m_{k+1}}}{x_{m_k}} \right)^{1-\varepsilon_{m_{k+1}}} \left(\frac{x_{m_k}}{x_i} \right)^{1-\varepsilon_{m_{k+1}}} \\ &= y_{m_{k+1}} \left(\frac{x_{m_{k+1}}}{x_i} \right)^{1-\varepsilon_{m_{k+1}}}. \end{aligned}$$

It follows that the inequality

$$y_i \leq y_{m_{k+1}} \left(\frac{x_{m_{k+1}}}{x_i} \right)^{1-\varepsilon_{m_{k+1}}}$$

holds for all i with $m_{k-1} < i \leq m_{k+1}$, which contradicts our choice of m_k. It is clear that the previous reasoning establishes the validity of (7.20). Therefore, the function g is monotonically decreasing. Moreover, it is not hard to see that g is right-continuous on the interval $[x_{m_1}, \infty)$.

For the sake of simplicity, put $\tilde{x}_k = x_{m_k}$, $\tilde{y}_k = y_{m_k}$, and $\tilde{\varepsilon}_k = \varepsilon_{m_k}$. The following estimates prove the integrability of the function g:

$$\begin{aligned} \int_0^\infty g(x)\,dx &= \sum_0^\infty \int_{\tilde{x}_k}^{\tilde{x}_{k+1}} g(x)\,dx = 2g(0) + \sum_{k+1}^\infty \int_{\tilde{x}_k}^{\tilde{x}_{k+1}} \tilde{y}_{k+1} \left(\frac{\tilde{x}_{k+1}}{x} \right)^{1-\tilde{\varepsilon}_{k+1}} dx \\ &\leq 2g(0) + \sum_{k+1}^\infty \int_0^{\tilde{x}_{k+1}} \tilde{y}_{k+1} \left(\frac{\tilde{x}_{k+1}}{x} \right)^{1-\tilde{\varepsilon}_{k+1}} dx \\ &= 2g(0) + \sum_{k=1}^\infty \frac{\tilde{y}_{k+1}\tilde{x}_{k+1}}{\tilde{\varepsilon}_{k+1}} < \infty. \end{aligned} \tag{7.22}$$

In the proof of (7.22), we use (7.17).

Our next goal is to construct a majorant h satisfying the conditions in Theorem 7.7. We will first define a sequence of integers $\{\tilde{m}_i\}_{i \geq 1}$ step by step. For the sake of simplicity, the following notation will be used in the sequel: $\hat{x}_k = \tilde{x}_{\tilde{m}_k}$, $\hat{y}_k = \tilde{y}_{\tilde{m}_k}$, and $\hat{\varepsilon}_k = \tilde{\varepsilon}_{\tilde{m}_k}$.

Put $\tilde{m}_1 = m_1 = 1$ and

$$z_1 = \min\left\{x > \hat{x}_1 : \hat{y}_1\left(\frac{\hat{x}_1}{x}\right)^{1+\hat{\varepsilon}_1} = g(x)\right\}.$$

Then z_1 may be finite, or infinite. For $z_1 = \infty$, we have

$$g(x) < \hat{y}_1\left(\frac{\hat{x}_1}{x}\right)^{1+\hat{\varepsilon}_1}, \quad x > \hat{x}_1. \tag{7.23}$$

In order to establish the inequality in (7.23), we observe that

$$g(\hat{x}_1) < \eta(\hat{x}_1) \tag{7.24}$$

where

$$\eta(x) = \hat{y}_1\left(\frac{\hat{x}_1}{x}\right)^{1+\hat{\varepsilon}_1}.$$

Inequality (7.24) follows from (7.20) with $k = 1$. Note that the function η is continuous, while the function g is right-continuous. Now it is not hard to see that (7.23) holds.

Under the assumption $z_1 = \infty$, set

$$h(x) = \frac{c}{x \log^2 x}, \quad x > \hat{x}_1,$$

where $c > 0$ is an appropriate constant. Then h is a nonincreasing continuous integrable majorant of the function f on the interval $[\hat{x}_1, \infty)$. In addition, we have $h \in R_{-1}$.

If $z_1 < \infty$, then we define the number $\tilde{m}_2$ by $\tilde{m}_2 = \min\{n : \tilde{x}_n > z_1\}$ and set

$$h(x) = \begin{cases} \hat{y}_1(\frac{\hat{x}_1}{x})^{1+\hat{\varepsilon}_1} & \text{if } \hat{x}_1 \le x < z_1, \\ g(x) & \text{if } z_1 \le x < \hat{x}_2. \end{cases}$$

Let us proceed by induction. Suppose the number $\tilde{m}_i$ is defined for all i with $1 \le i \le k$, and the function $h(x)$ is defined for all $x \in [\hat{x}_1, \hat{x}_k)$. Put

$$z_k = \min\left\{x > \hat{x}_k : \hat{y}_k\left(\frac{\hat{x}_k}{x}\right)^{1+\hat{\varepsilon}_k} = g(x)\right\}. \tag{7.25}$$

If $z_k = \infty$, then we can finish the proof exactly as in the case where $z_1 = \infty$. If $z_k < \infty$, then we set

$$\tilde{m}_{k+1} = \min\{n : \tilde{x}_n > z_k\}, \tag{7.26}$$

and define the function h on the interval $[\hat{x}_k, \hat{x}_{k+1})$ by

$$h(x) = \begin{cases} \hat{y}_k (\frac{\hat{x}_k}{x})^{1+\hat{\varepsilon}_k} & \text{if } \hat{x}_k \le x < z_k, \\ g(x) & \text{if } z_k \le x < \hat{x}_{k+1}. \end{cases}$$

It follows from (7.26) that

$$z_k \ge \tilde{x}_{\tilde{m}_{k+1}-1} \tag{7.27}$$

for all $k \ge 1$. Note that the right-continuity of the function g and formula (7.25) imply the following equality:

$$\hat{y}_k \left(\frac{\hat{x}_k}{z_k} \right)^{1+\hat{\varepsilon}_k} = \hat{y}_{k+1} \left(\frac{\hat{x}_{k+1}}{z_k} \right)^{1-\hat{\varepsilon}_{k+1}} \tag{7.28}$$

for all $k \ge 1$.

We will next prove the continuity of the function h on the interval $[\hat{x}_1, \infty)$. It suffices to show that h is continuous at the points z_k and $\hat{x}_{k+1}$ for all $k \ge 1$. It follows from the condition

$$\lim_{x \to z_k} h(x) = g(z_k) = \hat{y}_k \left(\frac{\hat{x}_k}{z_k} \right)^{1+\hat{\varepsilon}_k}$$

that the function h is continuous at z_k for all $k \ge 1$. Moreover, the function h is continuous at $\hat{x}_{k+1}$ for all $k \ge 1$, since

$$h(\hat{x}_{k+1}) = \lim_{x \to \hat{x}_{k+1}} g(x) = \hat{y}_{k+1}.$$

It is also clear that h is a nonincreasing function on $[\hat{x}_1, \infty)$.

We have

$$\begin{aligned} \int_{\hat{x}_1}^{\infty} h(x)\, dx &= \sum_{k=1}^{\infty} \int_{\hat{x}_k}^{\hat{x}_{k+1}} h(x)\, dx \\ &= \sum_{k=1}^{\infty} \int_{\hat{x}_k}^{z_k} \hat{y}_k \left(\frac{\hat{x}_k}{x} \right)^{1+\hat{\varepsilon}_k} dx + \sum_{k=1}^{\infty} \int_{z_k}^{\hat{x}_{k+1}} g(x)\, dx \\ &\le \int_{\hat{x}_1}^{\infty} g(x)\, dx + \sum_{k=1}^{\infty} \int_{\hat{x}_k}^{\infty} \hat{y}_k \left(\frac{\hat{x}_k}{x} \right)^{1+\hat{\varepsilon}_k} dx \\ &= \int_{\hat{x}_1}^{\infty} g(x)\, dx + \sum_{k=1}^{\infty} \frac{\hat{y}_k \hat{x}_k}{\hat{\varepsilon}_k} < \infty. \end{aligned} \tag{7.29}$$

In the proof of (7.29), we use (7.17).

It remains to show that $h \in R_{-1}$. Let us define a function on $[\hat{x}_1, \infty)$ by the following formula:

$$\tilde{h}(x) = \frac{\hat{x}_1 \hat{y}_1}{x} \exp\left\{ - \int_2^x \frac{\varepsilon(y)}{y} dy \right\},$$

where

$$\varepsilon(y) = \begin{cases} \hat{\varepsilon}_k & \text{if } \hat{x}_k \le x < z_k, \\ -\hat{\varepsilon}_{k+1} & \text{if } z_k \le x < \hat{x}_{k+1}. \end{cases}$$

It is not hard to see that $\tilde{h} \in R_{-1}$.

Our next goal is to prove that the functions $\tilde{h}$ and h coincide. Indeed, for $x \in [\hat{x}_k, z_k)$, we have

$$\tilde{h}(x) = \frac{\hat{x}_1 \hat{y}_1}{x} \left(\frac{x}{\hat{x}_k} \right)^{-\hat{\varepsilon}_k} \prod_{j=1}^{k-1} \left(\frac{z_j}{\hat{x}_j} \right)^{-\hat{\varepsilon}_j} \left(\frac{\hat{x}_{j+1}}{z_j} \right)^{\hat{\varepsilon}_{j+1}}. \tag{7.30}$$

Similarly, for $x \in [z_k, \hat{x}_{k+1})$, we get

$$\tilde{h}(x) = \frac{\hat{x}_1 \hat{y}_1}{x} \left(\frac{x}{\hat{x}_{k+1}} \right)^{\hat{\varepsilon}_{k+1}} \prod_{j=1}^{k} \left(\frac{z_j}{\hat{x}_j} \right)^{-\hat{\varepsilon}_j} \left(\frac{\hat{x}_{j+1}}{z_j} \right)^{\hat{\varepsilon}_{j+1}}. \tag{7.31}$$

Now using (7.30), (7.31), and (7.28), we see that

$$\tilde{h}(x) = \begin{cases} \hat{y}_k (\frac{\hat{x}_k}{x})^{1+\hat{\varepsilon}_k} & \text{if } \hat{x}_k \le x < z_k, \\ \hat{y}_{k+1} (\frac{\hat{x}_{k+1}}{x})^{1-\hat{\varepsilon}_{k+1}} & \text{if } z_k \le x < \hat{x}_{k+1}. \end{cases} \tag{7.32}$$

It follows from (7.32), (7.19), and (7.27) that $\tilde{h}(x) = h(x)$ for all $x \ge \hat{x}_1$.

The proof of Theorem 7.7 is thus completed. □

7.3 Fractional Integrals of Regularly Varying Functions

In this section, we prove an asymptotic formula for the fractional integral F_α of a regularly varying function. This formula can be found in [Sen76]. Recall that the operator F_α is defined by (5.20).

Theorem 7.10 *Let $\alpha > 0$ and $\beta < 0$ be such that $\alpha + \beta < 0$, and suppose g is a function on $[c, \infty)$ satisfying the following condition:*

$$g(\sigma) = f(\sigma)\big(1 + O(\rho(\sigma))\big) \quad \text{as } \sigma \to \infty,$$

where $f \in R_\beta$ and $\rho(\sigma) \downarrow 0$ as $\sigma \to \infty$. Then

$$F_\alpha g(\sigma) \sim \frac{\Gamma(|\alpha + \beta|)}{\Gamma(|\beta|)} \sigma^\alpha g(\sigma) \sim \frac{\Gamma(|\alpha + \beta|)}{\Gamma(|\beta|)} \sigma^\alpha f(\sigma) \tag{7.33}$$

as $\sigma \to \infty$.

Proof We have

$$F_\alpha g(\sigma) = F_\alpha f(\sigma)\big(1 + O\big(\rho(\sigma)\big)\big), \quad \sigma \to \infty. \tag{7.34}$$

Moreover

$$F_\alpha f(\sigma) = \frac{\sigma^\alpha}{\Gamma(\alpha)} \int_1^\infty (y-1)^{\alpha-1} f(\sigma y)\,dy.$$

It follows from $f(y) = y^\beta L(y)$ with $L \in R_0$ that

$$F_\alpha f(\sigma) = \frac{\sigma^\alpha f(\sigma)}{\Gamma(\alpha)} \int_1^\infty (y-1)^{\alpha-1} y^\beta \frac{L(\sigma y)}{L(\sigma)}\,dy. \tag{7.35}$$

Since for all $y \ge 1$, $\frac{L(\sigma y)}{L(\sigma)} \to 1$ as $\sigma \to \infty$, Theorem 7.5 gives

$$\begin{aligned} \sup_{\sigma > \sigma_0} \frac{L(\sigma y)}{L(\sigma)} &\le \exp\big\{\eta(\sigma y) - \eta(\sigma)\big\} \exp\left\{\int_\sigma^{\sigma y} \frac{\varepsilon(t)}{t}\,dt\right\} \\ &\le C \exp\left\{(\log y) \sup_{t \ge \sigma_0} \varepsilon(t)\right\} = C y^{\sup_{t \ge \sigma_0} \varepsilon(t)}. \end{aligned}$$

Using the fact that $\varepsilon(t) \to 0$ as $t \to \infty$, we obtain

$$\alpha + \beta + \sup_{t \ge \sigma_0} \varepsilon(t) < 0$$

for all sufficiently large σ_0. Next, applying the dominated convergence theorem to the integral in (7.35), we see that

$$F_\alpha f(\sigma) \sim \frac{\sigma^\alpha}{\Gamma(\alpha)} f(\sigma) \int_1^\infty (y-1)^{\alpha-1} y^\beta\,dy \tag{7.36}$$

as $\sigma \to \infty$.

Recall that we denoted by $B(r, s)$ the Beta function given by

$$B(r, s) = \int_0^1 (1-t)^{r-1} t^{s-1}\,dt$$

(see Definition 1.28). As has already been mentioned, the following formula holds:

$$B(r, s) = \frac{\Gamma(r)\Gamma(s)}{\Gamma(r+s)} \tag{7.37}$$

(see (1.39)). It is also true that

$$B(r, s) = \int_0^\infty \frac{u^{r-1}}{(1+u)^{r+s}}\,du \tag{7.38}$$

(see [SSh03], p. 175–176). Using (7.36) and (7.38), we obtain

$$F_\alpha f(\sigma) \sim \frac{B(\alpha, |\alpha + \beta|)}{\Gamma(\alpha)} \sigma^\alpha f(\sigma)$$

as $\sigma \to \infty$.

Now, it is clear that Theorem 7.10 follows from (7.34) and from (7.37) with $r = \alpha$ and $s = -\alpha - \beta$. □

7.4 Slowly Varying Functions with Remainder

In this section, we discuss slowly varying functions with remainder. Such functions appear in various asymptotic formulas with error estimates.

Definition 7.11 Let L be a slowly varying function, and let h be a positive measurable function on $(0, \infty)$ such that $h(x) \to 0$ as $x \to \infty$. Then the function L is called slowly varying with remainder h if for every $\lambda > 1$,

$$\frac{L(\lambda x)}{L(x)} - 1 = O\big(h(x)\big) \quad \text{as } x \to \infty.$$

For a fixed function h, we denote by $R_{0,h}$ the class of functions described in Definition 7.11, and let the symbol $R_{\beta,h}$ stand for the class of functions f satisfying the condition $f(x) = x^\beta L(x)$ with $L \in R_{0,h}$. Some examples of functions from the classes $R_{0,h}$ are as follows:

- $f(x) = (\log x)^\alpha$ with $\alpha \neq 0$. Here we have $h(x) = (\log x)^{-1}$.
- $f(x) = (\log\log x)^\alpha$ with $\alpha \neq 0$. Here we have $h(x) = (\log x \log\log x)^{-1}$.
- $f(x) = \exp\{\alpha(\log x)^\beta\}$ where $\alpha \neq 0$, $\beta \neq 0$, and $\beta < 1$. Here we can take $h(x) = (\log x)^{\beta-1}$.

Lemma 7.12 *Let $f_1 \in R_{0,g}$ and $f_2 \in R_{0,h}$. Then $f_1 + f_2 \in R_{0,g+h}$.*

Proof The following estimates imply Lemma 7.12:

$$\begin{aligned}\frac{f_1(\lambda x) + f_2(\lambda x)}{f_1(x) + f_2(x)} - 1 &= \frac{f_1(x) + f_2(x) + f_1(x)O(g(x)) + f_2(x)O(h(x))}{f_1(x) + f_2(x)} - 1 \\ &= \frac{f_1(x)O(g(x)) + f_2(x)O(h(x))}{f_1(x) + f_2(x)} \\ &= O\big(g(x) + h(x)\big)\end{aligned}$$

as $x \to \infty$. □

The next statement can be obtained from Lemma 7.12 by using the first and the third examples from the list above.

Corollary 7.13 *Let $a \in \mathbb{R}$ and $b > 0$. Then the function defined by*

$$h_{a,b}(x) = (\log x)^a e^{b\sqrt{\log x}}$$

belongs to the class $R_{0,g}$ with $g(x) = (\log x)^{-\frac{1}{2}}$.

The next assertion concerns the structure of slowly varying functions with remainder.

Theorem 7.14 *Let $L \in R_{0,h}$ and assume that the remainder h is a decreasing function. Then there exist constants C and $X > 0$ such that*

$$L(x) = \exp\left\{C + O\big(h(x)\big) + \int_X^x \frac{O(h(t))}{t}\, dt\right\} \tag{7.39}$$

as $x \to \infty$. In (7.39), the O functions are measurable.

The proof of Theorem 7.14 can be found in [GS87] and [BGT87].

We will next obtain a version of formula (7.33) for functions from the class $R_{\beta,h}$.

Theorem 7.15 *Let $\alpha > 0$ and $\beta < 0$ be such that $\alpha + \beta < 0$. Let g be a function on $[c, \infty)$ satisfying the condition*

$$g(\sigma) = f(\sigma)\big(1 + O\big(\rho(\sigma)\big)\big)$$

as $\sigma \to \infty$, where $f \in R_{\beta,h}$. Assume $\rho(\sigma) \downarrow 0$ and $h(\sigma) \downarrow 0$ as $\sigma \to \infty$. Then

$$F_\alpha g(\sigma) = \frac{\Gamma(|\alpha + \beta|)}{\Gamma(|\beta|)} \sigma^\alpha f(\sigma)\big[1 + O\big(\rho(\sigma)\big) + O\big(h(\sigma)\big)\big]$$

as $\sigma \to \infty$.

Proof The proof of Theorem 7.15 is similar to that of Theorem 7.10. Since $f(x) = x^\beta L(x)$, where L is a slowly varying function with remainder h, we can estimate the expression

$$I(\sigma) = F_\alpha f(\sigma) - \frac{B(\alpha, |\alpha + \beta|)}{\Gamma(\alpha)} \sigma^\alpha f(\sigma)$$

as follows. It is not hard to see using formulas (7.35) and (7.38) that

$$\begin{aligned} \big|I(\sigma)\big| &= \frac{\sigma^\alpha f(\sigma)}{\Gamma(\alpha)} \left| \int_1^\infty (y-1)^{\alpha-1} y^\beta \frac{L(\sigma y)}{L(\sigma)}\, dy - B\big(\alpha, |\alpha + \beta|\big) \right| \\ &\le \frac{\sigma^\alpha f(\sigma)}{\Gamma(\alpha)} \int_1^\infty (y-1)^{\alpha-1} y^\beta \left| \frac{L(\sigma y)}{L(\sigma)} - 1 \right| dy. \end{aligned} \tag{7.40}$$

To continue the proof of Theorem 7.15, we need the following result (see Proposition 2.5.1 in [GS87]):

Lemma 7.16 *Let $L \in R_{0,h}$, where h is a decreasing function, and let v be a positive function on $(1,\infty)$ such that there exists $\varepsilon > 0$ for which*

$$\int_1^\infty y^\varepsilon v(y)\,dy < \infty. \tag{7.41}$$

Then

$$\int_1^\infty v(y)\frac{L(\sigma y)}{L(\sigma)}\,dy = \int_1^\infty v(y)\,dy + O\big(h(\sigma)\big)$$

as $x \to \infty$.

Proof It suffices to prove that under the conditions in Lemma 7.16, the function

$$y \mapsto v(y)\frac{|L(\sigma y) - L(\sigma)|}{L(\sigma)h(\sigma)} \tag{7.42}$$

is dominated in a neighborhood of infinity by an integrable function. Using Theorem 7.14, we see that there exist constants $K > 0$ and $\sigma_0 > 0$ such that for all $\sigma > \sigma_0$ and $y > 1$,

$$\begin{aligned}\frac{|L(\sigma y) - L(\sigma)|}{L(\sigma)h(\sigma)} &\le \frac{1}{h(\sigma)}\big[\exp\big\{Kh(\sigma)(1+\log y)\big\} - 1\big]\\ &\le K(1+\log y)\exp\big\{Kh(\sigma)(1+\log y)\big\}\\ &\le K\exp\big\{Kh(\sigma_0)\big\}(1+\log y)y^{Kh(\sigma)}.\end{aligned} \tag{7.43}$$

In the proof of (7.43) we use the estimate $e^u - 1 \le ue^u$, $u > 0$. It follows from (7.43) that for every $\tau > 0$ there exists a constant $\sigma_\tau > \sigma_0$, depending on τ, and such that for all $\sigma > \sigma_\tau$ and $y > 1$,

$$\frac{|L(\sigma y) - L(\sigma)|}{L(\sigma)h(\sigma)} \le y^\tau. \tag{7.44}$$

Now we see from (7.44) that the function in (7.42) is dominated by the function $y \mapsto y^\tau v(y)$ uniformly with respect to $\sigma > \sigma_\tau$. Finally, using (7.41), we establish Lemma 7.16. □

Let us return to the proof of Theorem 7.15. Applying Lemma 7.16 with $v(y) = (y-1)^{\alpha-1}y^\beta$ and taking into account formula (7.40), we see that $I(\sigma) = O(h(\sigma))$ as $\sigma \to \infty$.

This completes the proof of Theorem 7.15. □

7.5 Smoothly Varying Functions

The class SR_α of smoothly varying functions with index α is an important subclass of the class R_α. Functions from the class SR_α are infinitely differentiable, and any regularly varying function is asymptotically equivalent to a smoothly varying function.

Definition 7.17 Let $\alpha \in \mathbb{R}$, and let f be a positive function defined on some neighborhood of infinity. The function f is called smoothly varying with index α if the function $h(x) = \log f(e^x)$ is infinitely differentiable and the following conditions hold:

1. $h'(x) \to \alpha$ as $x \to \infty$.
2. For any $n \geq 2$, $h^{(n)}(x) \to 0$ as $x \to \infty$.

An equivalent definition of the class SR_α is as follows:

$$f \in SR_\alpha \Leftrightarrow \lim_{x\to\infty} \frac{x^n f^{(n)}(x)}{f(x)} = \alpha(\alpha-1)\cdots(\alpha-n+1) \tag{7.45}$$

for all $n \geq 1$ (see [BGT87] for more information on smoothly varying functions).

Suppose $f \in SR_\alpha$ and put $h(x) = \log f(e^x)$. Then

$$\frac{f(\lambda x)}{f(x)} = \exp\{h(\log y + \log\lambda) - h(\log y)\}. \tag{7.46}$$

Next using (7.46), the mean value theorem, and the first condition in Definition 7.17, we see that $f \in R_\alpha$. This establishes the inclusion $SR_\alpha \subset R_\alpha$.

It is not hard to prove that a function from the class SR_α with $\alpha > 0$ is eventually increasing and tends to infinity, while a function from the class SR_α with $\alpha < 0$ is eventually decreasing and tends to zero. The next lemma shows how the differentiation operation acts on the classes of smoothly varying functions.

Lemma 7.18 *If $f \in SR_\alpha$ with $\alpha > 0$, then $f' \in SR_{\alpha-1}$. If $f \in SR_\alpha$ with $\alpha < 0$, then $|f'| \in SR_{\alpha-1}$.*

Proof It is clear that Lemma 7.18 follows from (7.45) and the equality

$$\frac{x^n f^{(n+1)}(x)}{f'(x)} = \frac{x^{n+1} f^{(n+1)}(x)}{f(x)} \frac{f(x)}{xf'(x)}.$$

□

The next definition introduces perfect proximate orders (see [Tar04]).

Definition 7.19 A proximate order ρ (see Definition 7.2)) is called perfect if

$$\lim_{x\to\infty} x^n \log x \rho^{(n)}(x) = 0$$

for all $n \geq 1$.

Let f be a positive function on (a,∞). Then it is clear that $f(x)=x^{\rho(x)}$ where

$$\rho(x)=\frac{\log f(x)}{\log x}. \tag{7.47}$$

Tarov obtained the following characterization of a smoothly varying function f in terms of the function ρ (see [Tar04]).

Theorem 7.20 *A positive function f belongs to the class SR_α if and only if the function ρ defined by* (7.47) *is a perfect proximate order with $\rho(x)\to\alpha$ as $x\to\infty$.*

The next assertion is called the smooth variation theorem (see [BGT87], Theorem 1.8.2).

Theorem 7.21 *Suppose $f\in R_\alpha$. Then there exist $f_1\in SR_\alpha$ and $f_2\in SR_\alpha$ such that $f_1(x)\le f(x)\le f_2(x)$ and $f_1(x)\sim f_2(x)$ as $x\to\infty$. In particular, for every $f\in R_\alpha$ there exists $g\in SR_\alpha$ such that $f(x)\sim g(x)$ as $x\to\infty$.*

Theorem 7.21 states that any regularly varying function can be squeezed between two asymptotically equivalent smoothly varying functions with the same index of regular variation.

Regularly varying functions appear in numerous asymptotic formulas. The following statement, often called Bingham's lemma, contains such a formula.

Theorem 7.22 *Let $\alpha>0$ and $f\in R_\alpha$. Then*

$$-\log\int_x^\infty e^{-f(y)}\,dy\sim f(x)\quad \text{as } x\to\infty. \tag{7.48}$$

Proof The proof of Theorem 7.22 given below is analogous to the proof of Theorem 4.12.10(i) in [BGT87]. However, for the functions from the class SR_α, we obtain sharper asymptotic formulas with error estimates.

Lemma 7.23 *Suppose $f\in SR_\alpha$ with $\alpha>0$. Then*

$$-\log\int_x^\infty e^{-f(y)}\,dy=f(x)+\log f'(x)+O\left(\frac{1}{f(x)}\right) \tag{7.49}$$

as $x\to\infty$.

Actually, a stronger result holds.

Lemma 7.24 *Suppose $f\in SR_\alpha$ with $\alpha>0$. Then*

$$\int_x^\infty e^{-f(y)}\,du=\frac{e^{-f(x)}}{f'(x)}\left(1+O\left(\frac{1}{f(x)}\right)\right) \tag{7.50}$$

as $x\to\infty$.

Proof Using the integration by parts formula, we see that

$$\int_x^\infty e^{-f(y)}\,dy = \frac{e^{-f(x)}}{f'(x)} - \int_x^\infty e^{-f(y)}\rho_1(y)\,dy \tag{7.51}$$

where

$$\rho_1(y) = \left(\frac{1}{f'(y)}\right)' = \frac{f''(y)}{f'(y)^2}.$$

It follows from (7.45) that

$$|\rho_1(y)| = o\big(f(y)^{-1}\big)$$

as $y \to \infty$. Using (7.45) again, we obtain

$$\begin{aligned}\int_x^\infty e^{-f(y)}\rho_1(y)\,dy &= o\left(\int_x^\infty f'(y)e^{-f(y)}\frac{1}{f(y)f'(y)}\,dy\right)\\ &= o\left(\int_x^\infty f'(y)e^{-f(y)}\frac{y}{f(y)^2}\,dy\right)\end{aligned} \tag{7.52}$$

as $x \to \infty$. Since for every $\varepsilon > 0$ the function $y \mapsto e^{-\varepsilon f(y)} y f(y)^{-2}$ is ultimately decreasing, formula (7.52) implies that

$$\int_x^\infty e^{-f(y)}\rho_1(y)\,dy = o\left(e^{-f(x)}\frac{x}{f(x)^2}\right) \tag{7.53}$$

as $x \to \infty$. Now Lemma 7.24 follows from (7.51), (7.53), and (7.45). □

Proof of Lemma 7.23 Taking the logarithm of the functions on both sides of formula (7.50), we obtain

$$-\log \int_x^\infty e^{-f(y)}\,dy = f(x) + \log f'(x) + \log\left(1 + O\left(\frac{1}{f(x)}\right)\right)$$

as $x \to \infty$. It is clear that the previous formula implies (7.49).

The proof of Lemma 7.23 is thus completed. □

Let us return to the proof of Theorem 7.22. Suppose $f \in R_\alpha$. Then there exist functions $f_1 \in SR_\alpha$ and $f_2 \in SR_\alpha$ for which the conclusions in Theorem 7.21 hold. Next, using Lemma 7.24, we obtain

$$\begin{aligned} f_1(x) + \log f_1'(x) + O\left(\frac{1}{f_1(x)}\right) &\le -\log \int_x^\infty e^{-f(x)}\,dx\\ &\le f_2(x) + \log f_2'(x) + O\left(\frac{1}{f_2(x)}\right)\end{aligned}$$

as $x \to \infty$. The previous estimates show that it suffices to prove the following statement. For every function $r \in SR_\alpha$,

$$\lim_{x\to\infty} \frac{\log r'(x)}{r(x)} = 0. \tag{7.54}$$

We have

$$\log \frac{xr'(x)}{r(x)} = \log x + \log r'(x) - \log r(x). \tag{7.55}$$

It follows from the definition of the class SR_α that

$$\lim_{x\to\infty} \log \frac{xr'(x)}{r(x)} = \log \alpha.$$

Next, taking into account that the function $r(x)$ grows faster than x^ε with some $\varepsilon > 0$, we see that (7.54) can be derived from (7.55).

This completes the proof of Theorem 7.22. □

7.6 Pareto-Type Distributions

Pareto-type functions are frequently encountered in the theory of stochastic volatility models. These functions are regularly varying functions in disguise. It is said that a function f is of Pareto type if f is asymptotically equivalent to a regularly varying function g. The previous definition clearly implies that the function f is also regularly varying. What really counts here is the explicit representation of the function g in the standard form $g(x) = x^\alpha l(x)$, where l is a slowly varying function. The Pareto type of f is determined by the couple (α, l). Various functions associated with stochastic volatility models, for instance, complementary cumulative distribution functions of the stock price, call pricing functions, and stock price distribution densities, are often of Pareto type.

Definition 7.25 Let F be a positive function on (c, ∞) with $c \geq 0$, and suppose

$$F(x) \sim x^\alpha l(x) \quad \text{as } x \to \infty,$$

where $\alpha \in \mathbb{R}$ and $l \in R_0$. Then we say that the function F is of Pareto type (α, l) near infinity.

Let G be a positive function on $(0, c)$ with $c > 0$, and suppose

$$G(x) \sim x^{-\alpha} l\left(x^{-1}\right) \quad \text{as } x \to 0,$$

where $\alpha \in \mathbb{R}$ and $l \in R_0$. Then we say that the function G is of Pareto type (α, l) near zero.

In the next definition, we introduce weak Pareto-type functions. Such functions can be squeezed between regularly varying functions with the same index of regular variation.

Definition 7.26 Let F be a positive function on (c, ∞) with $c \geq 0$. If there exist a number $y_0 > c$ and two positive functions $f_1 \in R_\alpha$ and $f_2 \in R_\alpha$, satisfying the condition

$$f_1(y) \leq F(y) \leq f_2(y) \quad \text{for all } y > y_0,$$

then we say that the function F is of weak Pareto type near infinity with index α.

Let G be a positive function on $(0, c)$. If there exist a number $0 < y_1 < c$ and two positive functions $g_1 \in R_\alpha$ and $g_2 \in R_\alpha$ such that

$$g_1\left(y^{-1}\right) \leq G(y) \leq g_2\left(y^{-1}\right) \quad \text{for all } 0 < y < y_1,$$

then we say that the function G is of weak Pareto type near zero with index α.

Let X be a random variable on a probability space $(\Omega, \mathcal{F}, \mathbb{P})$, and let F be the cumulative distribution function of X defined by

$$F(y) = \mathbb{P}[X \leq y], \quad y \in \mathbb{R}.$$

The complementary cumulative distribution function $\overline{F}$ of X is given by

$$\overline{F}(y) = 1 - F(y), \quad y \in \mathbb{R}.$$

Definition 7.27 It is said that the random variable X is distributed according a Pareto-type law if the complementary cumulative distribution function $\overline{F}$ of X is a function of Pareto type (α, l).

If $l(y) = A$, $A > 0$, in Definition 7.27, then F is called a Pareto distribution. Pareto distributions and Pareto-type distributions are examples of "heavy-tailed" distributions. It was observed by Mandelbrot that empirical distributions of price changes of a stock have heavy tails (see [Man63]).

For a Pareto distribution, the log-log plot of the function $f_\alpha(y) = Ax^{-\alpha}$, approximating $\overline{F}(y)$, is linear. Indeed, $\log f_\alpha(y) = -\alpha \log y + \log A$, and hence

$$\log f_\alpha\left(e^u\right) = -\alpha u + \log A.$$

For a Pareto-type distribution, we have $f_\alpha(y) = y^{-\alpha} l(y)$, and the log-log plot is as follows:

$$\log f_\alpha(u) = -\alpha u + \log l\left(e^u\right).$$

7.7 Pareto-Type Distributions in Stochastic Volatility Models

The tails of stock price distributions in the Hull–White, Stein–Stein, and Heston models are of Pareto type. This statement follows from sharp asymptotic formulas established in Sects. 6.1, 6.4, and 6.5. We will next provide more details. Let us recall that the following formulas hold for the stock price distribution density D_t in classical stochastic volatility models.

1. *Correlated Heston model* (*see Sect.* 6.1): There exist positive constants A_1, A_2, and A_3 such that

$$D_t(x) = A_1(\log x)^{-\frac{3}{4}+\frac{a}{c^2}} e^{A_2\sqrt{\log x}} x^{-A_3}\left(1 + O\left((\log x)^{-\frac{1}{2}}\right)\right) \tag{7.56}$$

as $x \to \infty$. The constant A_3 satisfies $A_3 > 2$. Explicit expressions for the constants A_1, A_2, and A_3 can be found in Sects. 6.1.6 and 6.1.7.

2. *Correlated Stein–Stein model* (*see Sect.* 6.4): There exist positive constants B_1, B_2, and B_3 such that

$$D_t(x) = B_1(\log x)^{-\frac{1}{2}} e^{B_2\sqrt{\log x}} x^{-B_3}\left(1 + O\left((\log x)^{-\frac{1}{2}}\right)\right) \tag{7.57}$$

as $x \to \infty$. The constant B_3 satisfies $B_3 > 2$. Explicit formulas for the constants B_1, B_2, and B_3 in uncorrelated Stein–Stein models can be found in Sect. 6.4.

3. *Uncorrelated Hull–White model* (*see Sect.* 6.5): There exist positive constants C, c_2, and c_3 such that

$$\begin{aligned} D_t(x) &= Cx^{-2}(\log x)^{\frac{c_2-1}{2}}(\log\log x)^{c_3} \\ &\quad\times \exp\left\{-\frac{1}{2t\xi^2}\left(\log\left[\frac{1}{y_0}\sqrt{\frac{2\log x}{t}}\right] + \frac{1}{2}\log\log\left[\frac{1}{y_0}\sqrt{\frac{2\log x}{t}}\right]\right)^2\right\} \\ &\quad\times \left(1 + O\left((\log\log x)^{-\frac{1}{2}}\right)\right) \end{aligned} \tag{7.58}$$

as $x \to \infty$. Explicit formulas for the constants C, c_2, and c_3 can be found in Sect. 6.5.

Equalities (7.56), (7.57), and (7.58) imply that the distribution densities of the stock price in the Heston, Stein–Stein, and Hull–White models, under the restrictions on the parameters imposed in Sects. 6.1, 6.4, and 6.5, are of Pareto type. Indeed, for the Heston model, we have

$$D_t(x) \sim x^{-\alpha_t} l_t(x), \quad x \to \infty, \tag{7.59}$$

with

$$\alpha_t = A_3 \quad \text{and} \quad l_t(x) = A_1(\log x)^{-\frac{3}{4}+\frac{a}{c^2}} e^{A_2\sqrt{\log x}}. \tag{7.60}$$

For the Stein–Stein model, (7.59) is valid with

$$\alpha_t = B_3 \quad \text{and} \quad l_t(x) = B_1(\log x)^{-\frac{1}{2}} e^{B_2\sqrt{\log x}}. \tag{7.61}$$

For the Hull–White model, (7.59) holds with $\alpha_t = 2$ and

$$l_t(x) = C(\log x)^{\frac{c_2-1}{2}}(\log\log x)^{c_3} \times \exp\left\{-\frac{1}{2t\xi^2}\left(\log\left[\frac{1}{y_0}\sqrt{\frac{2\log x}{t}}\right] + \frac{1}{2}\log\log\left[\frac{1}{y_0}\sqrt{\frac{2\log x}{t}}\right]\right)^2\right\}. \tag{7.62}$$

It is not hard to see that the functions l_t in (7.60), (7.61), and (7.62) are all slowly varying.

We will next simplify relation (7.59) for the Hull–White model using a weaker asymptotic equivalence.

Lemma 7.28 *The following formula holds for the stock price distribution density in the Hull–White model*:

$$D_t(x) \approx x^{-2}(\log x)^{V_1}(\log\log x)^{V_2} \times \exp\left\{-\frac{1}{8t\xi^2}(\log\log x + \log\log\log x)^2\right\} \tag{7.63}$$

as $x \to \infty$, *where*

$$V_1 = \frac{c_2 - 1}{2} + \frac{1}{2t\xi^2}\left(\frac{1}{2}\log t + \log y_0\right) \tag{7.64}$$

and

$$V_2 = c_3 + \frac{1}{2t\xi^2}\left(\frac{1}{2}\log t + \log y_0\right). \tag{7.65}$$

Proof We can rewrite (7.62) as follows:

$$h_t(x) = C(\log x)^{\frac{c_2-1}{2}}(\log\log x)^{c_3} \times \exp\left\{-\frac{1}{2t\xi^2}\left(v + \frac{1}{2}\log\log x + \frac{1}{2}\log\left[v + \frac{1}{2}\log\log x\right]\right)^2\right\},$$

where $v = \frac{1}{2}\log 2 - \frac{1}{2}\log t - \log y_0$. It is not hard to see that

$$\begin{aligned} h_t(x) &\approx (\log x)^{\frac{c_2-1}{2}}(\log\log x)^{c_3} \\ &\quad \times \exp\left\{-\frac{1}{2t\xi^2}\left(v + \frac{1}{2}\log\log x + \frac{1}{2}\log\left[\frac{1}{2}\log\log x\right]\right)^2\right\} \\ &\approx (\log x)^{\frac{c_2-1}{2}}(\log\log x)^{c_3} \\ &\quad \times \exp\left\{-\frac{1}{2t\xi^2}\left(v - \frac{1}{2}\log 2 + \frac{1}{2}\log\log x + \frac{1}{2}\log\log\log x\right)^2\right\} \end{aligned}$$

as $x \to \infty$. Next, using the previous equivalence, we see that Lemma 7.28 holds. □

Note that for the Heston and Stein–Stein models, the constant α_t in (7.59) depends on t, while for the Hull–White model, we have $\beta_t = 2$ for all $t > 0$.

The next theorem shows that the stock price X_t in the Stein–Stein, Heston, and Hull–White models is distributed according to a Pareto-type law.

Theorem 7.29 *The following statements are true:*

1. *Let $t > 0$ and let $\overline{F}_t$ be the complementary cumulative distribution function of the stock price X_t in the correlated Stein–Stein model. Then*

$$\overline{F}_t(y) \sim y^{-\beta_t} \tilde{l}_t(y) \tag{7.66}$$

as $y \to \infty$. In (7.66),

$$\beta_t = B_3 - 1 \quad \textit{and} \quad \tilde{l}_t(y) = \frac{1}{B_3 - 1} l_t(y)$$

where the constant B_3 is the same as in (7.57) *and l_t is defined in* (7.61).

2. *For the correlated Heston model, formula* (7.66) *holds with*

$$\beta_t = A_3 - 1 \quad \textit{and} \quad \tilde{l}(y) = \frac{1}{A_3 - 1} l_t(y)$$

where the constant A_3 is the same as in (7.56) *and l_t is defined in* (7.60).

3. *For the uncorrelated Hull–White model, the formula*

$$\overline{F}_t(y) \sim y^{-1} l_t(y)$$

holds, where l_t is defined in (7.62).

Theorem 7.29 can be established by integrating the equalities in (7.56), (7.57), and (7.58) on the interval $[x, \infty)$ and using part 1 of Theorem 7.6.

It follows from Theorem 7.29 that the tail index β_t of the stock price X_t in the Stein–Stein model is equal to $B_3 - 1$. For the Heston model, we have $\beta_t = A_3 - 1$. Note that for these models $\beta_t > 1$. For the Hull–White model, the tail index satisfies $\beta_t = 1$ for all $t > 0$.

7.8 Notes and References

- The basic definitions and results in the theory of regularly varying functions go back to J. Karamata (see [Tom01] and [Nik02] for short scientific biographies of Karamata). The monograph [BGT87] by N.H. Bingham, C.M. Goldie, and J.L. Teugels is a very rich source of information about regularly varying functions. We also recommend the following books: [Sen76, Res07] (the latter book contains a crash course on the theory of regular variation), and [EKM97]. For the multivariate case, the reader can consult [Res87, JM06, Res07].

- The book by W. Feller [Fel66] contains a section on regularly varying functions (Sect. 8 in Chap. VIII). This book played an important role in early dissemination of information on regularly varying functions in the probability theory community.
- The paper [HR83] studies regularly varying utility functions.
- Theorem 7.3 was obtained by M. Vuilleumier in [Vui63] (see also [BGT87]).
- In [Con01], Pareto-type functions are mentioned in the context of heavy tails of stock returns, market indices, and exchange rates. Weak Pareto-type functions were introduced in [Gul12].

Chapter 8
Asymptotic Analysis of Option Pricing Functions

This chapter is devoted to European style call and put options in general asset price models. In such models, the random behavior of the asset price is driven by an adapted positive stochastic process X defined on a filtered probability space $(\Omega, \mathcal{F}, \{\mathcal{F}_t\}, \mathbb{P}^*)$. We assume that the following conditions are satisfied:

- The interest rate r is a nonnegative constant.
- The process X starts at $x_0 > 0$.
- The process X is integrable. This means that $\mathbb{E}^*[X_t] < \infty$ for every $t \geq 0$.
- $\mathbb{P}^*$ is a risk-neutral measure. More precisely, the discounted stock price process $\{e^{-rt} X_t\}_{t \geq 0}$ is an $(\{\mathcal{F}_t\}, \mathbb{P}^*)$-martingale.

In the model described above, the asset price distributions are modeled by the marginal distributions of the process X with respect to the probability measure $\mathbb{P}^*$. Note that the integrability condition for X implies the existence of asset price moments only for the orders between zero and one, while the martingality condition for the discounted asset price process leads to fair pricing formulas for European call and put options.

The present chapter focuses first on definitions and general properties of call and put pricing functions. In Sect. 8.1, we prove a characterization theorem for call pricing functions (Theorem 8.3). Sects. 8.2 and 8.3 are devoted to the Black–Scholes model of call and put option pricing, while Sect. 8.4 lists partial derivatives of the Black–Scholes pricing function with respect to various model parameters. Finally, in Sect. 8.5 we establish asymptotic formulas with relative error estimates for call pricing functions in the Hull–White, Stein–Stein, and Heston models.

8.1 Call and Put Pricing Functions in Stochastic Asset Price Models

Let X be an asset price process under a risk-neutral measure $\mathbb{P}^*$. For every real number u set $u^+ = \max\{u, 0\}$.

A. Gulisashvili, *Analytically Tractable Stochastic Stock Price Models*,
Springer Finance, DOI 10.1007/978-3-642-31214-4_8,
© Springer-Verlag Berlin Heidelberg 2012

An European style call option on the underlying asset, with strike price K and maturity T, is a special contract, which gives its holder the right, but not the obligation, to buy one unit of the asset from the seller of the option, for the price K on the date T. An European style put option differs from the call option only in the right of the holder to sell a unit of the asset instead of buying it. The price that the buyer of the option pays for the contract is called the option premium. Note that call and put options can be exercised only on the expiration date. Buying options is less risky than buying units of underlying asset, because the holder of the option has no obligation to exercise it if things go wrong.

In a risk-neutral environment, a natural way to price an European style option is to choose the expected value of the discounted payoff of the option at maturity to be the option premium. For a call option, this payoff is given by $(X_T - K)^+$, while for the put option, the payoff is $(K - X_T)^+$. Combining the premiums for all maturities and strikes, we obtain the so-called pricing function associated with the option.

Definition 8.1 The European call and put option pricing functions C and P in a stochastic asset price model are defined as follows:

$$C(T,K) = e^{-rT}\mathbb{E}^*\big[(X_T - K)^+\big], \quad T \ge 0,\ K \ge 0, \tag{8.1}$$

and

$$P(T,K) = e^{-rT}\mathbb{E}^*\big[(K - X_T)^+\big], \quad T \ge 0,\ K \ge 0. \tag{8.2}$$

Let us denote by μ_T the distribution of the asset price X_T. Then we have

$$C(T,K) = e^{-rT}\int_K^\infty x\,d\mu_T(x) - e^{-rT}K\int_K^\infty d\mu_T(x) \tag{8.3}$$

and

$$P(T,K) = e^{-rT}K\int_0^K d\mu_T(x) - e^{-rT}\int_0^K x\,d\mu_T(x). \tag{8.4}$$

It is clear from (8.3) and (8.4) that

$$C(T,K) = P(T,K) + x_0 - e^{-rT}K.$$

The previous equality is called the put–call parity formula.

It is not hard to see, using the martingality condition, that

$$x_0 = e^{-rt}\mathbb{E}^*[X_t] \quad \text{for all } t \ge 0. \tag{8.5}$$

Therefore,

$$x_0 = C(T,0) = e^{-rT}\int_0^\infty x\,d\mu_T(x) \quad \text{for all } T \ge 0. \tag{8.6}$$

What are the most important properties of call pricing functions? This question will be answered below. Let $\widetilde{C}(T,K)$ be a positive function defined for all $T \ge 0$

and $K \ge 0$, and suppose we want to model call prices by the function $\widetilde{C}$. Let r and x_0 be the interest rate and the initial price of the underlying asset, respectively.

Definition 8.2 It is said that the model described by the function $\widetilde{C}$ is free from static arbitrage if there exist a stochastic process X and a risk-neutral measure $\mathbb{P}^*$ such that the function $\widetilde{C}$ coincides with the call pricing function C defined by (8.1).

Convex functions play an important role in this section. We will next briefly overview elements of the theory of convex functions on a half-line (a, ∞). The reader is referred to the appendix in [RY04] for more information concerning convex functions.

A function φ defined on the open half-line (a, ∞) is called convex if

$$\varphi\big(tx + (1-t)y\big) \le t\varphi(x) + (1-t)\varphi(y)$$

for all $0 \le t \le 1$ and $x, y \in (a, \infty)$. On the other hand, if a function φ is defined on the closed half-line $[b, \infty)$ for some $b \in \mathbb{R}$, then φ is called convex if there exist $a < b$ and a convex function $\widetilde{\varphi}$ on (a, ∞) such that $\widetilde{\varphi}(x) = \varphi(x)$ for all $x \in [b, \infty)$. A convex function on an open half-line is absolutely continuous on any closed subinterval of the half-line. For a convex function φ on (a, ∞), the right-hand derivative φ'_+ and the left-hand derivative φ'_- exist at every point in (a, ∞). Moreover, for all $a < x < y < \infty$, we have

$$\varphi'_+(x) \le \frac{\varphi(y) - \varphi(x)}{y - x} \le \varphi'_-(y). \tag{8.7}$$

The function φ'_+ is increasing and right-continuous on (a, ∞), while the function φ'_- is increasing and left-continuous on (a, ∞). In addition, the set $\{x : \varphi'_+(x) \ne \varphi'_-(x)\}$ is at most countable. If φ is a convex function on (a, ∞), then the second distributional derivative μ of the function φ is a locally finite Borel measure on $(0, \infty)$, and any such measure is the second derivative of a convex function, which is unique up to the addition of an affine function. It is also true that the measure μ is the Lebesgue–Stieltjes measure generated by the function φ'_+.

Our next goal is to prove an assertion that provides a characterization of a general call pricing function.

Theorem 8.3 *A nonnegative function $C(T, K)$, $T \ge 0$, $K \ge 0$, is a call pricing function with interest rate r and initial condition x_0 if and only if the following conditions hold*:

1. *For every $T \ge 0$ the function $K \to C(T, K)$ is convex on $[0, \infty)$.*
2. *For every $T \ge 0$ the second distributional derivative μ_T of the function $K \mapsto e^{rT} C(T, K)$ is a Borel probability measure on $[0, \infty)$ such that*

$$\int_{[0,\infty)} x \, d\mu_T(x) = x_0 e^{rT}. \tag{8.8}$$

3. *For every* $K \geq 0$ *the function* $T \to C(T, e^{rT} K)$ *is non-decreasing.*
4. *For every* $K \geq 0$, $C(0, K) = (x_0 - K)^+$.
5. *For every* $T \geq 0$, $\lim_{K \to \infty} C(T, K) = 0$.

Corollary 8.4 *Let* r *and* x_0 *be the interest rate and the initial price, respectively. The option pricing model described by a function* $\widetilde{C}$ *is free of static arbitrage if and only if the function* $\widetilde{C}$ *satisfies conditions* 1–5 *in Theorem* 8.3.

Proof of Theorem 8.3 Let C be a call pricing function and denote by μ_T the distribution of the random variable X_T. Then it is not hard to see that the second distributional derivative of the function $K \mapsto e^{rT} C(T, K)$ coincides with the measure μ_T. Our goal is to establish that conditions 1–5 in the formulation of Theorem 8.3 hold. It is clear that conditions 1 and 4 follow from the definitions. Condition 2 can be established using the equivalence of (8.5) and (8.8).

We will next prove that condition 3 holds. This condition is equivalent to the following inequality:

$$\int_0^\infty (x - K)^+ \, d\mu_S\left(e^{rS} x\right) \leq \int_0^\infty (x - K)^+ \, d\mu_T\left(e^{rT} x\right) \tag{8.9}$$

for all $K \geq 0$ and $0 \leq S \leq T < \infty$. The inequality in (8.9) can be established by taking into account the fact that the process $e^{-rt} X_t$, $t \geq 0$, is a martingale and applying Jensen's inequality. Finally, condition 5 follows from the estimate

$$C(T, K) \leq e^{-rT} \int_K^\infty x \, d\mu_T(x)$$

and (8.6). To finish the proof of the necessity part of Theorem 8.3, it suffices to show that the function $K \mapsto C(T, K)$ can be extended from $[0, \infty)$ to a convex function on $\mathbb{R}$. We already know that the function $K \mapsto C(T, K)$ is convex on $(0, \infty)$, and it follows from (8.3) that

$$C(T, 0) = e^{-rT} x_0 = \lim_{K \to 0} C(T, K).$$

Moreover, $C'_+(K) = -\mu_T[K, \infty)$ for all $K \geq 0$, and hence

$$-1 \leq \lim_{K \to 0} C'_+(K) < 0.$$

Now it is clear that the function $K \mapsto C(T, K)$ can be extended to a convex function on $\mathbb{R}$ by affine extrapolation.

We will next prove the sufficiency part of Theorem 8.3. Suppose C is a function such that conditions 1–5 in the formulation of Theorem 8.3 hold, and denote a convex extension of the function $K \mapsto C(T, K)$ to the half-line $(-a_T, \infty)$, $a_T > 0$, by $\widetilde{C}$. Denote the second distributional derivative of the function $\widetilde{C}(T, K)$ by $\tilde{\mu}_T$. Then the measure μ_T is the restriction of $\tilde{\mu}_T$ to $[0, \infty)$.

It follows from condition 2 that the following function is finite:

$$V(T,K)=\int_K^{\infty} x\,d\tilde{\mu}_T(x)-K\int_K^{\infty} d\tilde{\mu}_T(x), \quad T\ge 0,\ K\in(-a_T,\infty).$$

It is clear that the second distributional derivative of the function $K\mapsto V(T,K)$ coincides with the measure $\tilde{\mu}_T$. Therefore,

$$e^{rT}C(T,K)=V(T,K)+a(T)K+b(T)$$

for all $T\ge 0$ and $K\ge 0$, where the functions a and b do not depend on K. Since $C(T,K)\to 0$ as $K\to\infty$ (condition 5) and $V(T,K)\to 0$ as $K\to\infty$, we see that $a(T)=b(T)=0$, and hence

$$C(T,K)=e^{-rT}\int_0^{\infty}(x-K)^+\,d\mu_T(x) \tag{8.10}$$

for all $T\ge 0$ and $K\ge 0$.

Consider the family of Borel probability measures $\{\nu_T\}_{T\ge 0}$ on $\mathbb{R}$ defined as follows. For every $T\ge 0$, $\nu_T(A)=\mu_T(e^{rT}A)$ if A is a Borel subset of $[0,\infty)$, and $\nu_T((-\infty,0))=0$. Since condition 3 holds, we have

$$\int_{\mathbb{R}}(x-K)^+\,d\nu_S(x)\le\int_{\mathbb{R}}(x-K)^+\,d\nu_T(x) \tag{8.11}$$

for all $K\ge 0$ and $0\le S\le T<\infty$.

Let μ and ν be Borel probability measures on $\mathbb{R}$. It is said that ν dominates μ in the convex order if for every nonnegative convex function φ, which is integrable with respect to μ and ν, the inequality

$$\int_{\mathbb{R}}\varphi\,d\mu\le\int_{\mathbb{R}}\varphi\,d\nu$$

holds. Our next goal is to show that for all $0\le S\le T<\infty$ the measure ν_T dominates the measure ν_S in the convex order.

It is not hard to see that for any nonnegative convex function φ on $\mathbb{R}$ we have

$$\varphi(x)=\int_{\mathbb{R}}(x-u)^+\,d\eta(u)+ax+b \tag{8.12}$$

for all $x\in\mathbb{R}$, where a and b are some constants, and the symbol η stands for the second distributional derivative of the function φ. It follows from condition 2 in Theorem 8.3, (8.11), and (8.12) that for $0\le S\le T<\infty$,

$$\begin{aligned}\int_{\mathbb{R}}\varphi(x)\,d\nu_S(x)&=\int_0^{\infty}d\eta(u)\int_0^{\infty}(x-u)^+\,d\nu_S(x)+ax_0+b\\&\le\int_0^{\infty}d\eta(u)\int_0^{\infty}(x-u)^+\,d\nu_T(x)+ax_0+b=\int_{\mathbb{R}}\varphi(x)\,d\nu_T(x).\end{aligned}$$

Therefore, the measure ν_T dominates the measure μ_S in the convex order.

We will need the following result due to Kellerer (see [Kel72], Theorem 3). Kellerer's theorem answers the following question. What are the conditions under which there exists a martingale having prescribed marginal distributions?

Theorem 8.5 *Let ν_t, $0 \le t \le t_0$, be a family of Borel probability measures on $\mathbb{R}$. Suppose the first moment of the measure ν_t exists for every t, and for $s < t$ the measure ν_t dominates the measure ν_s in the convex order. Then there exist a filtered probability space $(\Omega, \mathcal{F}, \{\mathcal{F}_t\}, \mathbb{P}^*)$ and a Markov $(\{\mathcal{F}_t\}, \mathbb{P}^*)$-submartingale X such that the distribution of X_t coincides with ν_t for every t with $0 \le t \le t_0$. Furthermore, if the first moment $\int_{\mathbb{R}} x \, d\nu_t(x)$ is independent of t, then the process X is an $(\{\mathcal{F}_t\}, \mathbb{P}^*)$-martingale.*

Let us return to the proof of Theorem 8.3. By Kellerer's theorem, there exist a filtered probability space $(\Omega, \mathcal{F}, \{\mathcal{F}_T\}, \mathbb{P}^*)$ and a Markov $(\{\mathcal{F}_T\}, \mathbb{P}^*)$-martingale Y such that the distribution of Y_T coincides with the measure ν_T for every $T \ge 0$. Now put $X_T = e^{rT} Y_T$, $\quad T \ge 0$. It follows that the measure μ_T is the distribution of the random variable X_T for every $T \ge 0$. This produces a stock price process X such that the process $e^{-rt} X_t$ is a martingale. Using condition 4, we obtain $\mu_0 = \delta_{x_0}$, and hence $X_0 = x_0$ $\mathbb{P}^*$-a.s. Now it is clear that (8.10) implies that

$$C(T, K) = e^{-rT} \mathbb{E}^* \big[(X_T - K)^+ \big].$$

Therefore the function C is a call pricing function.

This completes the proof of Theorem 8.3. □

Our next goal is to characterize call pricing functions in the case where the maturity T is fixed. We will do it for $r = 0$.

Theorem 8.6 *Suppose $T \ge 0$ is fixed, and let ψ be a nonnegative function on $[0, \infty)$. Then the equality $\psi(K) = C(T, K)$, $K \ge 0$, holds for some call pricing function C with interest rate $r = 0$ and initial condition x_0 if and only if*

1. *The function ψ is convex.*
2. *The second distributional derivative μ of the function ψ is a Borel probability measure such that $\int_0^\infty x \, d\mu(x) = x_0$.*
3. *$\psi(K) \to 0$ as $K \to \infty$.*
4. *$\psi(K) \ge (x_0 - K)^+$ for all $K \ge 0$.*

Proof The sufficiency part of Theorem 8.6 follows from Theorem 8.3.

To prove the necessity part, let us assume that a function ψ, satisfying the conditions in the formulation of Theorem 8.6, is given. Define the following function on $[0, \infty)^2$:

$$C(T, K) = \begin{cases} (x_0 - K)^+, & \text{if } T = 0, \\ \psi(K), & \text{if } T > 0. \end{cases}$$

We only need to show that the function C is a call pricing function with $r = 0$. However, the previous statement can be easily established using Theorem 8.3.

This completes the proof of Theorem 8.6. □

8.2 The Black–Scholes Model

Without any doubt, the most famous examples of call and put pricing functions are the functions C_{BS} and P_{BS}, appearing in the Black–Scholes model of option pricing (see [BS73]). In this model, the asset is a stock, and the price process is a geometric Brownian motion, satisfying the Osborne–Samuelson equation

$$dX_t = X_t(\mu\, dt + \sigma\, dW_t) \tag{8.13}$$

(see Sect. 2.1). In (8.13), $\mu \in \mathbb{R}$ and $\sigma > 0$ are the drift and the volatility of the stock, respectively. The initial price of the stock is denoted by x_0. Note that we do not assume that the physical measure $\mathbb{P}$ is such that the discounted price process is a martingale under $\mathbb{P}$. However, using Girsanov's theorem, we can switch to a risk-neutral setting by introducing a new measure $\mathbb{P}^*$ defined by

$$d\mathbb{P}^* = \exp\left\{-\frac{1}{2}\left(\frac{\mu - r}{\sigma}\right)^2 T - \frac{\mu - r}{\sigma} W_T\right\}$$

and a new Brownian motion given by

$$W_t^* = W_t + \frac{\mu - r}{\sigma} t, \quad 0 \le t \le T.$$

In the previous formulas, the symbol T stands for a fixed time horizon.

Under the equivalent martingale measure $\mathbb{P}^*$, the model in (8.13) takes the following form:

$$dX_t = X_t\big(r\, dt + \sigma\, dW_t^*\big). \tag{8.14}$$

According to formulas (8.1) and (8.2), the call and put pricing functions C_{BS} and P_{BS} in the Black–Scholes model are given by

$$C_{\mathrm{BS}}(T, K, \sigma, r) = e^{-rT}\mathbb{E}^*\big[(X_T - K)^+\big] \tag{8.15}$$

and

$$P_{\mathrm{BS}}(T, K, \sigma, r) = e^{-rT}\mathbb{E}^*\big[(K - X_T)^+\big], \tag{8.16}$$

respectively. In (8.15) and (8.16), the process X satisfies (8.14).

8.3 Black–Scholes Formulas

In [BS73], F. Black and M. Scholes found explicit expressions for the call and put pricing functions C_{BS} and P_{BS}. We will next derive the Black–Scholes formulas.

Let $T > 0$ and $K > 0$. It follows from (8.15) that

$$C_{\mathrm{BS}}(T,K) = e^{-rT} \int_K^\infty u D_T(u)\,du - e^{-rT} K \int_K^\infty D_T(u)\,du, \tag{8.17}$$

where D_T is the distribution density of the stock price X_T at maturity with respect to the measure $\mathbb{P}^*$. This density is given by

$$D_T(u) = \frac{\sqrt{x_0 e^{rT}}}{\sqrt{2\pi T}\sigma} \exp\left\{-\frac{\sigma^2 T}{8}\right\} u^{-\frac{3}{2}} \exp\left\{-\frac{1}{2T\sigma^2}\left(\log \frac{u}{x_0 e^{rT}}\right)^2\right\}$$

(see formula (1.8)). Therefore,

$$\int_K^\infty u D_T(u)\,du$$
$$= \frac{\sqrt{x_0 e^{rT}}}{\sqrt{2\pi T}\sigma} \exp\left\{-\frac{\sigma^2 T}{8}\right\} \int_K^\infty u^{-\frac{1}{2}} \exp\left\{-\frac{1}{2T\sigma^2}\left(\log \frac{u}{x_0 e^{rT}}\right)^2\right\} du.$$

Making the change of variables

$$y = \frac{1}{\sigma\sqrt{T}} \log \frac{x_0 e^{rT}}{u} + \frac{1}{2}\sigma\sqrt{T},$$

we see that

$$\int_K^\infty u D_T(u)\,du = \frac{x_0 e^{rT}}{\sqrt{2\pi}} \int_{-\infty}^{d_1} \exp\left\{-\frac{y^2}{2}\right\} dy, \tag{8.18}$$

where

$$d_1 = \frac{\log x_0 - \log K + (r + \frac{1}{2}\sigma^2)T}{\sigma\sqrt{T}}. \tag{8.19}$$

Moreover,

$$\int_K^\infty D_T(u)\,du$$
$$= \frac{\sqrt{x_0 e^{rT}}}{\sqrt{2\pi T}\sigma} \exp\left\{-\frac{\sigma^2 T}{8}\right\} \int_K^\infty u^{-\frac{3}{2}} \exp\left\{-\frac{1}{2T\sigma^2}\left(\log \frac{u}{x_0 e^{rT}}\right)^2\right\} du.$$

Making another change of variables

$$y = \frac{1}{\sigma\sqrt{T}} \log \frac{x_0 e^{rT}}{u} - \frac{1}{2}\sigma\sqrt{T},$$

we obtain

$$\int_K^\infty D_T(u)\,du = \frac{1}{\sqrt{2\pi}} \int_{-\infty}^{d_2} \exp\left\{-\frac{y^2}{2}\right\} dy, \tag{8.20}$$

where

$$d_2 = \frac{\log x_0 - \log K + (r - \frac{1}{2}\sigma^2)T}{\sigma\sqrt{T}}. \tag{8.21}$$

Remark 8.7 It is not hard to see, using the equality

$$d_1^2 - d_2^2 = 2\log\frac{x_0 e^{rT}}{K},$$

that the functions d_1 and d_2 are related as follows:

$$\exp\left\{-\frac{d_2^2}{2}\right\} = \exp\left\{-\frac{d_1^2}{2}\right\}\frac{x_0 e^{rT}}{K}.$$

The formulas in the next assertion are the celebrated Black–Scholes formulas.

Theorem 8.8 *Let C_{BS} be the call pricing function in the Black–Scholes model. Then*

$$C_{\mathrm{BS}}(T,K) = \frac{x_0}{\sqrt{2\pi}} \int_{-\infty}^{d_1} e^{-\frac{y^2}{2}}\,dy - \frac{Ke^{-rT}}{\sqrt{2\pi}} \int_{-\infty}^{d_2} e^{-\frac{y^2}{2}}\,dy, \tag{8.22}$$

where d_1 and d_2 are given by (8.19) *and* (8.21), *respectively. Similarly, let P_{BS} be the put pricing function in the Black–Scholes model. Then*

$$P_{\mathrm{BS}}(T,K) = \frac{Ke^{-rT}}{\sqrt{2\pi}} \int_{-\infty}^{-d_2} e^{-\frac{y^2}{2}}\,dy - \frac{x_0}{\sqrt{2\pi}} \int_{-\infty}^{-d_1} e^{-\frac{y^2}{2}}\,dy. \tag{8.23}$$

Formula (8.22) follows from (8.17), (8.18), and (8.20), while formula (8.23) can be established using the put–call parity relation.

The function

$$\Phi(x) = \frac{1}{\sqrt{2\pi}} \int_{-\infty}^{x} e^{-\frac{y^2}{2}}\,dy$$

appearing in Theorem 8.8 is the cumulative distribution function of the standard normal distribution. The complementary cumulative distribution function of the standard normal distribution is given by

$$\Psi(x) = 1 - \Phi(x) = \frac{1}{\sqrt{2\pi}} \int_{x}^{\infty} e^{-\frac{y^2}{2}}\,dy.$$

The following asymptotic formula holds:

$$\Psi(x)=\frac{1}{\sqrt{2\pi}x}e^{-\frac{x^2}{2}}\left[1+\sum_{n=1}^{\infty}(-1)^n x^{-2n}\prod_{j=1}^{n}(2j-1)\right]. \tag{8.24}$$

It is not hard to derive formula (8.24) from a similar formula for the complementary error function, that is, the function given by

$$\operatorname{erfc}(x)=\frac{2}{\sqrt{\pi}}\int_x^{\infty}e^{-y^2}\,dy$$

(see [BH95], p. 96).

Formula (8.24) provides an asymptotic series representation for the function Ψ. This means that for every $N\geq 1$ we have

$$\Psi(x)=\frac{1}{\sqrt{2\pi}x}e^{-\frac{x^2}{2}}\left[1+\sum_{n=1}^{N}(-1)^n x^{-2n}\prod_{j=1}^{n}(2j-1)+O\left(x^{-2N-2}\right)\right]$$

as $x\to\infty$. The case where $N=0$ is as follows:

$$\Psi(x)=\frac{1}{\sqrt{2\pi}x}e^{-\frac{x^2}{2}}\left(1+O\left(x^{-2}\right)\right),\quad x\to\infty. \tag{8.25}$$

Remark 8.9 Formulas (8.22) and (8.23) are also meaningful for $K=0$ or $T=0$, provided that we define the functions d_1 and d_2 at the exceptional points as the corresponding limits (those limits may be infinite).

8.4 Derivatives of Option Pricing Functions

In finance, the derivatives of the pricing function C_{BS} with respect to the Black–Scholes model parameters are collectively called "the Greeks", because it is generally accepted to denote some of them by the letters of the Greek alphabet. For a general option pricing model, the Greeks measure the sensitivities of option prices to small changes in the model parameters.

In this section, we gather explicit formulas for certain derivatives of the function C_{BS}. We also include the "Greek" names of the derivatives when these names are known. Note that some of them do not have anything in common with the letters of the Greek alphabet. In the formulas listed below, d_1 and d_2 are the functions defined by (8.19) and (8.21), respectively, and we use the symbol x for the initial price of the stock. All the formulas on the list below can be checked by straightforward computations.

Explicit Formulas for the Greeks in the Black–Scholes Model

- The first order derivatives of the Black–Scholes pricing function C_{BS}
  - Delta:

$$\frac{\partial C}{\partial x} = \frac{1}{\sqrt{2\pi}} \int_{-\infty}^{d_1} \exp\left\{-\frac{y^2}{2}\right\} dy.$$

  - Without a name:

$$\frac{\partial C}{\partial K} = -e^{-rT} \frac{1}{\sqrt{2\pi}} \int_{-\infty}^{d_2} \exp\left\{-\frac{y^2}{2}\right\} dy.$$

  - Vega:

$$\frac{\partial C}{\partial \sigma} = \frac{x\sqrt{T}}{\sqrt{2\pi}} \exp\left\{-\frac{d_1^2}{2}\right\}.$$

  - Theta:

$$\frac{\partial C}{\partial T} = \frac{re^{-rT}K}{\sqrt{2\pi}} \int_{-\infty}^{d_2} \exp\left\{-\frac{y^2}{2}\right\} dy + \frac{\sigma x}{2\sqrt{T}\sqrt{2\pi}} \exp\left\{-\frac{d_1^2}{2}\right\}.$$

  - Rho:

$$\frac{\partial C}{\partial r} = \frac{KTe^{-rT}}{\sqrt{2\pi}} \int_{-\infty}^{d_2} \exp\left\{-\frac{y^2}{2}\right\} dy.$$

- The second order derivatives of the Black–Scholes pricing function C_{BS}
  - Gamma:

$$\frac{\partial^2 C}{\partial x^2} = \frac{1}{\sqrt{2\pi T}\sigma x} \exp\left\{-\frac{d_1^2}{2}\right\}.$$

  - Without a name:

$$\frac{\partial^2 C}{\partial K^2} = \frac{x}{\sqrt{2\pi T}K^2\sigma} \exp\left\{-\frac{d_1^2}{2}\right\}.$$

  - Without a name:

$$\frac{\partial^2 C}{\partial K \partial \sigma} = \frac{x}{\sqrt{2\pi}K\sigma} d_1 \exp\left\{-\frac{d_1^2}{2}\right\}.$$

  - Vanna:

$$\frac{\partial^2 C}{\partial x \partial \sigma} = -\frac{1}{\sqrt{2\pi}\sigma} d_2 \exp\left\{-\frac{d_1^2}{2}\right\}.$$

  - Volga:

$$\frac{\partial^2 C}{\partial \sigma^2} = \frac{x\sqrt{T}}{\sqrt{2\pi}\sigma} d_1 d_2 \exp\left\{-\frac{d_1^2}{2}\right\}.$$

- Charm:

$$\frac{\partial^2 C}{\partial x \partial T} = \frac{1}{\sqrt{2\pi}} \exp\left\{-\frac{d_1^2}{2}\right\} \frac{2rT - \sigma\sqrt{T}d_2}{2T\sigma\sqrt{T}}.$$

8.5 Asymptotic Behavior of Pricing Functions in Stochastic Volatility Models

This section studies call pricing functions in stochastic volatility models. The next statement provides a sharp asymptotic formula for such a function, under the assumption that the distribution density of the stock price is equivalent to a regularly varying function.

Theorem 8.10 *Let C be a call pricing function, and suppose the distribution of the stock price X_T admits a density D_T. Suppose also that*

$$D_T(x) = x^{\beta} h(x)\big(1 + O\big(\rho(x)\big)\big) \tag{8.26}$$

as $x \to \infty$, where $\beta < -2$, the function h is slowly varying with remainder g, and $\rho(x) \downarrow 0$ as $x \to \infty$. Then

$$C(K) = e^{-rT} \frac{1}{(\beta+1)(\beta+2)} K^{\beta+2} h(K)\big[1 + O\big(\rho(K)\big) + O\big(g(K)\big)\big] \tag{8.27}$$

as $K \to \infty$.

Proof It follows from (8.26) and Lemma 7.16 that

$$\begin{aligned}
C(K) &= e^{-rT} \int_K^{\infty} (x-K) D_T(x)\, dx \\
&= e^{-rT} \int_K^{\infty} (x-K) x^{\beta} h(x)\, dx \,\big(1 + O\big(\rho(K)\big)\big) \\
&= e^{-rT} K^{\beta+2} h(K) \int_1^{\infty} (y-1) y^{\beta} \frac{h(Ky)}{h(K)}\, dy \,\big(1 + O\big(\rho(K)\big)\big) \\
&= e^{-rT} K^{\beta+2} h(K) \int_1^{\infty} (y-1) y^{\beta}\, dy \big[1 + O\big(\rho(K)\big) + O\big(g(K)\big)\big]
\end{aligned}$$

as $K \to \infty$. Now it is clear that formula (8.27) holds, and the proof of Theorem 8.10 is thus completed. □

Theorem 8.10 allows us to characterize the asymptotic behavior of the call pricing function $C(K)$ in the Heston and the Stein–Stein models.

Theorem 8.11

(a) *The following formula holds for the call pricing function C in the correlated Heston model with $r = 0$, $x_0 = 1$, and $-1 < \rho \le 0$:*

$$C(K) = \frac{A_1}{(A_3 - 1)(A_3 - 2)} (\log K)^{-\frac{3}{4} + \frac{qm}{c^2}} e^{A_2 \sqrt{\log K}} K^{2 - A_3} \times \left(1 + O\left((\log K)^{-\frac{1}{2}}\right)\right) \tag{8.28}$$

as $K \to \infty$. The constants in (8.28) *are the same as in* (7.56).

(b) *The following formula holds for the call pricing function C in the correlated Stein–Stein model*:

$$C(K) = e^{-rT} \frac{B_1}{(1 - B_3)(2 - B_3)} (\log K)^{-\frac{1}{2}} e^{B_2 \sqrt{\log K}} K^{2 - B_3} \times \left(1 + O\left((\log K)^{-\frac{1}{2}}\right)\right) \tag{8.29}$$

as $K \to \infty$. The constants in (8.29) *are the same as in* (7.57).

It is not hard to see that Theorem 8.11 follows from (7.56), (7.57), Corollary 7.13, and Theorem 8.10.

Next we turn our attention to the uncorrelated Hull–White model. Note that Theorem 8.11 cannot be applied in this case since for the Hull–White model we have $\beta = -2$. Instead, we will employ the asymptotic formula for fractional integrals (see Theorem 5.3). A special case of this formula is as follows. Let $b(x) = B(\log x)$ be a positive increasing function on $[c, \infty)$ with $B''(x) \approx 1$ as $x \to \infty$. Then

$$\int_K^\infty \exp\{-b(x)\}\, dx = \frac{\exp\{-b(K)\}}{b'(K)} \left(1 + O\left((\log K)^{-1}\right)\right) \tag{8.30}$$

as $K \to \infty$.

Theorem 8.12 *Let C be a call pricing function, and suppose the distribution of the stock price X_T admits a density D_T. Suppose also that*

$$D_T(x) = x^{-2} \exp\{-b(\log x)\} \left(1 + O(\rho(x))\right) \tag{8.31}$$

as $x \to \infty$, where the function b is such as in formula (8.30), *and the function ρ satisfies $\rho(x) \downarrow 0$ as $x \to \infty$. Then*

$$C(K) = e^{-rT} \frac{\exp\{-b(\log K)\} \log K}{B'(\log \log K)} \times \left[1 + O\left((\log \log K)^{-1}\right) + O(\rho(K))\right] \tag{8.32}$$

as $K \to \infty$.

Proof We have

$$
\begin{aligned}
e^{rT}C(K) &= \int_K^\infty x D_T(x)\,dx - K\int_K^\infty D_T(x)\,dx \\
&= \left[\int_K^\infty x^{-1}\exp\{-b(\log x)\}\,dx - K\int_K^\infty x^{-2}\exp\{-b(\log x)\}\,dx\right] \\
&\quad \times (1+O(\rho(K)) \\
&= \int_K^\infty x^{-1}\exp\{-b(\log x)\}\,dx\bigl(1+O(\rho(K))\bigr)+O\bigl(\exp\{-b(\log K)\}\bigr) \\
&= \int_{\log K}^\infty \exp\{-b(u)\}\,dx\bigl(1+O(\rho(K))\bigr)+O\bigl(\exp\{-b(\log K)\}\bigr). \qquad (8.33)
\end{aligned}
$$

Using (8.30) we get

$$
\begin{aligned}
C(K) &= e^{-rT}\frac{\exp\{-b(\log K)\}}{b'(\log K)}\bigl(1+O((\log\log K)^{-1})\bigr)\bigl[1+O(\rho(K))\bigr] \\
&\quad + O\bigl(\exp\{-b(\log K)\}\bigr) \\
&= e^{-rT}\frac{\exp\{-b(\log K)\}\log K}{B'(\log\log K)}\bigl[1+O((\log\log K)^{-1})+O(\rho(K))\bigr] \\
&\quad + O\bigl(\exp\{-b(\log K)\}\bigr). \qquad (8.34)
\end{aligned}
$$

Since $B'(x)\approx x$ as $x\to\infty$, (8.34) implies (8.32). □

The next assertion characterizes the asymptotic behavior of a call pricing function in the uncorrelated Hull–White model.

Theorem 8.13 *Let C be the call pricing function in the uncorrelated Hull–White model. Then*

$$
\begin{aligned}
C(K) &= 4T\xi^2 C_0 e^{-rT}(\log K)^{\frac{c_2+1}{2}}(\log\log K)^{c_3-1} \\
&\quad \times \exp\left\{-\frac{1}{2T\xi^2}\left(\log\left[\frac{1}{y_0}\sqrt{\frac{2\log K}{T}}\right]+\frac{1}{2}\log\log\left[\frac{1}{y_0}\sqrt{\frac{2\log K}{T}}\right]\right)^2\right\} \\
&\quad \times \bigl(1+O((\log\log K)^{-\frac{1}{2}})\bigr) \qquad (8.35)
\end{aligned}
$$

as $K\to\infty$. The constants in (8.35) *are the same as in formula* (7.57).

Proof We will employ Theorem 8.12 in the proof. It is not hard to see using (7.58) that formula (8.31) holds for the distribution density D_T of the stock price in the Hull–White model. Here we choose the functions b, B, and ρ as follows:

$$b(u) = -\log C_0 - \frac{c_2 - 1}{2} \log u - c_3 \log\log u$$
$$+ \frac{1}{2T\xi^2}\left(\log\left[\frac{1}{y_0}\sqrt{\frac{2u}{T}}\right] + \frac{1}{2}\log\log\left[\frac{1}{y_0}\sqrt{\frac{2u}{T}}\right]\right)^2,$$

$$B(u) = -\log C_0 - \frac{c_2 - 1}{2} u - c_3 \log u$$
$$+ \frac{1}{2T\xi^2}\left[\log \frac{1}{y_0}\sqrt{\frac{2}{T}} + \frac{1}{2}u + \frac{1}{2}\log\left(\log \frac{1}{y_0}\sqrt{\frac{2}{T}} + \frac{1}{2}u\right)\right]^2,$$

and $\rho(x) = (\log\log x)^{-\frac{1}{2}}$.

It is clear that $B''(u) \approx 1$ and $B'(u) \approx u$ as $u \to \infty$. Moreover, using the mean value theorem, we obtain the following estimate:

$$\frac{1}{B'(\log\log K)} - \frac{4T\xi^2}{\log\log K} = O\big((\log\log K)^{-2}\big) \tag{8.36}$$

as $K \to \infty$. Next, taking into account (8.32) and (8.36), we see that (8.35) holds.

This completes the proof of Theorem 8.13. □

8.6 Notes and References

- The absence of arbitrage imposes certain restrictions on call pricing models (see Theorem 8.3 in Sect. 8.1). Similar results can be found in Proposition 3.2 in [Bue06], Theorem 2.1 in [Rop10], and Theorem 3.3 in [Gul10]. The reader can also consult Sect. 1 of [CN09], where necessary conditions for static no-arbitrage are given. Note that the conditions used in Theorem 8.3 are essentially the same as those in [CN09].
- Several results preceded Kellerer's theorem (Theorem 8.5 in this chapter), e.g., the Sherman–Stein–Blackwell theorem (see, e.g., [DH07]) and Strassen's theorem (see [Str65]). The Sherman–Stein–Blackwell theorem concerns martingale transition matrices, while Strassen's theorem provides necessary and sufficient conditions for the existence of a discrete time martingale with marginal distributions matching a given sequence of measures. Kellerer's theorem is a continuous time generalization of Strassen's result. These three theorems have become useful tools in the study of option pricing models reproducing observed option prices (see [CM05, Bue06, DH07, Cou07] and the references therein). More information on matching theorems for stochastic processes can be found in the book [HPRY11] by F. Hirsch, C. Profeta, B. Roynette, and M. Yor. We thank P. Embrechts for providing this reference.

- The Black–Scholes call and put option pricing formulas (see [BS73]) are without doubt the most famous quantitative results in theoretical and applied finance. We give analytical proofs of the Black–Scholes formulas in Sect. 8.3.
- The sharp asymptotic formulas for the call pricing functions in the Heston, Stein–Stein, and Hull–White models gathered in Sect. 8.5 were obtained in Sect. 7 of [Gul10].

Chapter 9
Asymptotic Analysis of Implied Volatility

The implied volatility was first introduced in the paper [LR76] of H.A. Latané and R.J. Rendleman under the name "the implied standard deviation". Latané and Rendleman studied standard deviations of asset returns, which are implied in actual call option prices when investors price options according to the Black–Scholes model. For a general model of call option prices, the implied volatility can be obtained by inverting the Black–Scholes call pricing function with respect to the volatility variable and composing the resulting inverse function with the original call pricing function.

This chapter mainly concerns the asymptotics of the implied volatility at extreme strikes. In Sect. 9.1, we define the implied volatility in general models of call option prices and discuss its elementary properties. Implied volatility models free of static arbitrage are characterized in Sect. 9.2 (see Theorem 9.6). The rest of the chapter is devoted to sharp asymptotic formulas with error estimates for the implied volatility. We discuss asymptotic formulas of various orders, and show how certain symmetries hidden in stochastic asset price models allow to analyze the asymptotic behavior of the implied volatility for small strikes, by using information about its behavior for large strikes. These symmetries become more explicit in the so-called symmetric models, which are also discussed in the present chapter.

9.1 Implied Volatility in General Option Pricing Models

Fix $K > 0$ and $T > 0$. Then the function $\rho(\sigma) = C_{BS}(T, K, \sigma)$ is increasing on $(0, \infty)$. This follows from the fact that the Greek vega is positive (see Sect. 8.4). If $0 < K < x_0 e^{rT}$, then the range of the function ρ coincides with the interval $(x_0 - Ke^{-rT}, x_0)$, while for $x_0 e^{rT} \geq K$, the range of ρ is the interval $(0, x_0)$.

Definition 9.1 Let C be a call pricing function. For $(T, K) \in (0, \infty)^2$, the implied volatility $I(T, K)$ associated with C is the value of the volatility σ in the Black–Scholes model for which $C(T, K) = C_{BS}(T, K, \sigma)$. The implied volatility $I(T, K)$ is defined only if such a number σ exists and is unique.

A. Gulisashvili, *Analytically Tractable Stochastic Stock Price Models*,
Springer Finance, DOI 10.1007/978-3-642-31214-4_9,

© Springer-Verlag Berlin Heidelberg 2012

It follows from the discussion above that if $0 < K < x_0 e^{rT}$, then the condition $x_0 - Ke^{-rT} < C(T,K) < x_0$ is necessary for the existence of the implied volatility $I(T,K)$. Similarly, if $x_0 e^{rT} \leq K$, then $I(T,K)$ is defined if and only if $0 < C(T,K) < x_0$. Note that the inequality $C(T,K) < x_0$ holds for all $T \geq 0$ and $K > 0$. Moreover, if $(T,K) \in [0,\infty)^2$, then

$$\left(x_0 - Ke^{-rT}\right)^+ \leq C(T,K).$$

In the next definitions, we introduce special classes of call pricing functions.

Definition 9.2 The class PF_∞ consists of all call pricing functions C, for which one of the following equivalent conditions holds:

1. $C(T,K) > 0$ for all $T > 0$ and $K > 0$ with $x_0 e^{rT} \leq K$.
2. $P(T,K) > e^{-rT} K - x_0$ for all $T > 0$ and $K > 0$ with $x_0 e^{rT} \leq K$.
3. For every $T > 0$ and all $a > 0$ the random variable X_T is such that $\mathbb{P}^*[X_T < a] < 1$.

Definition 9.3 The class PF_0 consists of all call pricing functions C, for which one of the following equivalent conditions holds:

1. $P(T,K) > 0$ for all $T > 0$ and $K > 0$ with $K < x_0 e^{rT}$.
2. $C(T,K) > x_0 - e^{-rT} K$ for all $T > 0$ and $K > 0$ with $K < x_0 e^{rT}$.
3. For every $T > 0$ and all $a > 0$ the random variable X_T is such that $0 < \mathbb{P}^*[X_T < a]$.

Remark 9.4 Suppose the maturity $T > 0$ is fixed, and consider the pricing function C and the implied volatility I as functions of the strike price K. If $C \in PF_\infty$, then the implied volatility $I(K)$ is defined for large values of K. This allows to study the asymptotic behavior of the implied volatility as $K \to \infty$. Similarly, if $C \in PF_0$, then $I(K)$ exists for small values of K. Finally, if $C \in PF_\infty \cap PF_0$, then the implied volatility $I(T,K)$ exists for all $T > 0$ and $K > 0$.

9.2 Implied Volatility Surfaces and Static Arbitrage

Let $I(T,K)$ with $(T,K) \in (0,\infty)^2$ be a positive function of two variables, and suppose we would like to model the implied volatility surface by this function. Then the function $\widetilde{C}$ defined on $[0,\infty)^2$ by

$$\widetilde{C}(T,K) = \begin{cases} C_{BS}(T,K,I(T,K)), & \text{if } (T,K) \in (0,\infty)^2, \\ (x_0 - K)^+, & \text{if } T = 0,\ K \geq 0, \\ x_0, & \text{if } T \geq 0,\ K = 0, \end{cases} \tag{9.1}$$

where x_0 is the initial price of the asset in the Black–Scholes model, should be a call pricing function. For the sake of simplicity, we will assume that $r = 0$. The next definition concerns the implied volatility in a no-arbitrage environment.

Definition 9.5 It is said that the function I modeling the implied volatility is free of static arbitrage if the model of call prices given by the function $\widetilde{C}$ in (9.1) is free of static arbitrage (see Definition 8.2).

Our next goal is to provide necessary and sufficient conditions for the absence of static arbitrage in a given implied volatility model.

Theorem 9.6 *Suppose the function I models the implied volatility. Suppose also that for every $T>0$ the function $K \mapsto I(T,K)$ is twice differentiable on $(0,\infty)$. Then I is free of static arbitrage if and only if the following conditions hold*:

1. *For all $(T,K)\in(0,\infty)^2$,*

$$\left(1-\frac{K}{I}\log\left(\frac{K}{x_0}\right)\frac{\partial I}{\partial K}\right)^2+TK^2I\frac{\partial^2 I}{\partial K^2}-\frac{1}{4}T^2K^2I^2\left(\frac{\partial I}{\partial K}\right)^2 + TKI\frac{\partial I}{\partial K}\geq 0. \tag{9.2}$$

2. *For every $K>0$ the function $T\mapsto \sqrt{T}I(T,K)$ is increasing on $(0,\infty)$.*
3. *For every $T>0$, $\lim_{K\to\infty} d_1(T,K,I(T,K))=-\infty$.*

Proof It suffices to prove that the conditions in Theorem 8.3, formulated for the function $\widetilde{C}$ given by (9.1), are equivalent to conditions 1–3 in Theorem 9.6.

Fix $T>0$, and differentiate the function $\widetilde{C}$ on $(0,\infty)$ with respect to K. This gives

$$\frac{\partial \widetilde{C}}{\partial K}=\frac{\partial C_{\mathrm{BS}}}{\partial K}\big(T,K,I(T,K)\big)+\frac{\partial C_{\mathrm{BS}}}{\partial \sigma}\big(T,K,I(T,K)\big)\frac{\partial I}{\partial K}.$$

Differentiating again, we obtain

$$\frac{\partial^2 \widetilde{C}}{\partial K^2}=\frac{\partial^2 C_{\mathrm{BS}}}{\partial K^2}\big(T,K,I(T,K)\big)+2\frac{\partial^2 C_{\mathrm{BS}}}{\partial K\partial \sigma}\big(T,K,I(T,K)\big)\frac{\partial I}{\partial K} + \frac{\partial C_{\mathrm{BS}}}{\partial \sigma^2}\big(T,K,I(T,K)\big)\left(\frac{\partial I}{\partial K}\right)^2+\frac{\partial C_{\mathrm{BS}}}{\partial \sigma}\big(T,K,I(T,K)\big)\frac{\partial^2 I}{\partial K^2}.$$

Next, taking into account explicit formulas for the Greeks (see Sect. 8.4), we see that for every $T>0$ the convexity of the function $K\mapsto \widetilde{C}(T,K)$ on $(0,\infty)$ is equivalent to the following inequality:

$$\frac{1}{\sqrt{T}K^2I}+\frac{2d_1(T,K,I(T,K))}{KI}\frac{\partial I}{\partial K} + \frac{\sqrt{T}d_1(T,K,I(T,K))d_2(T,K,I(T,K))}{I}\left(\frac{\partial I}{\partial K}\right)^2 + \sqrt{T}\frac{\partial^2 I}{\partial K^2}\geq 0,\quad K>0. \tag{9.3}$$

It is not hard to see, using the definition of d_1 and d_2, that (9.3) is equivalent to (9.2). Hence the convexity of the function $K \mapsto \widetilde{C}(0, K)$ on $(0, \infty)$ is equivalent to the validity of (9.1).

We will next turn our attention to the convexity conditions for the function $K \mapsto \widetilde{C}(T, K)$ on $[0, \infty)$. Let us assume that condition 1 in Theorem 9.6 holds, and put $\varphi(K) = \widetilde{C}(T, K)$. Then the function φ is twice differentiable and convex on $(0, \infty)$. Moreover, the function φ' is increasing on $(0, \infty)$, and it follows from (8.7) that for all $0 < x < y < \infty$,

$$\varphi'(x) \le \frac{\varphi(y) - \varphi(x)}{y - x} \le \varphi'(y). \tag{9.4}$$

Using the definition of the Black–Scholes call pricing function, we see that for all $T > 0$ and $K > 0$,

$$\begin{aligned} \varphi(K) = \frac{x_0}{\sqrt{2\pi}} \int_{-\infty}^{d_1(T,K,I(T,K))} e^{-\frac{y^2}{2}} dy \\ - \frac{K}{\sqrt{2\pi}} \int_{-\infty}^{d_2(T,K,I(T,K))} e^{-\frac{y^2}{2}} dy. \end{aligned} \tag{9.5}$$

It will be shown next that

$$\lim_{K \to 0} d_1\big(T, K, I(T, K)\big) = \infty. \tag{9.6}$$

Indeed, for small values of K we have

$$d_1(T, K, I) = \frac{\log \frac{x_0}{K} + \frac{1}{2} T I^2}{\sqrt{T} I} \ge \sqrt{2 \log \frac{x_0}{K}}, \tag{9.7}$$

and (9.6) follows. Using (9.5), we obtain the following equality:

$$\lim_{K \to 0} \widetilde{C}(T, K) = x_0. \tag{9.8}$$

Therefore, the function $K \mapsto \widetilde{C}(T, K)$ is continuous on $[0, \infty)$. Our next goal is to prove the differentiability of this function from the right at $K = 0$. It follows from (9.4) that there exists the limit $M = \lim_{K \to 0} \varphi'(K)$. In addition, (9.4) and (9.8) give

$$M \le \frac{\varphi(S) - \varphi(0)}{S} = \frac{\varphi(S) - x_0}{S} \le \varphi'(S)$$

for all $0 < K < S < \infty$. Therefore $M = \varphi'_+(0)$. Moreover, (9.5) and (9.7) imply

$$\begin{aligned} \varphi(K) = x_0 - K - \frac{x_0}{\sqrt{2\pi}} \int_{d_1(T,K,I(T,K))}^{\infty} e^{-\frac{y^2}{2}} dy \\ + \frac{K}{\sqrt{2\pi}} \int_{d_2(T,K,I(T,K))}^{\infty} e^{-\frac{y^2}{2}} dy \end{aligned}$$

$$= x_0 - K - \frac{x_0}{\sqrt{2\pi}} \int_{d_1(T,K,I(T,K))}^{\infty} e^{-\frac{y^2}{2}} dy + o(K)$$

$$= x_0 - K - \frac{x_0}{\sqrt{2\pi}} \int_{\sqrt{2\log\frac{x_0}{K}}}^{\infty} e^{-\frac{y^2}{2}} dy + o(K)$$

$$= x_0 - K + o(K)$$

as $K \to 0$. Therefore, $M = -1$, and it follows that for every $T > 0$ the function $K \mapsto \widetilde{C}(T, K)$ is convex on $[0, \infty)$ (use affine extrapolation).

The next step in the proof deals with condition 3 in Theorem 9.6. Our goal is to show that

$$\lim_{K\to\infty} d_1\big(T, K, I(T, K)\big) = -\infty \iff \lim_{K\to\infty} \widetilde{C}(T, K) = 0. \tag{9.9}$$

We will first prove the following equality:

$$\lim_{K\to\infty} d_2\big(T, K, I(T, K)\big) = -\infty. \tag{9.10}$$

Suppose $d_2(T, K, I(T, K))$ does not tend to $-\infty$ as $K \to \infty$. Then there exists a sequence $K_n \uparrow \infty$ such that

$$\int_{-\infty}^{d_2(T,K_n,I(T,K_n))} e^{-\frac{y^2}{2}} dy \geq c > 0$$

for all $n \geq 1$. It follows from (9.5) that $\widetilde{C}(T, K_n) < 0$ for $n > n_0$, which is impossible. Therefore, (9.10) holds.

It will be shown next that we always have

$$K \int_{-\infty}^{d_2(T,K,I(T,K))} e^{-\frac{y^2}{2}} dy \to 0 \tag{9.11}$$

as $K \to \infty$. Reasoning as in (9.7), we see that for large values of K,

$$d_2(T, K, I)^2 \geq 2\log\frac{K}{x_0}. \tag{9.12}$$

Using formula (8.25), we obtain

$$F(K) \sim \frac{K}{|d_2(T, K, I(T, K))|} \exp\left\{-\frac{d_2(T, K, I(T, K))^2}{2}\right\},$$

as $K \to \infty$, where F denotes the function on the left-hand side of (9.11). It follows from (9.10) and (9.12) that (9.11) holds. Now it is clear that (9.5) implies the equivalence in (9.9). Note that the condition on the right-hand side of (9.9) also holds for $T = 0$. This follows from the definition of the function $\widetilde{C}$.

Next, we turn our attention to condition 2 in Theorem 9.6. It is not hard to see that this condition is equivalent to the following:

$$\frac{1}{2\sqrt{T}}+\sqrt{T}\frac{\partial I}{\partial T}\geq 0, \quad T>0. \tag{9.13}$$

On the other hand,

$$\frac{\partial \widetilde{C}}{\partial T}=\frac{\partial C_{\mathrm{BS}}}{\partial T}(T,K,I)+\frac{\partial C_{\mathrm{BS}}}{\partial \sigma}(T,K,I)\frac{\partial I}{\partial T},$$

and using the formulas for the Greeks in Sect. 8.4, we obtain

$$\frac{\partial \widetilde{C}}{\partial T}=\frac{1}{\sqrt{2\pi}}\exp\left\{-\frac{d_1(T,K,I(T,K))^2}{2}\right\}\left[\frac{1}{2\sqrt{T}}+\sqrt{T}\frac{\partial I}{\partial T}\right].$$

Now (9.13) implies that condition 2 in Theorem 9.6 is equivalent to the following condition. For all $K>0$, the function $T\mapsto \widetilde{C}(T,K)$ is non-decreasing on $(0,\infty)$. For $K=0$, the same conclusion follows from the definition of the function $\widetilde{C}$. In addition, the function $T\mapsto \widetilde{C}(T,K)$ is also non-decreasing on $[0,\infty)$. Indeed, for any volatility parameter σ in the Black–Scholes model, we have $(x_0-K)^+\leq C_{\mathrm{BS}}(T,K,\sigma)$, $T\geq 0$, $K\geq 0$. Therefore, $\widetilde{C}(0,K)\leq \widetilde{C}(T,K)$.

Let us denote by μ_T the second distributional derivative of the function $K\mapsto \widetilde{C}(T,K)$, and suppose that the conditions in the formulation of Theorem 9.6 hold for the function $\widetilde{C}$. Recall that $\varphi_+(0)=x_0$, $\varphi'_+(0)=-1$, and $\lim_{K\to\infty}\varphi(K)=0$. The function φ' is non-decreasing (see (9.4)) and integrable on $[0,\infty)$. Therefore, φ' is non-positive. Our next goal is to prove that

$$\lim_{K\to\infty} K\left|\varphi'(K)\right|=0. \tag{9.14}$$

Using (9.4), we see that $K\left|\varphi'(2K)\right|\leq \varphi(K)-\varphi(2K)$, and it is clear that the previous estimate implies (9.14). Next, taking into account (9.14), we obtain

$$\mu_T\big([0,\infty)\big)=\lim_{K\to\infty}\varphi'(K)-\varphi'_+(0)=1.$$

Moreover, the integration by parts formula for Stieltjes integrals implies the following equality:

$$\int_{[0,\infty)} x\,d\mu_T(x)=\varphi_+(0)-\lim_{K\to\infty}\varphi(K)+\lim_{K\to\infty}K\varphi'(K)=x_0.$$

It follows that condition 2 in Theorem 8.3 is valid for the function $\widetilde{C}$, provided that the conditions in the formulation of Theorem 9.6 hold. Finally, it is not hard to see, taking into account what was said above and applying Theorem 8.3, that Theorem 9.6 holds. □

9.3 Asymptotic Behavior of Implied Volatility Near Infinity

In this section, we find sharp asymptotic formulas for the implied volatility $K \mapsto I(K)$ associated with a general call pricing function C. It is assumed that the maturity T is fixed and the implied volatility is considered as a function of the strike price. We also assume that $C \in PF_\infty$. This guarantees the existence of the implied volatility for large values of the strike price.

The next theorem provides an asymptotic formula for the implied volatility associated with a general call pricing function.

Theorem 9.7 *Let $C \in PF_\infty$. Then*

$$\begin{aligned} I(K) = {} & \frac{1}{\sqrt{T}}\sqrt{2\log K + 2\log\frac{1}{C(K)} - \log\log\frac{1}{C(K)}} \\ & - \frac{1}{\sqrt{T}}\sqrt{2\log\frac{1}{C(K)} - \log\log\frac{1}{C(K)}} \\ & + O\left(\left(\log\frac{1}{C(K)}\right)^{-\frac{1}{2}}\right) \end{aligned} \tag{9.15}$$

as $K \to \infty$.

Theorem 9.7 and the mean value theorem imply the following statement:

Corollary 9.8 *For any call pricing function $C \in PF_\infty$,*

$$\begin{aligned} I(K) = {} & \frac{\sqrt{2}}{\sqrt{T}}\left[\sqrt{\log K + \log\frac{1}{C(K)}} - \sqrt{\log\frac{1}{C(K)}}\right] \\ & + O\left(\left(\log\frac{1}{C(K)}\right)^{-\frac{1}{2}} \log\log\frac{1}{C(K)}\right) \end{aligned} \tag{9.16}$$

as $K \to \infty$.

Proof of Theorem 9.7 The next lemma will be needed in the proof of Theorem 9.7.

Lemma 9.9 *Let C be a call pricing function, and fix a positive continuous increasing function ψ, satisfying $\psi(K) \to \infty$ as $K \to \infty$. Suppose ϕ is a positive function such that $\phi(K) \to \infty$ as $K \to \infty$ and*

$$C(K) \approx \frac{\psi(K)}{\phi(K)} \exp\left\{-\frac{\phi(K)^2}{2}\right\}. \tag{9.17}$$

Then the following asymptotic formula holds:

$$I(K)=\frac{1}{\sqrt{T}}\left(\sqrt{2\log\frac{K}{x_0e^{rT}}+\phi(K)^2}-\phi(K)\right)+O\left(\frac{\psi(K)}{\phi(K)}\right) \tag{9.18}$$

as $K\to\infty$.

Remark 9.10 It is easy to see that if (9.17) holds, then $C\in PF_\infty$.

Proof of Lemma 9.9 Let us compare the implied volatility I with a function $\widetilde{I}$ such that

$$0<\widetilde{I}(K)<I(K),\quad K>K_0. \tag{9.19}$$

Our goal is to prove that

$$I(K)=\widetilde{I}(K)+O\left(C(K)\exp\left\{\frac{1}{2}d_1\left(K,\widetilde{I}(K)\right)^2\right\}\right) \tag{9.20}$$

as $K\to\infty$, where $d_1(K,\sigma)$ is defined in (8.19).

It is not hard to see that the function ρ given by

$$\rho(K)=\frac{1}{\sqrt{T}}\sqrt{2\log\frac{K}{x_0e^{rT}}}$$

satisfies the equalities

$$d_1\bigl(K,\rho(K)\bigr)=0 \tag{9.21}$$

and

$$d_2\bigl(K,\rho(K)\bigr)=\sqrt{T}\rho(K). \tag{9.22}$$

Plugging (9.21) and (9.22) into the Black–Scholes formula (formula (8.22)), we obtain

$$C_{\mathrm{BS}}\bigl(K,\rho(K)\bigr)=\frac{x_0}{2}-Ke^{-rT}\frac{1}{\sqrt{2\pi}}\int_{\sqrt{T}\rho(K)}^{\infty}\exp\left\{-\frac{y^2}{2}\right\}dy\to\frac{x_0}{2} \tag{9.23}$$

as $K\to\infty$. Next, taking into account (9.23) and the fact that

$$C_{\mathrm{BS}}\bigl(K,I(K)\bigr)=C(K)\to 0$$

as $K\to\infty$, we see that $C_{\mathrm{BS}}(K,I(K))<C_{\mathrm{BS}}(K,\rho(K))$ for all $K>K_0$. Therefore,

$$I(K)<\rho(K),\quad K>K_0. \tag{9.24}$$

Here we use the fact that for every fixed $K>0$ and $T>0$ the vega is a strictly increasing function of σ.

It is easy to see that for sufficiently large values of K, the function

$$\sigma \mapsto d_1(K,\sigma) \tag{9.25}$$

increases. It follows from (9.21) and (9.24) that

$$d_1\big(K, I(K)\big) < 0, \quad K > K_1. \tag{9.26}$$

Moreover, using the explicit expression for the vega (see Sect. 8.4) and the mean value theorem, we get

$$\begin{aligned} &C_{\mathrm{BS}}\big(K, I(K)\big) - C_{\mathrm{BS}}\big(K, \widetilde{I}(K)\big) \\ &\quad = \frac{x_0\sqrt{T}}{\sqrt{2\pi}}\big(I(K) - \widetilde{I}(K)\big)\exp\left\{-\frac{d_1^2(K,\lambda)}{2}\right\}, \quad K > K_1, \end{aligned} \tag{9.27}$$

where $\widetilde{I}(K) < \lambda < I(K)$. Since the function in (9.25) increases and (9.26) holds,

$$d_1\big(K, \tilde{I}(K)\big) < d_1(K,\lambda) < d_1\big(K, I(K)\big) < 0, \quad K > K_1. \tag{9.28}$$

Now, using (9.27) and (9.28), we establish the validity of formula (9.20).

Let us continue the proof of Lemma 9.9. Suppose $\widetilde{I}$ is a function satisfying the equality

$$d_1\big(K, \tilde{I}(K)\big) = -\phi(K), \quad K > K_0. \tag{9.29}$$

Such a function exists, since for large values of K the function $\sigma \mapsto d_1(K,\sigma)$ increases from $-\infty$ to ∞. It follows from (9.29) and from the definition of d_1 that

$$\widetilde{I}(K) = \frac{1}{\sqrt{T}}\left(\sqrt{2\log\frac{K}{x_0 e^{rT}} + \phi(K)^2} - \phi(K)\right). \tag{9.30}$$

Our next goal is to use formula (9.20) with $\widetilde{I}$ defined in (9.30). However, we have to first prove inequality (9.19). Using (8.22), (8.25), and (9.29), we see that there exist constants $c_1 > 0$ and $c_2 > 0$ such that

$$\begin{aligned} &C_{\mathrm{BS}}\big(K, I(K)\big) - C_{\mathrm{BS}}\big(K, \tilde{I}(K)\big) \\ &\quad = C(K) - C_{\mathrm{BS}}\big(K, \tilde{I}(K)\big) \\ &\quad \geq c_1\frac{\psi(K)}{\phi(K)}\exp\left\{-\frac{\phi(K)^2}{2}\right\} - c_2\frac{1}{d_1(K,\widetilde{I}(K))}\exp\left\{-\frac{d_1(K,\widetilde{I}(K))^2}{2}\right\} \\ &\quad = c_1\frac{\psi(K)}{\phi(K)}\exp\left\{-\frac{\phi(K)^2}{2}\right\} - c_2\frac{1}{\phi(K)}\exp\left\{-\frac{\phi(K)^2}{2}\right\}, \quad K > K_2. \end{aligned} \tag{9.31}$$

Since $\psi(K) \to \infty$ as $K \to \infty$ and (9.31) holds, we get

$$C_{\mathrm{BS}}\big(K, I(K)\big) > C_{\mathrm{BS}}\big(K, \tilde{I}(K)\big)$$

for sufficiently large values of K. Using the fact that the vega is an increasing function of σ, we obtain inequality (9.19). Now it is clear that (9.18) follows from (9.17), (9.20), and (9.29).

The proof of Lemma 9.9 is thus completed. □

Let us return to the proof of Theorem 9.7. Let ψ be a positive increasing function such that $\psi(K) \to \infty$ as $K \to \infty$. We also assume that the function $\psi(K)$ tends to infinity slower than the function $K \mapsto \log\log\frac{1}{C(K)}$. Put

$$\phi(K) = \left[2\log\frac{1}{C(K)} - \log\log\frac{1}{C(K)} + 2\log\psi(K)\right]^{\frac{1}{2}}.$$

Then we have

$$\phi(K) \approx \sqrt{2\log\frac{1}{C(K)}}$$

as $K \to \infty$. It follows that

$$\psi(K)\exp\left\{-\frac{\phi(K)^2}{2}\right\}\phi(K)^{-1} \approx C(K)$$

as $K \to \infty$. Using formula (9.18), we obtain

$$\begin{aligned} I(K) &= \frac{1}{\sqrt{T}}\left(\sqrt{2\log\frac{K}{x_0e^{rT}} + \phi(K)^2} - \phi(K)\right) \\ &\quad + O\left(\left(\log\frac{1}{C(K)}\right)^{-\frac{1}{2}}\psi(K)\right) \end{aligned} \tag{9.32}$$

as $K \to \infty$. Now, it is not hard to see that (9.15) can be derived from (9.32), the mean value theorem, and Lemma 3.1.

This completes the proof of Theorem 9.7. □

9.4 Corollaries

Our objective in this section is to replace the function C in formula (9.15) by another function $\widetilde{C}$.

Corollary 9.11 *Let $C \in PF_\infty$, and suppose $\widetilde{C}$ is a positive function such that $\widetilde{C}(K) \approx C(K)$ as $K \to \infty$. Then*

$$I(K) = \frac{1}{\sqrt{T}}\sqrt{2\log K + 2\log\frac{1}{\widetilde{C}(K)} - \log\log\frac{1}{\widetilde{C}(K)}}$$

$$- \frac{1}{\sqrt{T}} \sqrt{2 \log \frac{1}{\widetilde{C}(K)} - \log \log \frac{1}{\widetilde{C}(K)}} + O\left(\left(\log \frac{1}{\widetilde{C}(K)}\right)^{-\frac{1}{2}}\right) \tag{9.33}$$

as $K \to \infty$. Therefore,

$$I(K) = \frac{\sqrt{2}}{\sqrt{T}} \left[\sqrt{\log K + \log \frac{1}{\widetilde{C}(K)}} - \sqrt{\log \frac{1}{\widetilde{C}(K)}}\right] + O\left(\left(\log \frac{1}{\widetilde{C}(K)}\right)^{-\frac{1}{2}} \log \log \frac{1}{\widetilde{C}(K)}\right) \tag{9.34}$$

as $K \to \infty$.

Formula (9.33) can be established exactly as (9.15). Formula (9.34) follows from (9.33) and the mean value theorem.

We can also replace a call pricing function C in (9.15) by a function $\widetilde{C}$ under more general conditions. However, this may lead to a weaker error estimate. For instance, put

$$\tau(K) = \left| \log \frac{1}{C(K)} - \log \frac{1}{\widetilde{C}(K)} \right|. \tag{9.35}$$

Then the following theorem holds:

Theorem 9.12 *Let $C \in PF_\infty$, and suppose $\widetilde{C}$ is a positive function satisfying the following condition. There exist $K_1 > 0$ and c with $0 < c < 1$ such that*

$$\tau(K) < c \log \frac{1}{\widetilde{C}(K)} \tag{9.36}$$

for all $K > K_1$, where τ is defined by (9.35). Then

$$I(K) = \frac{1}{\sqrt{T}} \sqrt{2 \log K + 2 \log \frac{1}{\widetilde{C}(K)} - \log \log \frac{1}{\widetilde{C}(K)}} - \frac{1}{\sqrt{T}} \sqrt{2 \log \frac{1}{\widetilde{C}(K)} - \log \log \frac{1}{\widetilde{C}(K)}} + O\left(\left(\log \frac{1}{\widetilde{C}(K)}\right)^{-\frac{1}{2}} [1 + \tau(K)]\right) \tag{9.37}$$

as $K \to \infty$.

Proof It is not hard to check that (9.36) implies the formula

$$\log \frac{1}{\widetilde{C}(K)} \approx \log \frac{1}{C(K)}$$

as $K \to \infty$. Now using (9.15), (9.35), and the mean value theorem, we obtain (9.37). □

The next statement follows from Theorem 9.12 and the mean value theorem.

Corollary 9.13 *Let $C \in PF_\infty$, and suppose $\widetilde{C}$ is a positive function satisfying the following condition. There exist $\nu > 0$ and $K_0 > 0$ such that*

$$\left| \log \frac{1}{\widetilde{C}(K)} - \log \frac{1}{C(K)} \right| \leq \nu \log \log \frac{1}{\widetilde{C}(K)} \tag{9.38}$$

for all $K > K_0$. Then

$$I(K) = \frac{\sqrt{2}}{\sqrt{T}} \left[\sqrt{\log K + \log \frac{1}{\widetilde{C}(K)}} - \sqrt{\log \frac{1}{\widetilde{C}(K)}} \right] + O\left(\left(\log \frac{1}{\widetilde{C}(K)} \right)^{-\frac{1}{2}} \log \log \frac{1}{\widetilde{C}(K)} \right)$$

as $K \to \infty$.

Remark 9.14 It is not hard to see that if $C(K) \approx \widetilde{C}(K)$ as $K \to \infty$, or if (9.38) holds, then $\log \frac{1}{C(K)} \sim \log \frac{1}{\widetilde{C}(K)}$ as $K \to \infty$.

Corollary 9.15 *Let $C \in PF_\infty$, and suppose $\widetilde{C}$ is a positive function satisfying the condition*

$$\log \frac{1}{C(K)} \sim \log \frac{1}{\widetilde{C}(K)} \tag{9.39}$$

as $K \to \infty$. Then

$$I(K) \sim \frac{\sqrt{2}}{\sqrt{T}} \left[\sqrt{\log K + \log \frac{1}{\widetilde{C}(K)}} - \sqrt{\log \frac{1}{\widetilde{C}(K)}} \right] \tag{9.40}$$

as $K \to \infty$.

Proof It follows from (9.16) that

$$I(K) \sim \frac{\sqrt{2}}{\sqrt{T}} \left[\sqrt{\log K + \log \frac{1}{\widetilde{C}(K)}} - \sqrt{\log \frac{1}{\widetilde{C}(K)}} \right] \Lambda(K) \tag{9.41}$$

where

$$\Lambda(K)=\frac{\sqrt{\log K+\log\frac{1}{\widetilde{C}(K)}}+\sqrt{\log\frac{1}{\widetilde{C}(K)}}}{\sqrt{\log K+\log\frac{1}{C(K)}}+\sqrt{\log\frac{1}{C(K)}}}.$$

We will next prove that $\Lambda(K)\to 1$ as $K\to\infty$. We have

$$\Lambda(K)=\frac{\sqrt{\Lambda_1(K)+\Lambda_2(K)}+\sqrt{\Lambda_2(K)}}{\sqrt{\Lambda_1(K)+1}+1}$$

where

$$\Lambda_1(K)=\frac{\log K}{\log\frac{1}{C(K)}}\quad\text{and}\quad\Lambda_2(K)=\frac{\log\frac{1}{\widetilde{C}(K)}}{\log\frac{1}{C(K)}}.$$

It is not hard to show that for all positive numbers a and b,

$$\left|\sqrt{a+b}-\sqrt{a+1}\right|\le\left|\sqrt{b}-1\right|.$$

Therefore,

$$\begin{aligned}\left|\Lambda(K)-1\right|&=\frac{|\sqrt{\Lambda_1(K)+\Lambda_2(K)}-\sqrt{\Lambda_1(K)+1}|+|\sqrt{\Lambda_2(K)}-1|}{\sqrt{\Lambda_1(K)+1}+1}\\&\le\left|\sqrt{\Lambda_2(K)}-1\right|\end{aligned}\tag{9.42}$$

for $K>K_0$. It follows from (9.39) and (9.42) that $\Lambda(K)\to 1$ as $K\to\infty$. Next using (9.41) we see that (9.40) holds.

This completes the proof of Corollary 9.15. □

9.5 Extra Terms: First-Order Asymptotic Formulas for Implied Volatility

Formula (9.15) characterizes the asymptotic behavior of the implied volatility in terms of the call pricing function C, while in formula (9.33), the function C is replaced by a function $\widetilde{C}$, equivalent to C in a certain sense. We call these formulas zero-order asymptotic formulas for the implied volatility. In an important recent paper [GL11], K. Gao and R. Lee obtained a hierarchy of higher-order asymptotic formulas generalizing formula (9.15). Note that formula (9.33) cannot be generalized in a similar way.

In the present section we establish a first-order asymptotic formula, which is different from similar first-order formulas obtained in [GL11]. Higher-order asymptotic formulas from [GL11] are discussed in Sect. 9.6. Our proofs of above-mentioned formulas are refinements of the proof of Theorem 9.7 given in Sect. 9.3, and they differ from the proofs given in [GL11].

For the sake of simplicity, we assume $x_0=1$ and $r=0$.

Theorem 9.16 *Let $C \in PF_\infty$, and suppose there exist a number $\lambda > 0$ and a continuous function Λ satisfying the following conditions:*

$$\Lambda(K) = o\left(\log \frac{1}{C(K)}\right)$$

and

$$\log \frac{1}{C(K)} = \lambda \log K + O\big(\Lambda(K)\big) \tag{9.43}$$

as $K \to \infty$. Then

$$\begin{aligned} I(K) = {} & \frac{\sqrt{2}}{\sqrt{T}} \sqrt{\log K + \log \frac{1}{C(K)} - \frac{1}{2} \log\log \frac{1}{C(K)} + \log \frac{\sqrt{\lambda+1} - \sqrt{\lambda}}{2\sqrt{\pi}\sqrt{\lambda+1}}} \\ & - \frac{\sqrt{2}}{\sqrt{T}} \sqrt{\log \frac{1}{C(K)} - \frac{1}{2} \log\log \frac{1}{C(K)} + \log \frac{\sqrt{\lambda+1} - \sqrt{\lambda}}{2\sqrt{\pi}\sqrt{\lambda+1}}} \\ & + O\left(\Lambda(K)\left(\log \frac{1}{C(K)}\right)^{-\frac{3}{2}}\right) \\ & + O\left(\log\log \frac{1}{C(K)} \left(\log \frac{1}{C(K)}\right)^{-\frac{3}{2}}\right) \end{aligned} \tag{9.44}$$

as $K \to \infty$.

Proof Suppose ψ is a positive slowly increasing function such that $\psi(K) \to \infty$ and

$$\psi(K) \frac{\Lambda(K) + \log\log \frac{1}{C(K)}}{\log \frac{1}{C(K)}} \to 0$$

as $K \to \infty$. Put

$$\begin{aligned} \varphi^2(K) = {} & 2 \log \frac{1}{C(K)} - \log\log \frac{1}{C(K)} + 2 \log A \\ & + 2 \log\left[1 + \psi(K) \frac{\Lambda(K) + \log\log \frac{1}{C(K)}}{\log \frac{1}{C(K)}}\right]. \end{aligned} \tag{9.45}$$

Here $A > 0$ is a constant that will be chosen later. We have

$$\begin{aligned} & \exp\left\{\frac{\varphi^2(K)}{2}\right\} C(K) \\ & = A \left(\log \frac{1}{C(K)}\right)^{-\frac{1}{2}} + A\psi(K) \left[\Lambda(K) + \log\log \frac{1}{C(K)}\right] \left(\log \frac{1}{C(K)}\right)^{-\frac{3}{2}}. \end{aligned} \tag{9.46}$$

Lemma 9.17 *Let $\widetilde{I}$ be the function, for which* (9.29) *holds with φ given by* (9.45). *Set*

$$A = \frac{\sqrt{\lambda+1}-\sqrt{\lambda}}{2\sqrt{\pi}\sqrt{\lambda+1}}. \tag{9.47}$$

Then $\widetilde{I}(K) \le I(K)$.

Proof It follows from (8.22) and (8.25) that

$$\exp\left\{\frac{\varphi^2(K)}{2}\right\} C_{\mathrm{BS}}\bigl(K, \widetilde{I}(K)\bigr) = \frac{1}{\sqrt{2\pi}}\frac{1}{\varphi(K)} - \frac{1}{\sqrt{2\pi}}\frac{1}{\sqrt{2\log K + \varphi^2(K)}} + O\bigl(\varphi^{-3}(K)\bigr) \tag{9.48}$$

as $K \to \infty$. Using (9.43), (9.45), and the formula

$$(1+h)^{-\frac{1}{2}} = 1 + O(h), \quad h \to 0,$$

we obtain

$$\frac{1}{\varphi(K)} = \frac{1}{\sqrt{2}}\left(\log\frac{1}{C(K)}\right)^{-\frac{1}{2}} + O\left(\log\log\frac{1}{C(K)}\left(\log\frac{1}{C(K)}\right)^{-\frac{3}{2}}\right) \tag{9.49}$$

as $K \to \infty$. Moreover, we have

$$\frac{1}{\sqrt{2\log K + \varphi^2(K)}} = \frac{1}{\sqrt{2}}\left(\frac{\lambda}{\lambda+1}\right)^{\frac{1}{2}}\left(\log\frac{1}{C(K)}\right)^{-\frac{1}{2}} + O\left(\Lambda(K)\left(\log\frac{1}{C(K)}\right)^{-\frac{3}{2}}\right) + O\left(\log\log\frac{1}{C(K)}\left(\log\frac{1}{C(K)}\right)^{-\frac{3}{2}}\right) \tag{9.50}$$

as $K \to \infty$.

Our next goal is to combine formulas (9.46)–(9.50). It is not hard to see that there exists $K_0 > 0$ such that

$$C_{\mathrm{BS}}\bigl(K, I(K)\bigr) - C_{\mathrm{BS}}\bigl(K, \widetilde{I}(K)\bigr) = C(K) - C_{\mathrm{BS}}\bigl(K, \widetilde{I}(K)\bigr) > 0$$

for all $K > K_0$. Now Lemma 9.17 follows from the fact that the vega is an increasing function of σ. □

Let us return to the proof of Theorem 9.16. Since formula (9.27) holds, we have

$$I(K) - \widetilde{I}(K) = O\left(\exp\left\{\frac{\varphi^2(K)}{2}\right\}\left[C(K) - C_{\mathrm{BS}}\left(K, \widetilde{I}(K)\right)\right]\right)$$

as $K \to \infty$. Now using formulas (9.46)–(9.50) again, we obtain

$$\begin{aligned} I(K) &= \widetilde{I}(K) \\ &\quad + O\left(\psi(K)\left[\Lambda(K) + \log\log\frac{1}{C(K)}\right]\left(\log\frac{1}{C(K)}\right)^{-\frac{3}{2}}\right) \end{aligned} \tag{9.51}$$

as $K \to \infty$. It follows from (9.30) and (9.45) that

$$\begin{aligned} \widetilde{I}(K) &= \sqrt{2\log K + 2\log\frac{1}{C(K)} - \log\log\frac{1}{C(K)} + 2\log A + V(K)} \\ &\quad - \sqrt{2\log\frac{1}{C(K)} - \log\log\frac{1}{C(K)} + 2\log A + V(K)}, \end{aligned} \tag{9.52}$$

where

$$\begin{aligned} V(K) &= 2\log\left(1 + \psi(K)\frac{\Lambda(K) + \log\log\frac{1}{C(K)}}{\log\frac{1}{C(K)}}\right) \\ &= O\left(\psi(K)\frac{\Lambda(K) + \log\log\frac{1}{C(K)}}{\log\frac{1}{C(K)}}\right) \end{aligned} \tag{9.53}$$

as $K \to \infty$. Applying the mean value theorem to (9.52) and taking into account (9.47), (9.51), and (9.53), we obtain (9.44) with an extra factor $\psi(K)$ in the error term. Finally, using Lemma 3.1, we get rid of the extra factor.

This completes the proof of Theorem 9.16. □

Formula (9.44) will be used in Sect. 10.5 to study the asymptotic behavior of the implied volatility in the correlated Heston model.

9.6 Extra Terms: Higher-Order Asymptotic Formulas for Implied Volatility

In this section, we discuss higher-order asymptotic formulas for the implied volatility obtained in [GL11]. We restrict ourselves to second- and third-order formulas, since the higher-order cases can be treated similarly. Note that when the order grows, the formulas become more and more complicated. That is why we decided to use simpler formulas from Sects. 9.3 and 9.4 in the rest of the present book.

Let us begin with a second-order formula (see [GL11], formula (6.2) in Corollary 6.1). Our presentation of this result of Gao and Lee is different from that in [GL11]. The main idea is to replace the constant λ in Theorem 9.16 by the function

$$\lambda(K) = (\log K)^{-1} \log \frac{1}{C(K)}$$

and put $\Lambda(K) = 0$. Then formula (9.47) takes the following form:

$$A(K) = \frac{\sqrt{\log K + \log \frac{1}{C(K)}} - \sqrt{\log \frac{1}{C(K)}}}{2\sqrt{\pi}\sqrt{\log K + \log \frac{1}{C(K)}}}. \tag{9.54}$$

This choice of the function A leads to the cancellation of all the terms in the upper estimate for the function $C(K) - C_{\mathrm{BS}}(K, \widetilde{I}(K))$, except for the higher-order error terms (see the proof of Theorem 9.16). To justify the previous statement, we will need the estimate

$$\begin{aligned} 0 &\le \log \frac{1}{A(K)} \\ &= \log(2\sqrt{\pi}) \\ &\quad + \log \frac{\sqrt{\log K + \log \frac{1}{C(K)}}(\sqrt{\log K + \log \frac{1}{C(K)}} + \sqrt{\log \frac{1}{C(K)}})}{\log K} \\ &= O\left(\log\log \frac{1}{C(K)}\right). \end{aligned} \tag{9.55}$$

Taking into account the previous remarks, we see that the following assertion holds.

Theorem 9.18 *Let $C \in PF_{\infty}$. Then*

$$\begin{aligned} I(K) &= \frac{\sqrt{2}}{\sqrt{T}}\sqrt{\log K + \log \frac{1}{C(K)} - \frac{1}{2}\log\log \frac{1}{C(K)} + \log A(K)} \\ &\quad - \frac{\sqrt{2}}{\sqrt{T}}\sqrt{\log \frac{1}{C(K)} - \frac{1}{2}\log\log \frac{1}{C(K)} + \log A(K)} \\ &\quad + O\left(\log\log \frac{1}{C(K)} \left(\log \frac{1}{C(K)}\right)^{-\frac{3}{2}}\right) \end{aligned} \tag{9.56}$$

as $K \to \infty$, where the function A is defined by (9.54).

Our next goal is to establish a third-order asymptotic formula for the implied volatility (see formula (9.68) below). The proof of this formula is similar to that of formula (9.56), but is more involved. Put

$$\varphi(K)^2 = 2\log\frac{1}{C(K)} - \log\log\frac{1}{C(K)} + 2\log A(K) + 2\log\left[1 + \frac{B(K)}{\log\frac{1}{C(K)}} + \psi(K)\frac{\log\log^2\frac{1}{C(K)}}{\log^2\frac{1}{C(K)}}\right]. \tag{9.57}$$

In (9.57), ψ is a positive continuous function such that $\psi(K) \to \infty$ and

$$\psi(K)\left(\log\frac{1}{C(K)}\right)^{-1} \to 0 \tag{9.58}$$

as $K \to \infty$. The function B, appearing in (9.57), will be chosen later. This function should satisfy the following condition:

$$B(K) = O\left(\log\log\frac{1}{C(K)}\right) \tag{9.59}$$

as $K \to \infty$. We have

$$\exp\left\{\frac{\varphi(K)^2}{2}\right\}C(K) = \left(\log\frac{1}{C(K)}\right)^{-\frac{1}{2}} A(K) \times \left[1 + \frac{B(K)}{\log\frac{1}{C(K)}} + \psi(K)\frac{\log\log^2\frac{1}{C(K)}}{\log^2\frac{1}{C(K)}}\right]. \tag{9.60}$$

On the other hand, using (8.22) and (8.25), we obtain

$$\exp\left\{\frac{\varphi(K)^2}{2}\right\}C_{BS}(K, \widetilde{I}(K)) = \frac{1}{\sqrt{2\pi}}\left[\frac{1}{\varphi(K)} - \frac{1}{(2\log K + \varphi(K)^2)^{\frac{1}{2}}} - \frac{1}{\varphi(K)^3} + \frac{1}{(2\log K + \varphi(K)^2)^{\frac{3}{2}}}\right] + O\left(\left(\log\frac{1}{C(K)}\right)^{-\frac{5}{2}}\right) \tag{9.61}$$

as $K \to \infty$. Set

$$h(K) = -\frac{\log\log\frac{1}{C(K)}}{2\log\frac{1}{C(K)}} + \frac{\log A(K)}{\log\frac{1}{C(K)}} + \left(\log\frac{1}{C(K)}\right)^{-1}\log\left[1 + \frac{B(K)}{\log\frac{1}{C(K)}} + \psi(K)\frac{(\log\log\frac{1}{C(K)})^2}{\log^2\frac{1}{C(K)}}\right].$$

Using (9.55), (9.58), and (9.59), we obtain

$$h(K)=O\left(\frac{\log\log\frac{1}{C(K)}}{\log\frac{1}{C(K)}}\right)$$

as $K\to\infty$. Therefore,

$$\begin{aligned}\frac{1}{\sqrt{2\pi}\varphi(K)}&=\frac{1}{2\sqrt{\pi}}\left(\log\frac{1}{C(K)}\right)^{-\frac{1}{2}}(1+h(K))^{-\frac{1}{2}}\\&=\frac{1}{2\sqrt{\pi}}\left(\log\frac{1}{C(K)}\right)^{-\frac{1}{2}}\left(1-\frac{1}{2}h(K)+O\big(h(K)^2\big)\right)\\&=\frac{1}{2\sqrt{\pi}}\left(\log\frac{1}{C(K)}\right)^{-\frac{1}{2}}+\frac{\log\log\frac{1}{C(K)}-2\log A(K)}{8\sqrt{\pi}(\log\frac{1}{C(K)})^{\frac{3}{2}}}\\&\quad+O\left(\left(\log\frac{1}{C(K)}\right)^{-\frac{5}{2}}\left(\log\log\frac{1}{C(K)}\right)^2\right)\end{aligned}\tag{9.62}$$

as $K\to\infty$. Similarly,

$$\begin{aligned}&\frac{1}{\sqrt{2\pi}\sqrt{2\log K+\varphi^2(K)}}\\&=\frac{1}{2\sqrt{\pi}}\left(\frac{\log\frac{1}{C(K)}}{\log K+\log\frac{1}{C(K)}}\right)^{\frac{1}{2}}\left(\log\frac{1}{C(K)}\right)^{-\frac{1}{2}}+\frac{\log\log\frac{1}{C(K)}-2\log A(K)}{8\sqrt{\pi}(\log K+\log\frac{1}{C(K)})^{\frac{3}{2}}}\\&\quad+O\left(\left(\log\frac{1}{C(K)}\right)^{-\frac{5}{2}}\left(\log\log\frac{1}{C(K)}\right)^2\right)\end{aligned}\tag{9.63}$$

as $K\to\infty$. Moreover,

$$\begin{aligned}\frac{1}{\sqrt{2\pi}\varphi(K)^3}&=\frac{1}{4\sqrt{\pi}}\left(\log\frac{1}{C(K)}\right)^{-\frac{3}{2}}\\&\quad+O\left(\left(\log\frac{1}{C(K)}\right)^{-\frac{5}{2}}\log\log\frac{1}{C(K)}\right)\end{aligned}\tag{9.64}$$

and

$$\begin{aligned}&\frac{1}{\sqrt{2\pi}(2\log K+\varphi^2(K))^{\frac{3}{2}}}\\&=\frac{1}{4\sqrt{\pi}}\left(\frac{\log\frac{1}{C(K)}}{\log K+\log\frac{1}{C(K)}}\right)^{\frac{3}{2}}\left(\log\frac{1}{C(K)}\right)^{-\frac{3}{2}}\end{aligned}$$

$$+O\left(\left(\log\frac{1}{C(K)}\right)^{-\frac{5}{2}}\log\log\frac{1}{C(K)}\right) \tag{9.65}$$

as $K\to\infty$.

Our next goal is to combine formulas (9.60)–(9.65). Recalling the cancellation properties of the function A, we see that the correct choice of the function B is as follows:

$$B(K)=\frac{1}{8\sqrt{\pi}} \times\frac{(\log\log\frac{1}{C(K)}-2\log A(K)-2)[(\log K+\log\frac{1}{C(K)})^{\frac{3}{2}}-(\log\frac{1}{C(K)})^{\frac{3}{2}}]}{A(K)(\log K+\log\frac{1}{C(K)})^{\frac{3}{2}}}. \tag{9.66}$$

Indeed, it is not hard to see that with this choice of B all the terms in the estimate for the difference $C(K)-C_{\mathrm{BS}}(K,\widetilde{I}(K))$, containing the factor $(\log\frac{1}{C(K)})^{-\frac{3}{2}}$, cancel out. It follows that formula (9.66) can be rewritten in the following form:

$$B(K)=\frac{\log\log\frac{1}{C(K)}-2\log A(K)-2}{4(\log K+\log\frac{1}{C(K)})} \times\left(\log K+2\log\frac{1}{C(K)}+\sqrt{\left(\log K+\log\frac{1}{C(K)}\right)\log\frac{1}{C(K)}}\right). \tag{9.67}$$

Here we take into account (9.54).

It remains to prove that the function B satisfies condition (9.59). It is not hard to see that this condition follows from formulas (9.55) and (9.67). Analyzing the proof sketched above, we see that the following assertion holds.

Theorem 9.19 *Let $C\in PF_{\infty}$. Then*

$$I(K) =\frac{\sqrt{2}}{\sqrt{T}} \times\sqrt{\log K+\log\frac{1}{C(K)}-\frac{1}{2}\log\log\frac{1}{C(K)}+\log A(K)+\log\left[1+\frac{B(K)}{\log\frac{1}{C(K)}}\right]} -\frac{\sqrt{2}}{\sqrt{T}}\sqrt{\log\frac{1}{C(K)}-\frac{1}{2}\log\log\frac{1}{C(K)}+\log A(K)+\log\left[1+\frac{B(K)}{\log\frac{1}{C(K)}}\right]} +O\left(\left(\log\log\frac{1}{C(K)}\right)^{2}\left(\log\frac{1}{C(K)}\right)^{-\frac{5}{2}}\right) \tag{9.68}$$

as $K \to \infty$, *where the function* A *is defined by* (9.54).

Formula (9.68) is a third-order asymptotic formula for the implied volatility in a general model of call prices.

9.7 Symmetries and Asymptotic Behavior of Implied Volatility Near Zero

In this section, we turn our attention to the asymptotic behavior of the implied volatility as $K \to 0$. It is interesting to mention that one can derive asymptotic formulas for the implied volatility at small strikes from similar results at large strikes, by taking into account certain symmetries existing in the world of stochastic asset price models. We will next describe those symmetries and explain what follows from them.

Let C be a general call pricing function, and let X be the corresponding stock price process. This process is defined on a filtered probability space $(\Omega, \mathcal{F}, \{\mathcal{F}_t\}, \mathbb{P}^*)$, where $\mathbb{P}^*$ is a risk-neutral probability measure. We assume that the interest rate r and the initial condition x_0 are fixed, and denote by μ_T the distribution of the random variable X_T. Put

$$\eta_T(K) = \left(x_0 e^{rT}\right)^2 K^{-1}.$$

We call η_T a symmetry transformation. It is easy to see that the Black–Scholes pricing function C_{BS} satisfies the following condition:

$$C_{\mathrm{BS}}(T, K, \sigma) = x_0 - K e^{-rT} + \frac{K e^{-rT}}{x_0} C_{\mathrm{BS}}\left(T, \eta_T(K), \sigma\right). \tag{9.69}$$

On the other hand, the put–call parity formula implies that

$$C(T, K) = x_0 - K e^{-rT} + \frac{K e^{-rT}}{x_0} G\left(T, \eta_T(K)\right), \tag{9.70}$$

where G is given by

$$G(T, K) = \frac{K}{x_0 e^{rT}} P\left(T, \eta_T(K)\right). \tag{9.71}$$

It follows from (8.4) and (9.71) that

$$G(T, K) = x_0 \int_0^{\eta_T(K)} d\mu_T(x) - \frac{K}{x_0 e^{2rT}} \int_0^{\eta_T(K)} x \, d\mu_T(x). \tag{9.72}$$

Define a family of Borel measures $\{\tilde{\mu}_T\}_{T \geq 0}$ on $(0, \infty)$ as follows. For every Borel subset A of $(0, \infty)$ put

$$\tilde{\mu}_T(A) = \frac{1}{x_0 e^{rT}} \int_{\eta_T(A)} x \, d\mu_T(x). \tag{9.73}$$

It is not hard to see that $\{\tilde{\mu}_T\}_{T\geq 0}$ is a family of probability measures. Moreover, for all $K > 0$ and $T \geq 0$, we have

$$\int_K^\infty d\tilde{\mu}_T(x) = \frac{1}{x_0 e^{rT}} \int_0^{\eta_T(K)} x\, d\mu_T(x) \tag{9.74}$$

and

$$\int_K^\infty x\, d\tilde{\mu}_T(x) = x_0 e^{rT} \int_0^{\eta_T(K)} d\mu_T(x). \tag{9.75}$$

It follows from (9.72), (9.74), and (9.75) that

$$G(T,K) = e^{-rT} \int_K^\infty x\, d\tilde{\mu}_T(x) - e^{-rT} K \int_K^\infty d\tilde{\mu}_T(x). \tag{9.76}$$

Remark 9.20 Suppose for every $T > 0$ the measure μ_T is absolutely continuous with respect to the Lebesgue measure on $(0,\infty)$. Denote the Radon–Nikodym derivative of μ_T with respect to the Lebesgue measure by D_T. Then, for every $T > 0$ the measure $\tilde{\mu}_T$ admits a density $\widetilde{D}_Y$ given by

$$\widetilde{D}_T(x) = \frac{(x_0 e^{rT})^3}{x^3} D_T\left(\frac{(x_0 e^{rT})^2}{x}\right), \quad x > 0.$$

The next theorem has important consequences. For example, it will allow us to establish a link between the asymptotic behavior of the implied volatility at large and small strikes.

Theorem 9.21 *Let C be a call pricing function and let P be the corresponding put pricing function. Then the function G defined by* (9.71) *is a call pricing function with the same interest rate r and the initial condition x_0 as the pricing function C. Moreover, if $\widetilde{X}$ is the stock price process associated with G, then for every $T > 0$ the measure $\tilde{\mu}_T$ defined by* (9.73) *is the distribution of the random variable $\widetilde{X}_T$.*

Proof According to Theorem 8.3, it suffices to prove that conditions 1–5 in the formulation of this theorem are valid for the function G. We have already shown that for every $T \geq 0$, $\tilde{\mu}_T$ is a probability measure. In addition, equality (8.8) holds for $\tilde{\mu}_T$, by (9.75). Put

$$V(T,K) = \int_K^\infty x\, d\tilde{\mu}_T(x) - K \int_K^\infty d\tilde{\mu}_T(x).$$

Then $G(T,K) = e^{-rT} V(T,K)$. Moreover, the function $K \mapsto V(T,K)$ is convex on $[0,\infty)$, since its second distributional derivative coincides with the measure $\tilde{\mu}_T$. This establishes conditions 1 and 2 in Theorem 8.3. The equality $G(0,K) = (x_0 - K)^+$ can be obtained using (9.71). Thus condition 4 holds. Next, we see that (9.76) implies

$$G(T,K) \le e^{-rT} \int_K^\infty x\, d\tilde{\mu}_T(x),$$

and hence $\lim_{K\to\infty} G(T,K) = 0$. This establishes condition 5. In order to prove the validity of condition 3 for G, we notice that (9.70) gives the following:

$$G\left(T, e^{rT} K\right) = \frac{K}{x_0} C\left(T, e^{rT} \frac{x_0^2}{K}\right) + x_0 - K. \tag{9.77}$$

Now it is clear that condition 3 for G follows from the same condition for C. Therefore, G is a call pricing function.

This completes the proof of Theorem 9.21. □

Remark 9.22 It is not hard to see that if the call pricing function C in Theorem 9.21 satisfies $C \in PF_\infty$, then $G \in PF_0$. Similarly, if $C \in PF_0$, then $G \in PF_\infty$.

Let C be a call pricing function such that $C \in PF_\infty \cap PF_0$. Then $G \in PF_\infty \cap PF_0$, and hence the implied volatilities I_C and I_G associated with the pricing functions C and G, respectively, exist for all $T > 0$ and $K > 0$. Replacing σ by $I_C(K)$ in (9.69) and taking into account (9.70) and the equality

$$C_{\mathrm{BS}}\left(T, K, I_C(T,K)\right) = C(T,K),$$

we see that

$$C_{\mathrm{BS}}\left(T, \eta_T(K), I_C(T,K)\right) = G\left(T, \eta_T(K)\right).$$

Therefore, the following lemma holds.

Lemma 9.23 *Let $C \in PF_\infty \cap PF_0$, and let G be defined by* (9.76). *Then*

$$I_C(T,K) = I_G\left(T, \eta_T(K)\right) \tag{9.78}$$

for all $T > 0$ and $K > 0$.

Lemma 9.23 shows that the implied volatility associated with C can be obtained from the implied volatility associated with G by applying the symmetry transformation.

9.8 Symmetric Models

The notion of a symmetric model is based on the symmetry properties of stochastic models discussed in the previous section.

Definition 9.24 A stochastic asset price model is called symmetric if, for every $T > 0$ the distributions μ_T and $\tilde{\mu}_T$ coincide.

Lemma 9.25 *The following statements hold*:

1. *Suppose for every $T > 0$ the measure μ_T admits a density D_T. Then the model is symmetric if and only if for all $T > 0$,*
$$D_T(x) = \left(x_0 e^{rT}\right)^3 x^{-3} D_T\left(\left(x_0 e^{rT}\right)^2 x^{-1}\right) \tag{9.79}$$
almost everywhere with respect to the Lebesgue measure on $(0, \infty)$.
2. *Suppose the asset price process X is strictly positive and for every $T > 0$ the measure μ_T admits a density D_T. Define the log-price process by $X^{\log} = \log X$ and denote by $D_T^{\log}$ the distribution density of $X_T^{\log}$, $T > 0$. Then the model is symmetric if and only if*
$$D_T^{\log}(x) = x_0 e^{rT} e^{-x} D_T^{\log}\left(-x + 2\log\left(x_0 e^{rT}\right)\right)$$
almost everywhere with respect to the Lebesgue measure on $\mathbb{R}$.
3. *The model is symmetric if and only if for all $T > 0$ and $K > 0$, $G(T, K) = C(T, K)$.*
4. *The model is symmetric if and only if for all $T > 0$ and $K > 0$,*
$$C(T, K) = \frac{K}{x_0 e^{rT}} C\left(T, \left(x_0 e^{rT}\right)^2 K^{-1}\right) + x_0 - e^{-rT} K.$$
5. *Let $C \in PF_\infty \cap PF_0$. Then the model is symmetric if and only if for all $T > 0$ and $K > 0$,*
$$I(T, K) = I\left(T, \left(x_0 e^{rT}\right)^2 K^{-1}\right).$$

Proof Part 3 of Lemma 9.25 follows from (8.3), (9.76), and from the fact that the measures μ_T and $\tilde{\mu}_T$ are the second distributional derivatives of the functions $K \mapsto C(T, K)$ and $K \mapsto G(T, K)$, respectfully. Part 4 can be easily derived from (9.77). As for part 5 of Lemma 9.25, it can be established using part 3 and Lemma 9.23. In addition, part 1 follows from Definition 9.24 and Remark 9.20. Finally, the equivalence $1 \Leftrightarrow 2$ follows from the standard equalities $D_T^{\log}(x) = e^x D_T(e^x)$ and $D_T(y) = y^{-1} D_T^{\log}(\log y)$.

This completes the proof of Lemma 9.25. □

Special examples of symmetric models are uncorrelated stochastic volatility models in a risk-neutral setting. Let us consider a stochastic model defined by
$$\begin{cases} dX_t = rX_t\,dt + f(Y_t)X_t\,dW_t, \\ dY_t = b(Y_t)\,dt + \sigma(Y_t)\,dZ_t, \end{cases} \tag{9.80}$$
where W and Z are independent Brownian motions on $(\Omega, \mathcal{F}, \{\mathcal{F}_t\}, \mathbb{P}^*)$, and suppose that the measure $\mathbb{P}^*$ is risk-neutral. Suppose also that the solvability conditions

discussed in Sect. 2.1 hold. It is clear that for such a model, formula (9.79) follows from formula (3.6). Therefore, part 1 of Lemma 9.25 shows that the model in (9.80) is symmetric.

Remark 9.26 The symmetry condition for the implied volatility in part 4 of Lemma 9.25 becomes especially simple if the strike K is replaced by the log-moneyness k defined by

$$k = \log \frac{K}{x_0 e^{rT}}, \quad K > 0.$$

In terms of the log-moneyness, the symmetry condition can be rewritten as follows: $I(k) = I(-k)$ for all $-\infty < k < \infty$. For uncorrelated stochastic volatility models, the previous equality was first obtained in [RT96].

In [CL09], P. Carr and R. Lee established that under certain restrictions, stochastic volatility models are symmetric if and only if $\rho = 0$. We will next prove this result of Carr and Lee. We restrict ourselves to models with time-homogeneous volatility equation. However, Theorem 9.27 also holds when volatility equations are inhomogeneous (see [CL09]).

Let us consider the stochastic model given by

$$\begin{cases} dX_t = rX_t\,dt + \sqrt{Y_t}X_t\,dW_t, \\ dY_t = b(Y_t)\,dt + \sigma(Y_t)\,dZ_t. \end{cases} \tag{9.81}$$

It is assumed in (9.81) that $Z = \sqrt{1-\rho^2}\widetilde{Z} + \rho W$, where $\widetilde{Z}$ is a standard Brownian motion independent of W, and the correlation coefficient ρ is such that $-1 \le \rho \le 1$. It is also assumed that the functions b and σ in (9.81) satisfy the linear growth condition and the Lipschitz condition, the function σ is positive, and for every ρ and every positive initial condition y_0 the solution Y to the second equation in (9.81) is a positive process.

Theorem 9.27 *Suppose the model in* (9.81) *satisfies the conditions formulated above. In addition, suppose the discounted price process is a martingale. Then the model is symmetric if and only if* $\rho = 0$.

Remark 9.28 It is worth mentioning that the conditions in Theorem 9.27 are rather restrictive. For example, this theorem is not applicable to the Stein–Stein model, or the Heston model. Indeed, in the Stein–Stein model the volatility process is not positive, while in the Heston model the function σ does not satisfy the Lipschitz condition. On the other hand, Theorem 9.27 can be used to prove that a negatively correlated Hull–White model cannot be symmetric. Indeed, in such a model the volatility process is a geometric Brownian motion, and hence it is a positive process. Moreover, the stock price process is a martingale (use Theorem 2.33). Note also that if a geometric Brownian motion Y is the solution to the equation

$$dY_t = \nu Y_t\,dt + \xi Y_t\,dZ_t$$

with the initial condition $y_0 > 0$, then the process $\widetilde{Y} = \sqrt{Y}$ is also a geometric Brownian motion satisfying the equation

$$d\widetilde{Y}_t = \left(\frac{\nu}{2} - \frac{\xi^2}{8}\right)\widetilde{Y}_t\,dt + \frac{\xi}{2}\widetilde{Y}_t\,dZ_t$$

with the initial condition $\sqrt{y_0}$. Summarizing what was said above, we see that Theorem 9.27 can be applied to the negatively correlated Hull–White model. If the Hull–White model is positively correlated, then Theorem 2.33 implies that the stock price process is not a martingale. Therefore, Theorem 9.27 cannot be applied to such a model. It would be interesting to extend Theorem 9.27 to a larger class of stochastic volatility models.

Proof It has already been established that for $\rho = 0$, the model is symmetric. We will next prove the converse statement. With no loss of generality, we can assume $r = 0$. Fix $\rho > 0$, and suppose the symmetry condition holds for the model given by

$$\begin{cases} dX_t = \sqrt{Y_t}X_t\,dW_t, \\ dY_t = b(Y_t)\,dt + \sqrt{1-\rho^2}\sigma(Y_t)\,d\widetilde{Z}_t + \rho\sigma(Y_t)\,dW_t. \end{cases}$$

Using the Itô formula, we can rewrite the model above in terms of the log-price process defined by $X^{\log} = \log X$ and $X_0^{\log} = \log x_0$. This gives

$$\begin{cases} dX_t^{\log} = -\dfrac{1}{2}Y_t\,dt + \sqrt{Y_t}\,dW_t, \\ dY_t = b(Y_t)\,dt + \sqrt{1-\rho^2}\sigma(Y_t)\,d\widetilde{Z}_t + \rho\sigma(Y_t)\,dW_t. \end{cases} \tag{9.82}$$

Let us fix $T > 0$. Since the process X is a martingale, the measure $\widetilde{\mathbb{P}}$ determined from $d\widetilde{\mathbb{P}} = x_0^{-1}X_T\,d\mathbb{P}$ is a probability measure. Define a new process by

$$\widehat{W}_t = W_t - \int_0^t \sqrt{Y_s}\,ds, \quad 0 \le t \le T.$$

It follows from Girsanov's theorem that the process $(\widehat{W}_t, \widetilde{Z}_t)$, $t \in [0, T]$, is a two-dimensional standard Brownian motion under the measure $\widetilde{\mathbb{P}}$. Therefore, the same is true for the process $(\widetilde{W}_t, \widetilde{Z}_t)$, $0 \le t \le T$, where $\widetilde{W}_t = -\widehat{W}_t$ for all $t \in [0, T]$. It is easy to see that under the measure $\widetilde{\mathbb{P}}$, the system in (9.82) can be rewritten as follows:

$$\begin{cases} d\left(-X_t^{\log}\right) = -\dfrac{1}{2}Y_t\,dt + \sqrt{Y_t}\,d\widetilde{W}_t, \\ dY_t = \Phi(Y_t)\,dt + \sqrt{1-\rho^2}\sigma(Y_t)\,d\widetilde{Z}_t + \rho\sigma(Y_t)\,d\widetilde{W}_t \end{cases} \tag{9.83}$$

where

$$\Phi(u) = b(u) + \rho\sigma(u)\sqrt{u}.$$

Recall that, by our assumption, the model described by (9.82) is symmetric. Using part 2 of Lemma 9.25, we obtain

$$\begin{aligned}\mathbb{E}\left[X_T^{\log}\right] &= \int_{-\infty}^{\infty} x D_T^{\log}(x)\,dx \\ &= -\frac{1}{x_0}\int_{-\infty}^{\infty} ue^u D_T^{\log}(u)\,du + \frac{2\log x_0}{x_0}\int_{-\infty}^{\infty} e^u D_T^{\log}(u)\,du \\ &= -\frac{1}{x_0}\mathbb{E}\left[X_T X_T^{\log}\right] + \frac{2\log x_0}{x_0}\mathbb{E}[X_T].\end{aligned}$$

Next, using the fact that the process X is a martingale, we see that

$$\mathbb{E}\left[X_T^{\log}\right] = -\widetilde{\mathbb{E}}\left[X_T^{\log}\right] + 2\log x_0. \tag{9.84}$$

The next step in the proof is to take the expectation $\mathbb{E}$ in the first stochastic differential equation in (9.82), written in the integral form. This gives

$$\mathbb{E}\left[X_T^{\log}\right] = -\frac{1}{2}\int_0^T \mathbb{E}[Y_t]\,dt + \log x_0. \tag{9.85}$$

Similarly, applying $\widetilde{\mathbb{E}}$ to the first equation in (9.83), we obtain

$$\widetilde{\mathbb{E}}\left[X_T^{\log}\right] = \frac{1}{2}\int_0^T \widetilde{\mathbb{E}}[Y_t]\,dt + \log x_0. \tag{9.86}$$

It follows from (9.84), (9.85), and (9.86) that

$$\int_0^T \mathbb{E}[Y_t]\,dt = \int_0^T \widetilde{\mathbb{E}}[Y_t]\,dt. \tag{9.87}$$

We will next use a coupling argument. Consider the following processes: $X^{\log}$, Y, W, $\widetilde{Z}$ under the measure $\mathbb{P}$ and $-X^{\log}$, Y, $\widetilde{W}$, $\widetilde{Z}$ under the measure $\widetilde{P}$. Applying the lemma formulated on p. 24 of [IW77], we see that there exist a filtered measure space $(\widetilde{\Omega}, \mathcal{F}, \mathcal{F}_t, \widehat{\mathbb{P}})$ and adapted stochastic processes $X^{(1)}$, Y^1, $X^{(2)}$, $Y^{(2)}$, $W^{(1)}$, and $Z^{(1)}$ on $\widetilde{\Omega}$ such that the following conditions hold:

- The processes $(X^{\log}, Y, W, \widetilde{Z})$ and $(X^{(1)}, Y^1, W^{(1)}, Z^{(1)})$ have the same law under the measures $\mathbb{P}$ and $\widehat{\mathbb{P}}$, respectively.
- The processes $(-X^{\log}, Y, \widetilde{W}, \widetilde{Z})$ and $(X^{(2)}, Y^2, W^{(1)}, Z^{(1)})$ have the same law under the measures $\widetilde{\mathbb{P}}$ and $\widehat{\mathbb{P}}$, respectively.
- The process $(W^{(1)}, Z^{(1)})$ is a two-dimensional $\mathcal{F}_t$-Brownian motion under the measure $\widehat{\mathbb{P}}$.

It follows (9.82), (9.83), and the previous statements that under the measure $\widehat{P}$,

$$\begin{cases} dX_t^{(1)} = -\frac{1}{2}Y_t^{(1)}\,dt + \sqrt{Y_t^{(1)}}\,dW_t^{(1)}, \\ dY_t^{(1)} = b\left(Y_t^{(1)}\right)dt + \sqrt{1-\rho^2}\sigma\left(Y_t^{(1)}\right)dZ_t^{(1)} + \rho\sigma\left(Y_t^{(1)}\right)dW_t^{(1)} \end{cases} \tag{9.88}$$

and

$$\begin{cases} dX_t^{(2)} = -\frac{1}{2} Y_t^{(2)}\, dt + \sqrt{Y_t^{(2)}}\, dW_t^{(1)}, \\ dY_t^{(2)} = \Phi\big(Y_t^{(2)}\big)\, dt + \sqrt{1-\rho^2}\sigma\big(Y_t^{(2)}\big)\, dZ_t^{(1)} + \rho\sigma(Y_t)\, dW_t^1. \end{cases} \tag{9.89}$$

Moreover, (9.87) implies that

$$\int_0^T \widehat{\mathbb{E}}\big[Y_t^{(1)}\big]\, dt = \int_0^T \widehat{\mathbb{E}}\big[Y_t^{(2)}\big]\, dt. \tag{9.90}$$

Now we are ready to finish the proof. Applying the strong comparison theorem for stochastic differential equations (Theorem 54 in [Pro04]) to (9.88) and (9.89), we see that

$$Y_t^{(2)} > Y_t^{(1)} \quad \text{for all } 0 < t < T. \tag{9.91}$$

Here we take into account that $b(u) < \Phi(u)$ and the initial condition (x_0, y_0) is the same for the processes $(X^{(1)}, Y^{(1)})$ and $(X^{(2)}, Y^{(2)})$. However, (9.91) contradicts (9.90). It follows that if $\rho > 0$, then the model cannot be symmetric. The case where $\rho < 0$ is similar.

This completes the proof of Theorem 9.27. □

9.9 Asymptotic Behavior of Implied Volatility for Small Strikes

Lemma 9.23 and the results obtained in Sect. 9.3 imply sharp asymptotic formulas for the implied volatility as $K \to 0$.

Theorem 9.29 *Let $C \in PF_0$, and let P be the corresponding put pricing function. Suppose*

$$P(K) \approx \widetilde{P}(K) \quad \text{as } K \to 0, \tag{9.92}$$

where $\widetilde{P}$ is a positive function. Then the following asymptotic formula holds:

$$\begin{aligned} I(K) = {} & \frac{\sqrt{2}}{\sqrt{T}} \sqrt{\log \frac{1}{\widetilde{P}(K)} - \frac{1}{2} \log\log \frac{K}{\widetilde{P}(K)}} \\ & - \frac{\sqrt{2}}{\sqrt{T}} \sqrt{\log \frac{K}{\widetilde{P}(K)} - \frac{1}{2} \log\log \frac{K}{\widetilde{P}(K)}} \\ & + O\bigg(\bigg(\log \frac{K}{\widetilde{P}(K)}\bigg)^{-\frac{1}{2}}\bigg) \end{aligned} \tag{9.93}$$

as $K \to 0$.

Corollary 9.30 *The following asymptotic formula holds*:

$$I(K) = \frac{\sqrt{2}}{\sqrt{T}}\left[\sqrt{\log\frac{1}{\widetilde{P}(K)}} - \sqrt{\log\frac{K}{\widetilde{P}(K)}}\right] + O\left(\left(\log\frac{K}{\widetilde{P}(K)}\right)^{-\frac{1}{2}}\log\log\frac{K}{\widetilde{P}(K)}\right)$$

as $K \to 0$.

An important special case of Theorem 9.29 is as follows:

Corollary 9.31 *Let* $C \in PF_0$, *and let* P *be the corresponding put pricing function. Then*

$$I(K) = \frac{\sqrt{2}}{\sqrt{T}}\left[\sqrt{\log\frac{1}{P(K)}} - \sqrt{\log\frac{K}{P(K)}}\right] + O\left(\left(\log\frac{K}{P(K)}\right)^{-\frac{1}{2}}\log\log\frac{K}{P(K)}\right)$$

as $K \to 0$.

Proof of Theorem 9.29 Formulas (9.71) and (9.92) imply that

$$G(K) \approx \widetilde{G}(K) \quad \text{as } K \to \infty$$

where

$$\widetilde{G}(K) = K\widetilde{P}\big(\eta_T(K)\big). \tag{9.94}$$

Next, applying Corollary 9.11 to G and $\widetilde{G}$, we get

$$\frac{\sqrt{T}}{\sqrt{2}}I_G(K) = \sqrt{\log K + \log\frac{1}{\widetilde{G}(K)} - \frac{1}{2}\log\log\frac{1}{\widetilde{G}(K)}} - \sqrt{\log\frac{1}{\widetilde{G}(K)} - \frac{1}{2}\log\log\frac{1}{\widetilde{G}(K)}} + O\left(\left(\log\frac{1}{\widetilde{G}(K)}\right)^{-\frac{1}{2}}\right) \tag{9.95}$$

as $K \to \infty$. It follows from (9.78), (9.94), (9.95), and from the mean value theorem that

$$\frac{\sqrt{T}}{\sqrt{2}}I(K) = \sqrt{\log\frac{(x_0e^{rT})^2}{K} + \log\frac{K}{(x_0e^{rT})^2\widetilde{P}(K)} - \frac{1}{2}\log\log\frac{K}{(x_0e^{rT})^2\widetilde{P}(K)}}$$

$$- \sqrt{\log \frac{K}{(x_0 e^{rT})^2 \widetilde{P}(K)} - \frac{1}{2} \log\log \frac{K}{(x_0 e^{rT})^2 \widetilde{P}(K)}}$$

$$+ O\left(\left(\log \frac{K}{\widetilde{P}(K)}\right)^{-\frac{1}{2}}\right)$$

$$= \sqrt{\log \frac{1}{\widetilde{P}(K)} - \frac{1}{2} \log\log \frac{K}{\widetilde{P}(K)}} - \sqrt{\log \frac{K}{\widetilde{P}(K)} - \frac{1}{2} \log\log \frac{K}{\widetilde{P}(K)}}$$

$$+ O\left(\left(\log \frac{K}{\widetilde{P}(K)}\right)^{-\frac{1}{2}}\right)$$

as $K \to 0$.

This completes the proof of Theorem 9.29. □

9.10 Notes and References

- The books [FPS00, Reb04, Haf04, Fen05, Gat06, H-L09], the dissertations [Dur04, Rop09], the surveys [Ski01, CL10], and the papers [SP99, SHK99, Lee01, CdF02, Lee04a, CGLS09, Fri10] are useful sources of information on the implied volatility.
- Section 9.2 is mostly adapted from [Rop10]. However, the conditions in Theorem 9.6 are not exactly the same as in the similar result (Theorem 2.9) in [Rop10]. Moreover, Theorem 9.6 is formulated in terms of the strike price, while the log-moneyness is used in [Rop10].
- The asymptotic formulas for the implied volatility included in Sects. 9.3, 9.4, and 9.9 are taken from [Gul10].
- The material in Sects. 9.7 and 9.8 (symmetries and symmetric models) comes mostly from [Gul10]. We send the interested reader to [CL09, Teh09a, DM10, DMM10] for more information on symmetric models.
- The paper [GL11] of K. Gao and R. Lee is an important recent work on smile asymptotics. In Sects. 9.5 and 9.6 of this chapter, several theorems from [GL11] are presented. These theorems provide higher-order approximations for the implied volatility at extreme strikes. However, we have not touched upon the results in [GL11] characterizing the asymptotic behavior of the implied volatility with respect to the maturity, or in certain combined regimes.

Chapter 10
More Formulas for Implied Volatility

This chapter deals primarily with applications of the asymptotic formulas for the implied volatility established in Chap. 9. We will show below that these formulas imply the following well-known results: R. Lee's moment formulas and the tail-wing formulas due to S. Benaim and P. Friz (see Sects. 10.1 and 10.2). We will also obtain sharp asymptotic formulas for the implied volatility in several special stochastic volatility models. These models include, on the one hand, the Hull–White, Stein–Stein, and Heston models (see Sect. 10.5), and on the other hand, a special Heston model with jumps (see Sect. 10.8).

The remainder of the chapter is devoted to "volatility smile" and to J. Gatheral's SVI parameterization of the implied variance. The expression "volatility smile" was coined to describe an observed feature of at-the-money options to have a smaller implied volatility than in-the-money or out-of-the-money options. In Sect. 10.9, we discuss an interesting result of E. Renault and N. Touzi concerning the existence of volatility smile in uncorrelated stochastic volatility models. The last section of this chapter (Sect. 10.10) discusses the SVI parameterization of the implied variance.

10.1 Moment Formulas

It will be demonstrated in this section that R. Lee's moment formulas for the implied volatility (see [Lee04b]) can be derived from Corollaries 9.8 and 9.31. The moment formulas provide certain relations between the implied volatility for large or small strikes and the orders of extreme moments of the stock price.

Definition 10.1 Let X be a nonnegative random variable on a probability space $(\Omega, \mathcal{F}, \mathbb{P})$. The moment $m_p(X)$ of order $p \in \mathbb{R}$ of the random variable X is defined as follows:

$$m_p(X) = \mathbb{E}^*\left[X^p\right].$$

A. Gulisashvili, *Analytically Tractable Stochastic Stock Price Models*,
Springer Finance, DOI 10.1007/978-3-642-31214-4_10,

© Springer-Verlag Berlin Heidelberg 2012

It is clear that

$$m_p(X) = \int_0^\infty x^p \, d\mu(x),$$

where the symbol μ stands for the distribution of X. It will be assumed in the present section that C is a call pricing function satisfying the condition $C \in PF_\infty \cap PF_0$. Let X be the corresponding stock price process. Then it is clear that $m_p(X_T) < \infty$ for all $p \in [0, 1]$ and $T > 0$. However, if $p \notin [0, 1]$, then $m_p(X_T)$ may be finite or infinite. Note that the moments of X_T are computed using the risk-neutral expectation $\mathbb{E}^*$.

We will next formulate Lee's theorems.

Theorem 10.2 *Let C be a call pricing function, and let I be the implied volatility associated with C. Fix $T > 0$, and define the number $\tilde{p}$ by*

$$\tilde{p} = \sup\{p \geq 0 : m_{1+p}(X_T) < \infty\}. \tag{10.1}$$

Then the following equality holds:

$$\limsup_{K\to\infty} \frac{TI(K)^2}{\log K} = \psi(\tilde{p}) \tag{10.2}$$

where the function ψ is given by

$$\psi(u) = 2 - 4\left(\sqrt{u^2 + u} - u\right), \quad u \geq 0. \tag{10.3}$$

Theorem 10.3 *Under the conditions in Theorem 10.2, define the number $\tilde{q}$ by*

$$\tilde{q} = \sup\{q \geq 0 : m_{-q}(X_T) < \infty\}. \tag{10.4}$$

Then the following formula holds:

$$\limsup_{K\to 0} \frac{TI(K)^2}{\log \frac{1}{K}} = \psi(\tilde{q}). \tag{10.5}$$

Formulas (10.2) and (10.5) are called Lee's moment formulas. For the asset price X_T, the numbers $1 + \tilde{p}$ and $\tilde{q}$ characterize the tail behavior of its distribution. These numbers are called the right-tail index and the left-tail index of the asset price distribution, respectively. The function ψ in (10.3) is strictly decreasing on the interval $[0, \infty]$ and maps this interval onto the interval $[0, 2]$.

Our next goal is to prove Lee's moment formulas. Since for every $a > 0$,

$$\sqrt{2}(\sqrt{1+a} - \sqrt{a}) = \sqrt{2}\left(1 - 2\left(\sqrt{a^2+a} - a\right)\right)^{\frac{1}{2}} = \sqrt{\psi(a)}, \tag{10.6}$$

formulas (10.2) and (10.5) can be rewritten as follows:

$$\limsup_{K\to\infty} \frac{\sqrt{T}I(K)}{\sqrt{\log K}} = \sqrt{2}(\sqrt{1+\tilde{p}} - \sqrt{\tilde{p}}) \tag{10.7}$$

and

$$\limsup_{K\to 0}\frac{\sqrt{T}I(K)}{\sqrt{\log\frac{1}{K}}}=\sqrt{2}(\sqrt{1+\tilde{q}}-\sqrt{\tilde{q}}). \tag{10.8}$$

We will first prove formula (10.7). The following lemma will be used in the proof.

Lemma 10.4 *Let $C\in PF_\infty\cap PF_0$ and put*

$$l=\liminf_{K\to\infty}(\log K)^{-1}\log\frac{1}{C(K)}. \tag{10.9}$$

Then

$$\limsup_{K\to\infty}\frac{\sqrt{T}I(K)}{\sqrt{\log K}}=\sqrt{2}(\sqrt{1+l}-\sqrt{l})=\sqrt{\psi(l)}. \tag{10.10}$$

Proof of Lemma 10.4 Observe that Corollary 9.8 implies

$$\begin{aligned}\frac{\sqrt{T}I(K)}{\sqrt{\log K}}&=\sqrt{2}\left(\sqrt{1+\frac{\log\frac{1}{C(K)}}{\log K}}-\sqrt{\frac{\log\frac{1}{C(K)}}{\log K}}\right)\\&\quad+O\left((\log K)^{-\frac{1}{2}}\left(\log\frac{1}{C(K)}\right)^{-\frac{1}{2}}\log\log\frac{1}{C(K)}\right)\\&=\sqrt{2}\left[\sqrt{1+\frac{\log\frac{1}{C(K)}}{\log K}}+\sqrt{\frac{\log\frac{1}{C(K)}}{\log K}}\right]^{-1}\\&\quad+O\left((\log K)^{-\frac{1}{2}}\left(\log\frac{1}{C(K)}\right)^{-\frac{1}{2}}\log\log\frac{1}{C(K)}\right)\end{aligned} \tag{10.11}$$

as $K\to\infty$. Now it is clear that (10.10) follows from (10.11).

Let us continue the proof of formula (10.7). Denote by $\overline{F}=\overline{F}_T$ the complementary cumulative distribution function of X_T defined by

$$\overline{F}(y)=\mathbb{P}[X_T>y],\quad y>0.$$

Then we have

$$C(K)=e^{-rT}\int_K^\infty\overline{F}(y)\,dy,\quad K>0. \tag{10.12}$$

Set

$$r^*=\sup\left\{r\geq 0:C(K)=O\left(K^{-r}\right)\text{ as }K\to\infty\right\}, \tag{10.13}$$

and

$$s^*=\sup\left\{s\geq 0:\overline{F}(y)=O\left(y^{-(1+s)}\right)\text{ as }y\to\infty\right\}. \tag{10.14}$$

□

Lemma 10.5 *The numbers $\tilde{p}$, l, r^*, and s^* given by* (10.1), (10.9), (10.13), *and* (10.14), *respectively, are equal.*

Proof If $s^* = 0$, then the inequality $s^* \le r^*$ is trivial. If $s > 0$ is such that $\overline{F}(y) = O(y^{-(1+s)})$ as $y \to \infty$, then

$$C(K) = O\left(\int_K^\infty y^{-(1+s)}\, dy\right) = O\left(K^{-s}\right)$$

as $K \to \infty$. Hence $s^* \le r^*$.

Next let $r \ge 0$ be such that $C(K) = O(K^{-r})$ as $K \to \infty$. Then (10.12) shows that there exist $c > 0$ and $K_0 > 0$ such that for all $K > K_0$,

$$cK^{-r} \ge e^{-rT} \int_K^\infty \overline{F}(y)\, dy \ge e^{-rT} \int_K^{2K} \overline{F}(y)\, dy \ge e^{-rT} \overline{F}(2K) K.$$

Therefore, $\overline{F}(K) = O(K^{-(r+1)})$ as $K \to \infty$. It follows that $r^* \le s^*$. This proves the equality $r^* = s^*$.

Suppose $0 < l < \infty$. Then for every $\varepsilon > 0$ there exists $K_\varepsilon > 0$ such that for all $K > K_\varepsilon$,

$$\log \frac{1}{C(K)} \ge (l - \varepsilon) \log K.$$

Therefore $C(K) \le K^{-l+\varepsilon}$ for all $K > K_\varepsilon$. It follows that $l - \varepsilon \le r^*$ for all $\varepsilon > 0$, and hence $l \le r^*$. The inequality $l \le r^*$ also holds if $l = 0$ or $l = \infty$. This fact can be established similarly.

To prove the inequality $r^* \le l$, suppose $r^* \ne 0$ and $r < r^*$. Then $C(K) = O(K^{-r})$ as $K \to \infty$, and hence $C(K)^{-1} \ge cK^r$ for some $c > 0$ and all $K > K_0$. It follows that

$$\log \frac{1}{C(K)} \ge \log c + r \log K, \quad K > K_0$$

and

$$\frac{\log \frac{1}{C(K)}}{\log K} \ge \frac{\log c}{\log K} + r.$$

Therefore

$$\liminf_{K \to \infty} \frac{\log \frac{1}{C(K)}}{\log K} \ge r. \tag{10.15}$$

Using (10.15), we see that $l \ge r^*$. If $r^* = 0$, then the inequality $l \ge r^*$ is trivial. This proves that $l = r^* = s^*$.

It is clear that for all $p \ge 0$,

$$m_{1+p}(X_T) = (1 + p) \int_0^\infty y^p \overline{F}(y)\, dy. \tag{10.16}$$

Suppose $s^* = 0$. Then the inequality $s^* \le \tilde{p}$ is trivial. If for some $s > 0$, $\overline{F}(y) = O(y^{-(1+s)})$ as $y \to \infty$, then it is not hard to see using (10.16) that $m_{1+p}(X_T) < \infty$ for all $p < s$. It follows that $s^* \le \tilde{p}$.

On the other hand, if $m_{1+p}(X_T) < \infty$ for some $p \ge 0$, then there exists a number $M > 0$ such that

$$M > \int_K^\infty y^p \overline{F}(y)\,dy \ge K^p \int_K^\infty \overline{F}(y)\,dy = e^{rT} K^p C(K). \tag{10.17}$$

In the proof of (10.17), we used (10.16) and (10.12). It follows from (10.17) that $C(K) = O(K^{-p})$ as $K \to \infty$, and hence $\tilde{p} \le r^*$.

This completes the proof of Lemma 10.5. □

To finish the proof of formula (10.2), we observe that (10.10) and the equality $l = \tilde{p}$ in Lemma 10.5 imply formula (10.7).

It will be explained next how to prove formula (10.8). The following lemma follows from Corollary 9.31:

Lemma 10.6 *Let $C \in PF_\infty \cap PF_0$ and define a number by*

$$m = \liminf_{K \to 0} \left(\log \frac{1}{K} \right)^{-1} \log \frac{1}{P(K)}. \tag{10.18}$$

Then

$$\limsup_{K \to 0} \frac{\sqrt{T} I(K)}{\sqrt{\log \frac{1}{K}}} = \sqrt{2}(\sqrt{m} - \sqrt{m-1}) = \sqrt{\psi(m-1)}. \tag{10.19}$$

It is not hard to see that $m \ge 1$, where m is defined by (10.18). Put

$$F(y) = \mathbb{P}[X_T \le y] = 1 - \overline{F}(y), \quad y \ge 0.$$

Then

$$P(K) = e^{-rT} \int_0^K F(y)\,dy$$

and

$$m_{-q}(X_T) = q \int_0^\infty y^{-q-1} F(y)\,dy$$

for all $q > 0$. Note that $F(0) = \mathbb{P}[X_T = 0]$.

Consider the following numbers:

$$u^* = \sup\{u \ge 1 : P(K) = O(K^u) \text{ as } K \to 0\} \tag{10.20}$$

and

$$v^* = \sup\{v \ge 0 : F(y) = O(y^v) \text{ as } y \to 0\}. \tag{10.21}$$

It is not hard to see, using the same ideas as in the proof of Lemma 10.5, that the following lemma holds:

Lemma 10.7 *The numbers* $\tilde{q}$, m, u^*, *and* v^* *defined by* (10.4), (10.18), (10.20), *and* (10.21), *respectively, satisfy the equalities*

$$\tilde{q} + 1 = m = u^* = v^* + 1.$$

Now it is clear that formula (10.8) follows from (10.19) and Lemma 10.7.

10.2 Tail-Wing Formulas

This section focuses on tail-wing formulas. Such formulas characterize the asymptotics of the implied volatility at large and small strikes (the wing asymptotics) in terms of the tail behavior of the stock price density.

Our first goal is to formulate and discuss the tail-wing formulas established by S. Benaim and P. Friz in [BF09] (see also [BF08]). Recall that we denoted by $\overline{F} = \overline{F}_T$ the complementary cumulative distribution function of the stock price X_T. This function is given by $\overline{F}(y) = \mathbb{P}[X_t > y]$, $y > 0$. We will also need the function ψ defined by (10.3).

The next statement is a part of Theorem 1 in [BF09] adapted to the style of the present book. Note that in [BF09] a different normalization is used in the Black–Scholes formula, and the normalized implied volatility is considered as a function of the log-strike k.

Theorem 10.8 *Let* C *be a call pricing function, and suppose the stock price* X_T *satisfies the condition*

$$m_{1+\varepsilon}(X_T) < \infty \quad \textit{for some } \varepsilon > 0. \tag{10.22}$$

Then the following are true:

1. *If* $C(K) = \exp\{-\eta(\log K)\}$ *with* $\eta \in R_\alpha$, $\alpha > 0$, *then*

$$I(K) \sim \frac{\sqrt{\log K}}{\sqrt{T}} \sqrt{\psi\left(-\frac{\log C(K)}{\log K}\right)} \quad \textit{as } K \to \infty. \tag{10.23}$$

2. *If* $\overline{F}(y) = \exp\{-\rho(\log y)\}$ *with* $\rho \in R_\alpha$, $\alpha > 0$, *then*

$$I(K) \sim \frac{\sqrt{\log K}}{\sqrt{T}} \sqrt{\psi\left(-\frac{\log[K\overline{F}(K)]}{\log K}\right)} \quad \textit{as } K \to \infty. \tag{10.24}$$

3. *If the distribution* μ_T *of the stock price* X_T *admits a density* D_T *and if*

$$D_T(x) = \frac{1}{x} \exp\{-h(\log x)\}$$

as $x \to \infty$, *where* $h \in R_\alpha$, $\alpha > 0$, *then*

$$I(K) \sim \frac{\sqrt{\log K}}{\sqrt{T}} \sqrt{\psi\left(-\frac{\log[K^2 D_T(K)]}{\log K}\right)} \quad \text{as } K \to \infty. \tag{10.25}$$

It is easy to see that the conditions in Theorem 10.8 imply the condition $C \in PF_\infty$. The formulas contained in Theorem 10.8 are called right-tail-wing formulas.

Remark 10.9 The functions $V(k)$ and $c(k)$ used in [BF09] correspond in our notation to the functions $\sqrt{T} I(K)$ and $e^{rT} C(K)$, respectively. We also take into account that for a strictly positive price process X, the distribution density $D_T^{\log}$ of the asset return $\log X_T$ is related to the density D_T by the formula $f(y) = e^y D_T(e^y)$.

Our next goal is to explain how to derive Theorem 10.8 from Corollary 9.15. The following statement is nothing else but Corollary 9.15 in disguise.

Corollary 10.10 *Let* $C \in PF_\infty$. *Then*

$$I(K) = \frac{\sqrt{\log K}}{\sqrt{T}} \sqrt{\psi\left(-\frac{\log C(K)}{\log K}\right)} + O\left(\left(\log \frac{1}{C(K)}\right)^{-\frac{1}{2}} \log\log \frac{1}{C(K)}\right) \tag{10.26}$$

as $K \to \infty$.

The equivalence of formulas (9.16) and (10.26) can be easily shown using (10.6).

Remark 10.11 It follows from Corollary 10.10 that formula (10.23) holds for any call pricing function from the class PF_∞, and hence no restrictions are needed in part 1 of Theorem 10.8. Moreover, formula (10.26) contains an error term, which is absent in formula (10.23). Note that the equality in Lee's moment formula (10.2) as well as the tail-wing formulas given in (10.24) and (10.25) do not hold without certain restrictions (see Sect. 2.3 of [BFL09]). We will provide necessary and sufficient conditions for the validity of the equality in formula (10.2) in Sect. 10.6.

It will be briefly explained next how to obtain (10.24) and (10.25). More precisely, a slightly more general statement will be obtained. We assume that

$$\overline{F}(y) \approx \exp\{-\rho(\log y)\} \tag{10.27}$$

as $y \to \infty$ in part 2 of Theorem 10.8 and

$$D_T(x) \approx x^{-1} \exp\{-h(\log x)\} \tag{10.28}$$

as $x \to \infty$ in part 3 of this theorem. Some of the ideas used in the proof below are borrowed from [BF09] (see, for instance, the proofs in Sect. 3 of [BF09]). With no loss of generality, we may suppose that $\alpha \geq 1$.

Our proof of the tail-wing formulas is based on Theorem 7.22, (10.12), and the equality

$$\overline{F}(y) = \int_y^\infty D_T(x)\,dx. \tag{10.29}$$

If $\alpha > 1$ in parts 2 or 3 of Theorem 10.8, then the moment condition (10.22) holds and we have $\rho(u) - u \in R_\alpha$ in part 2 and $h(u) - u \in R_\alpha$ in part 3. If $\alpha = 1$, then the moment condition gives $\rho(u) - u \in R_1$ in part 2 and $h(u) - u \in R_1$ in part 3 (see Sect. 3 in [BF09]).

Suppose (10.27) holds and put $\lambda(u) = \rho(u) - u$. Then we have $C(K) \approx \widehat{C}(K)$ as $K \to \infty$, where

$$\widehat{C}(K) = \int_{\log K}^\infty \exp\{-\lambda(u)\}\,du.$$

Applying formula (7.48) to the function λ, we obtain

$$\log \frac{1}{\widehat{C}(K)} \sim \lambda(\log K) = \log \frac{1}{K\overline{F}(K)}$$

as $K \to \infty$. Since $C(K) \approx \widehat{C}(K)$, we also have

$$\log \frac{1}{C(K)} \sim \log \frac{1}{\widehat{C}(K)},$$

and hence

$$\log \frac{1}{C(K)} \sim \log \frac{1}{K\overline{F}(K)}$$

as $K \to \infty$. Now it clear that formula (10.24) follows from (9.40) and (10.6).

Next assume that equality (10.28) holds. Then (10.29) implies (10.27) with

$$\rho(y) = -\log \int_y^\infty e^{-h(u)}\,du.$$

Applying Theorem 7.22, we see that $\rho \in R_\alpha$. This reduces the case of the distribution density D_T of the stock price in Theorem 10.8 to that of the complementary cumulative distribution function $\overline{F}$.

Remark 10.12 The tail-wing formula (10.24) also holds provided that $\alpha = 1$ and $\rho(u) - u \in R_\beta$ with $0 < \beta \leq 1$. A similar statement is true in the case of formula (10.25). The proof of these assertions does not differ much from the proof given above. Interesting examples here are $\rho(u) = u + u^\beta$ for $\beta < 1$ and $\rho(u) = u + \frac{u}{\log u}$ for $\beta = 1$. Note that condition (10.22) does not hold in these cases.

10.3 Tail-Wing Formulas with Error Estimates

Formulas (10.24) and (10.25) do not contain error estimates. In this section, we obtain tail-wing formulas with error estimates. Note that classes of smoothly varying functions play an important role in the formulations of the next theorems.

Theorem 10.13 *Let $\overline{F}$ be the complementary cumulative distribution function of the stock price X_T, and suppose*

$$\overline{F}(y) \approx \exp\{-\rho(\log y)\} \tag{10.30}$$

as $y \to \infty$, where ρ is a function such that either $\rho \in SR_\alpha$ with $\alpha > 1$, or $\rho \in SR_1$ and $\lambda(u) = \rho(u) - u \in R_\beta$ for some $0 < \beta \le 1$. Then

$$I(K) = \frac{\sqrt{2}}{\sqrt{T}}\left(\sqrt{\rho(\log K)} - \sqrt{\rho(\log K) - \log K}\right) + O\left(\frac{\log[\rho(\log K)]}{\sqrt{\rho(\log K)}}\right) \tag{10.31}$$

as $K \to \infty$.

Theorem 10.14 *Let D_T be the distribution density of the stock price X_T. Suppose*

$$D_T(x) \approx \frac{1}{x}\exp\{-h(\log x)\} \tag{10.32}$$

as $x \to \infty$, where h is a function such that either $h \in SR_\alpha$ with $\alpha > 1$, or $h \in SR_1$ and $g(u) = h(u) - u \in SR_\beta$ for some $0 < \beta \le 1$. Then

$$I(K) = \frac{\sqrt{2}}{\sqrt{T}}\left(\sqrt{h(\log K)} - \sqrt{h(\log K) - \log K}\right) + O\left(\frac{\log[h(\log K)]}{\sqrt{h(\log K)}}\right) \tag{10.33}$$

as $K \to \infty$.

Remark 10.15 Formulas (10.31) and (10.33) are equivalent to the formulas

$$I(K) = \frac{\sqrt{\log K}}{\sqrt{T}}\sqrt{\psi\left(\frac{\rho(\log K) - \log K}{\log K}\right)} + O\left(\frac{\log[\rho(\log K)]}{\sqrt{(\rho(\log K))}}\right)$$

and

$$I(K) = \frac{\sqrt{\log K}}{\sqrt{T}}\sqrt{\psi\left(\frac{h(\log K) - \log K}{\log K}\right)} + O\left(\frac{\log[h(\log K)]}{\sqrt{(h(\log K))}}\right),$$

respectively, where the function ψ is defined by (10.3).

If we replace the symbol $\approx$ in (10.30) and (10.32) by the equality sign, then the following tail-wing formulas with error estimates hold:

$$I(K)=\frac{\sqrt{\log K}}{\sqrt{T}}\sqrt{\psi\left(-\frac{\log[K\overline{F}(K)]}{\log K}\right)}$$
$$+O\left(\left(\log\frac{1}{[K\overline{F}(K)]}\right)^{-\frac{1}{2}}\log\log\frac{1}{[K\overline{F}(K)]}\right)$$

and

$$I(K)=\frac{\sqrt{\log K}}{\sqrt{T}}\sqrt{\psi\left(-\frac{\log[K^2D_T(K)]}{\log K}\right)}$$
$$+O\left(\left(\log\frac{1}{[K^2D_T(K)]}\right)^{-\frac{1}{2}}\log\log\frac{1}{[K^2D_T(K)]}\right)$$

as $K\to\infty$.

We will next prove Theorem 10.14. The proof of Theorem 10.13 is similar, but less complicated. We leave it as an exercise for the reader.

Proof of Theorem 10.14 For $h\in SR_\alpha$ with $\alpha>1$, we have $g\in SR_\alpha$. On the other hand, if $\alpha=1$, then we assume that $g\in SR_\beta$ with $0<\beta\le 1$. Consider the following functions:

$$\widetilde{D}_T(x)=\frac{1}{x}\exp\{-h(\log x)\}$$

and

$$\widehat{C}(K)=K^2\widetilde{D}_T(K)=\exp\{-g(\log K)\}.$$

We have

$$C(K)\approx\int_{\log K}^{\infty}e^{-g(u)}\,du-K\int_{\log K}^{\infty}e^{-h(u)}\,du \tag{10.34}$$

as $K\to\infty$. Now, applying Lemma 7.24 we get

$$\int_{\log K}^{\infty}e^{-g(u)}\,du=\frac{Ke^{-h(\log K)}}{g'(\log K)}\left(1+O\left(\frac{1}{g(\log K)}\right)\right)$$

and

$$K\int_{\log K}^{\infty}e^{-h(u)}\,du=\frac{Ke^{-h(\log K)}}{h'(\log K)}\left(1+O\left(\frac{1}{h(\log K)}\right)\right)$$

as $K \to \infty$. It follows that

$$\int_{\log K}^{\infty} e^{-g(u)}\,du - K\int_{\log K}^{\infty} e^{-h(u)}\,du$$
$$= \frac{Ke^{-h(\log K)}}{h'(\log K)g'(\log K)}\left(1 + O\left(\frac{h(\log K)}{g(\log K)\log K}\right)\right) \qquad (10.35)$$

as $K \to \infty$. The proof of (10.35) uses (7.45). Next we see that (10.34), (10.35), and (7.45) give

$$C(K) \approx \widetilde{C}(K) \quad \text{where } \widetilde{C}(K) = \frac{K(\log K)^2 e^{-h(\log K)}}{h(\log K)g(\log K)}.$$

Therefore

$$\log\frac{1}{\widetilde{C}(K)} - \log\frac{1}{\widehat{C}(K)} = \log\frac{h(\log K)g(\log K)}{(\log K)^2}. \qquad (10.36)$$

It follows from (10.36) that there exists $a > 0$ such that

$$\left|\log\frac{1}{\widetilde{C}(K)} - \log\frac{1}{\widehat{C}(K)}\right| \le a\log\log\frac{1}{\widehat{C}(K)}, \qquad K > K_1. \qquad (10.37)$$

Indeed, if $\alpha > 1$, we can take $a > \frac{2\alpha-2}{\alpha}$ in (10.37), and if $\alpha = 1$ and $0 < \beta \le 1$, we take $a > \frac{1-\beta}{\beta}$. It is not hard to see that an estimate similar to (10.37) is valid with C instead of $\widetilde{C}$. Now it follows from Corollary 9.13 that formula (10.33) holds.

The proof of Theorem 10.14 is thus completed. □

Formulas similar to those established in Sect. 10.2 and 10.3 also hold when $K \to 0$. We will next formulate one of such results. It is equivalent to Corollary 9.31.

Corollary 10.16 *Let $C \in PF_0$, and let P be the corresponding put pricing function. Then*

$$I(K) = \frac{\log\frac{1}{K}}{\sqrt{T}}\sqrt{\psi\left(\frac{\log P(K)}{\log K} - 1\right)} + O\left(\left(\log\frac{K}{P(K)}\right)^{-\frac{1}{2}}\log\log\frac{K}{P(K)}\right)$$

as $K \to 0$, where $\psi(u) = 2 - 4(\sqrt{u^2+u} - u)$, $u \ge 0$.

The equivalence of Corollary 9.31 and Corollary 10.16 can be shown using (10.6) with $a = (\log K)^{-1}\log P(K) - 1$.

10.4 Regularly Varying Stock Price Densities and Tail-Wing Formulas

In this section, we obtain tail-wing formulas for the implied volatility in the cases where the stock price density is equivalent to a regularly varying function.

Theorem 10.17 *Suppose that the distribution of the stock price X_T admits a density D_T such that*

$$D_T(x) \approx x^{\beta} h(x) \tag{10.38}$$

as $x \to \infty$, where $\beta < -2$, and h is a slowly varying function. Then

$$\begin{aligned} I(K) = {} & \frac{\sqrt{2}}{\sqrt{T}} \sqrt{\log K + \log \frac{1}{K^2 D_T(K)} - \frac{1}{2} \log \log \frac{1}{K^2 D_T(K)}} \\ & - \frac{\sqrt{2}}{\sqrt{T}} \sqrt{\log \frac{1}{K^2 D_T(K)} - \frac{1}{2} \log \log \frac{1}{K^2 D_T(K)}} \\ & + O\big((\log K)^{-\frac{1}{2}}\big) \\ = {} & \frac{\sqrt{2}}{\sqrt{T}} \sqrt{\log K + \log \frac{1}{K^{\beta+2} h(K)} - \frac{1}{2} \log \log \frac{1}{K^{\beta+2} h(K)}} \\ & - \frac{\sqrt{2}}{\sqrt{T}} \sqrt{\log \frac{1}{K^{\beta+2} h(K)} - \frac{1}{2} \log \log \frac{1}{K^{\beta+2} h(K)}} \\ & + O\big((\log K)^{-\frac{1}{2}}\big) \end{aligned}$$

as $K \to \infty$.

Proof It is not hard to see that (10.38) and part 1 of Karamata's theorem (Theorem 7.6) imply the following formula:

$$C(K) \approx K^{\beta+2} h(K) \quad \text{as } K \to \infty.$$

Therefore $C \in PF_{\infty}$, and we can apply Corollary 9.11 with $\widetilde{C}(K) = K^2 D_T(K)$ or $\widetilde{C}(K) = K^{\beta+2} h(K)$ to finish the proof of Theorem 10.17. □

A similar theorem holds for small values of the strike price.

Theorem 10.18 *Suppose that the distribution of the stock price X_T admits a density D_T such that*

$$D_T(x) \approx x^{\gamma} h\big(x^{-1}\big) \tag{10.39}$$

as $x \to 0$, where $\gamma > -1$, and h is a slowly varying function. Then

$$
\begin{aligned}
I(K) &= \frac{\sqrt{2}}{\sqrt{T}}\sqrt{\log\frac{1}{K^2 D_T(K)} - \frac{1}{2}\log\log\frac{1}{K D_T(K)}} \\
&\quad - \frac{\sqrt{2}}{\sqrt{T}}\sqrt{\log\frac{1}{K D_T(K)} - \frac{1}{2}\log\log\frac{1}{K D_T(K)}} \\
&\quad + O\left(\left(\log\frac{1}{K}\right)^{-\frac{1}{2}}\right) \\
&= \frac{\sqrt{2}}{\sqrt{T}}\sqrt{\log\frac{1}{K^{\gamma+2}h(K^{-1})} - \frac{1}{2}\log\log\frac{1}{K^{\gamma+1}h(K^{-1})}} \\
&\quad - \frac{\sqrt{2}}{\sqrt{T}}\sqrt{\log\frac{1}{K^{\gamma+1}h(K^{-1})} - \frac{1}{2}\log\log\frac{1}{K^{\gamma+1}h(K^{-1})}} \\
&\quad + O\left(\left(\log\frac{1}{K}\right)^{-\frac{1}{2}}\right)
\end{aligned}
$$

as $K \to 0$.

Proof Theorem 10.18 can be derived from Theorem 9.29. Indeed, using (10.39) and part 1 of Karamata's theorem (Theorem 7.6), we obtain the following formula: $P(K) \approx K^{\gamma+2}h(K^{-1})$ as $K \to 0$. Next, applying Theorem 9.29 with $\widetilde{P}(K) = K^2 D_T(K)$ or $\widetilde{P}(K) = K^{\gamma+2}h(K^{-1})$, we see that Theorem 10.18 holds. □

10.5 Implied Volatility in Stochastic Volatility Models

We begin this section by comparing two asymptotic formulas for the implied volatility in the Heston model. One formula will be derived from Theorem 10.17, while the other one follows from Theorem 9.16. Let us suppose that $r = 0$, $x_0 = 1$, and $-1 < \rho \le 0$. Then, using (7.59) and (7.60), we see that condition (10.38) holds for the Heston density. Next, applying Theorem 10.17, and using the mean value theorem to simplify the resulting asymptotic formula, we obtain

$$
I(K) = \beta_1\sqrt{\log K} + \beta_2 + \beta_3\frac{\log\log K}{\sqrt{\log K}} + O\left(\frac{1}{\sqrt{\log K}}\right) \tag{10.40}
$$

as $K \to \infty$, where

$$
\beta_1 = \frac{\sqrt{2}}{\sqrt{T}}(\sqrt{A_3 - 1} - \sqrt{A_3 - 2}), \tag{10.41}
$$

$$\beta_2 = \frac{A_2}{\sqrt{2T}}\left(\frac{1}{\sqrt{A_3 - 2}} - \frac{1}{\sqrt{A_3 - 1}}\right), \tag{10.42}$$

and

$$\beta_3 = \frac{1}{\sqrt{2T}}\left(\frac{1}{4} - \frac{a}{c^2}\right)\left(\frac{1}{\sqrt{A_3 - 1}} - \frac{1}{\sqrt{A_3 - 2}}\right). \tag{10.43}$$

Explicit expressions for the constants A_1, A_2, and A_3 can be found in (6.41) and (6.46).

A sharper formula than formula (10.40) can be derived from Theorem 9.16. Note that for the call pricing function in the Heston model, condition (9.43) is satisfied with $\lambda = A_3 - 2$ and $\Lambda(K) = (\log K)^{\frac{1}{2}}$ (use (8.28)). The next statement follows from Theorem 9.16 and the mean value theorem.

Theorem 10.19 *The following asymptotic formula holds for the implied volatility in the Heston model with $r = 0$, $x_0 = 1$, and $-1 < \rho \le 0$:*

$$I(K) = \beta_1\sqrt{\log K} + \beta_2 + \beta_3\frac{\log\log K}{\sqrt{\log K}} + \beta_4\frac{1}{\sqrt{\log K}} + O\big((\log K)^{-1}\big) \tag{10.44}$$

as $K \to \infty$, where β_1, β_2, and β_3 are given by (10.41), (10.42), *and* (10.43), *respectively, and where*

$$\beta_4 = \frac{1}{\sqrt{2\pi}}\left(\frac{1}{\sqrt{A_3 - 1}} - \frac{1}{\sqrt{A_3 - 2}}\right)\log\frac{(A_3 - 1)\sqrt{A_3 - 2}}{A_1}. \tag{10.45}$$

Remark 10.20 The constants β_1 and β_3 in (10.40) and (10.44) depend on the constant A_3, the constant β_2 depends on A_2 and A_3, while the constant β_4 depends on all the constants A_1, A_2, and A_3. Note that the constant A_1 affects only the fourth term in the asymptotic expansion of the implied volatility in the Heston model. This is one more manifestation of a better flexibility of formula (9.33) in comparison with formula (9.44).

We will next formulate a counterpart of formula (10.40) for the implied volatility in the Stein–Stein model. Put

$$\gamma_1 = \frac{\sqrt{2}}{\sqrt{T}}\left(\sqrt{B_3 - 1} - \sqrt{B_3 - 2}\right)$$

and

$$\gamma_2 = \frac{B_2}{\sqrt{2T}}\left(\frac{1}{\sqrt{B_3 - 2}} - \frac{1}{\sqrt{B_3 - 1}}\right),$$

where B_2 and B_3 are defined in Lemma 6.18. Then

$$I(K) = \gamma_1\sqrt{\log K} + \gamma_2 + O\left(\frac{1}{\sqrt{\log K}}\right) \tag{10.46}$$

as $K \to \infty$. Note that the term with $\frac{\log\log K}{\sqrt{\log K}}$ that is present in formula (10.40) is absent in formula (10.46). This happens because the coefficient in front of the expression $\frac{\log\log K}{\sqrt{\log K}}$ in the case of the Stein–Stein model is equal to zero.

Formula (10.46) follows from Theorem 10.17 and the mean value theorem. Here we take into account (7.61). It is also possible to add an additional term in (10.46) by applying Theorem 9.16.

For the uncorrelated Hull–White model, Theorem 10.17 cannot be used to characterize the asymptotic behavior of the implied volatility because for this model we have $\beta = -2$. The next assertion provides a special asymptotic formula, which can be applied to the implied volatility in the uncorrelated Hull–White model.

Theorem 10.21 *Suppose the distribution of the stock price X_T admits a density D_T. Suppose also that*

$$D_T(x) \approx x^{-2}\exp\{-b(\log x)\}$$

as $x \to \infty$, where the function b is positive, increasing on $[c, \infty)$ for some $c > 0$, and such that $b(x) = B(\log x)$ with $B''(x) \approx 1$ as $x \to \infty$. Then

$$\begin{aligned} I(K) &= \frac{\sqrt{2}}{\sqrt{T}}\sqrt{\log K} \\ &\quad - \frac{1}{\sqrt{T}}\sqrt{2\log\frac{1}{K^2 D_T(K)} - \log\log\frac{1}{K^2 D_T(K)} + 2\log\left(\frac{\log\log K}{\log K}\right)} \\ &\quad + O\big((\log\log K)^{-1}\big) \\ &= \frac{\sqrt{2}}{\sqrt{T}}\sqrt{\log K} - \frac{1}{\sqrt{T}}\sqrt{2b(\log K) - \log\big(b(\log K)\big) + 2\log\left(\frac{\log\log K}{\log K}\right)} \\ &\quad + O\big((\log\log K)^{-1}\big) \end{aligned}$$

as $K \to \infty$.

Proof To prove Theorem 10.21, we first reason as in (8.33) and (8.34) to get the relation

$$C(K) \approx \frac{\exp\{-b(\log K)\}\log K}{\log\log K} \approx K^2 D_T(K)\frac{\log K}{\log\log K}$$

as $K \to \infty$. It follows that $C \in PF_\infty$, and hence Corollary 9.11 can be applied. Simplifying the resulting expressions, using the mean value theorem, and taking into account that $b(u) \approx (\log u)^2$ as $u \to \infty$, we complete the proof of Theorem 10.21. □

The asymptotic behavior of the implied volatility in the Hull–White model is characterized in the next assertion.

Theorem 10.22 *The following formula holds for the implied volatility in the Hull–White model*:

$$
\begin{aligned}
I(K) = {} & \frac{\sqrt{2}}{\sqrt{T}}\sqrt{\log K} - \frac{1}{\sqrt{T}} \\
& \times \sqrt{\frac{1}{4T\xi^2}(\log\log K + \log\log\log K)^2 + a_1 \log\log K + a_2 \log\log\log K} \\
& + O\left(\frac{1}{\log\log K}\right)
\end{aligned}
$$

as $K \to \infty$, *where* $a_1 = -2V_1 - 2$, $a_2 = -2V_2$, *and the constants* V_1 *and* V_2 *are defined in* (7.64) *and* (7.65).

Proof It follows from formula (7.63) that the density D_T in the Hull–White model satisfies the conditions in Theorem 10.21 with the function b given by

$$
b(x) = \frac{1}{8t\xi^2}(\log x + \log\log x)^2 - V_1 \log x - V_2 \log\log x, \tag{10.47}
$$

where V_1 and V_1 are defined in (7.64) and (7.65). Note that

$$
B(u) = \frac{1}{8t\xi^2}(u + \log u)^2 - V_1 u - V_2 \log u.
$$

Applying Theorem 10.21 with b defined by (10.47) and making simplifications using the mean value theorem, we see that Theorem 10.22 holds. □

The expression $\frac{\sqrt{2}}{\sqrt{T}}\sqrt{\log K}$ is the leading term in the asymptotic expansion of the implied volatility in the Hull–White model. This expression also appears in Lee's moment formula formulated for the Hull–White model.

Remark 10.23 In [ZA98], Y. Zhu and M. Avellaneda introduced a special risk-neutral modification of the correlated Hull–White model, and obtained estimates for the implied volatility at large strikes in the new model. As far as we know, the problem of characterizing the asymptotic behavior of the implied volatility at extreme strikes in the correlated Hull–White model is still open.

10.6 Asymptotic Equivalence and Moment Formulas

Let C be a call pricing function, and let X be the corresponding asset price process. In this section, we use the same notation as in Sect. 10.1. The maturity T is fixed, the numbers $\tilde{p}$ and $\tilde{q}$ are defined by (10.1) and (10.4), respectively, and the function ψ is given by $\psi(u) = 2 - 4(\sqrt{u^2 + u} - u)$, $u \geq 0$.

Our next goal is to explain when the upper limit in Lee's moment formulas (10.2) and (10.5) can be replaced by the ordinary limit. To solve the previous problem, it suffices to find conditions, under which the following statements hold for the implied volatility I associated with C.

- Let the stock price X_T be such that $\tilde{p} < \infty$. Then

$$I(K) \sim \left(\frac{\psi(\tilde{p})}{T}\right)^{\frac{1}{2}} \sqrt{\log K}, \quad K \to \infty. \tag{10.48}$$

- Let the stock price X_T be such that $\tilde{q} < \infty$. Then

$$I(K) \sim \left(\frac{\psi(\tilde{q})}{T}\right)^{\frac{1}{2}} \sqrt{\log \frac{1}{K}}, \quad K \to 0. \tag{10.49}$$

The next statement provides a solution to the problem formulated above.

Theorem 10.24 *Let $C \in PF_\infty$ be a call pricing function for which $0 \le \tilde{p} < \infty$. Then formula* (10.48) *holds if and only if the function C is of weak Pareto type near infinity with index $\alpha = -\tilde{p}$.*

Proof It follows from Lemma 10.4, and Lemma 10.5 that formula (10.48) holds if and only if

$$\lim_{K\to\infty} (\log K)^{-1} \log \frac{1}{C(K)} = \tilde{p}. \tag{10.50}$$

Note that if the limit on the left-hand side of (10.50) exists, then it necessarily equals $\tilde{p}$ (see Lemma 10.5 and the definition of l in Lemma 10.4).

Let us suppose that formula (10.50) holds. Then for every $\varepsilon > 0$ there exists $K_\varepsilon > 0$ such that

$$K^{-\tilde{p}-\varepsilon} \le C(K) \le K^{-\tilde{p}+\varepsilon}$$

for all $K > K_\varepsilon$. Next, applying Theorem 7.3 to the functions $K^{\tilde{p}} C(K)$ and $(K^{\tilde{p}} C(K))^{-1}$, we see that there exist f_1 and f_2 such as in Definition 7.26 with $F = C$.

To prove the converse statement, we assume that for the function C there exist f_1 and f_2 such as in Definition 7.26 and put

$$\tau(K) = (\log K)^{-1} \log \frac{1}{C(K)}.$$

Then we have

$$(\log K)^{-1} \log \frac{1}{f_2(K)} \le \tau(K) \le (\log K)^{-1} \log \frac{1}{f_1(K)}, \quad K > K_0.$$

Since $f_1 \in R_{-\tilde{p}}$ and $f_2 \in R_{-\tilde{p}}$, we see that there exist slowly varying functions l_1 and l_2 such that

$$\frac{\tilde{p}\log K + \log l_1(K)}{\log K} \leq \tau(K) \leq \frac{\tilde{p}\log K + \log l_2(K)}{\log K}, \quad K > K_0. \tag{10.51}$$

It follows from the representation theorem for slowly varying functions (Theorem 7.5) that for every $l \in R_0$,

$$\lim_{K\to\infty} \frac{\log l(K)}{\log K} = 0.$$

Now it is clear that (10.51) implies (10.50).

This completes the proof of Theorem 10.24. □

The next lemma concerns the case where $\tilde{p} = 0$ in Theorem 10.24.

Lemma 10.25 *The following are equivalent*:

1. *The call pricing function C is of weak Pareto type near infinity with index $\alpha = 0$.*
2. *There exist a function $g_1 \in R_0$ and a number $K_0 > 0$ such that $g_1(K) \leq C(K)$ for all $K > K_0$.*

Proof Recall that we denoted by $\overline{F} = \overline{F}_T$ the complementary cumulative distribution function of X_T given by $\overline{F}_T(y) = \mathbb{P}[X_T > y]$, $y > 0$. It is clear that $\overline{F}$ is a nonnegative nonincreasing integrable function. It follows from (10.12) and Corollary 7.8 that there exists a function $f_2 \in R_0$ such that $f_2(K) \to 0$ as $K \to \infty$ and $C(K) \leq f_2(K)$, $K > K_0$. Now it is clear that Lemma 10.25 holds. □

Our next goal is to establish an assertion similar to Theorem 10.24 in the case where $K \to 0$. We will need the results obtained in Sect. 9.7. Let C be a call pricing function, X the corresponding asset price process, and μ_T the distribution of the asset price X_T. Let P be the put pricing function associated with X. Recall that in Sect. 9.7, we defined a function G by

$$G(T,K) = \frac{K}{x_0 e^{rT}} P\big(T, \eta_T(K)\big), \tag{10.52}$$

where $\eta_T(K) = (x_0 e^{rT})^2 K^{-1}$. We also proved that G is a call pricing function, and denoted by $\widetilde{X}$ an asset price process corresponding to G. The distribution $\tilde{\mu}_T$ of $\widetilde{X}_T$ is given by

$$\tilde{\mu}_T(A) = \frac{1}{x_0 e^{rT}} \int_{\eta_T(A)} x \, d\mu_T(x)$$

for all Borel sets A. Denote by $\widehat{F}_T$ the complementary cumulative distribution function of $\widetilde{X}_T$. We have

$$\widehat{F}_T(y) = \tilde{\mu}_T\big((y,\infty)\big) = \frac{1}{x_0 e^{rT}} \int_0^{\eta_T(y)} x \, d\mu_T(x), \quad y > 0. \tag{10.53}$$

Moreover, the asset price distribution densities D_T and $\widehat{D}_T$ associated with the pricing functions C and G, respectively, are related by the following formula:

$$\widehat{D}_T(x) = \frac{(x_0e^{rT})^3}{x^3} D_T\big(\eta_T(x)\big), \quad x > 0.$$

Furthermore, the equality $I_C(T,K) = I_G(T,\eta_T(K))$ holds for all $T > 0$ and $K > 0$.

The next statement provides a relation between the moments of the asset price processes X and $\widetilde{X}$.

Lemma 10.26 *For fixed $T > 0$ and $p \neq 0$, the following formula holds:*

$$m_p(\widetilde{X}_T) = \big(x_0e^{rT}\big)^{2p-1} m_{1-p}(X_T). \tag{10.54}$$

Proof For every $p > 0$ we have

$$m_p(\widetilde{X}_T) = p\int_0^\infty y^{p-1}\widehat{F}_T(y)\,dy. \tag{10.55}$$

It follows from (10.55) that

$$\begin{aligned} m_p(\widetilde{X}_T) &= \frac{p}{x_0e^{rT}}\int_0^\infty y^{p-1}\,dy\int_0^{\eta_T(y)} x\,d\mu_T(x) \\ &= \frac{p}{x_0e^{rT}}\int_0^\infty x\,d\mu_T(x)\int_0^{\eta_T(x)} y^{p-1}\,dy \\ &= \big(x_0e^{rT}\big)^{2p-1}\int_0^\infty x^{1-p}\,d\mu_T(x) \\ &= \big(x_0e^{rT}\big)^{2p-1} m_{1-p}(X_T). \end{aligned}$$

Now let $p < 0$. Then

$$m_p(\widetilde{X}_T) = \int_0^\infty \big[1 - \widehat{F}\big(y^{\frac{1}{p}}\big)\big]\,dy. \tag{10.56}$$

Using (10.53) and (10.56), we see that

$$\begin{aligned} m_p(\widetilde{X}_T) &= \frac{1}{x_0e^{rT}}\int_0^\infty dy\int_{\eta_T(y^{\frac{1}{p}})}^\infty x\,d\mu_T(x) \\ &= \frac{|p|}{x_0e^{rT}}\int_0^\infty u^{p-1}\,du\int_{\eta_T(u)}^\infty x\,d\mu_T(x) \\ &= \frac{|p|}{x_0e^{rT}}\int_0^\infty x\,d\mu_T(x)\int_{\eta_T(x)}^\infty u^{p-1}\,du \\ &= \big(x_0e^{rT}\big)^{2p-1}\int_0^\infty x^{1-p}\,d\mu_T(x) \end{aligned}$$

$$= \left(x_0 e^{rT}\right)^{2p-1} m_{1-p}(X_T).$$

This completes the proof of Lemma 10.26. □

Recall the definition of the critical orders $\tilde{p}$ and $\tilde{q}$ (see Theorems 10.2 and 10.3). In order to distinguish among the critical orders associated with the call pricing functions C and G, we will use the symbols $\tilde{p}_C$, $\tilde{q}_C$, $\tilde{p}_G$, and $\tilde{q}_G$.

The next assertion can be easily obtained from Lemma 10.26.

Corollary 10.27 *The following equalities hold*: $\tilde{p}_G = \tilde{q}_C$ *and* $\tilde{q}_G = \tilde{p}_C$.

Theorem 10.28 *Let* $C \in PF_0$ *and let* P *be the corresponding put pricing function. Suppose* $0 \le \tilde{q} < \infty$. *Then formula* (10.49) *holds if and only if the function* P *is of weak Pareto type near zero with index* $\alpha = -\tilde{q} - 1$.

Proof We use the notation $\tilde{p} = \tilde{p}_G$ and $\tilde{q} = \tilde{q}_C$ in the proof. By Corollary 10.27, we have $\tilde{p} = \tilde{q}$. It follows from (9.78) that formula (10.49) is equivalent to the following:

$$I_G\left(\eta_T(K)\right) \sim \left(\frac{\psi(\tilde{p})}{T}\right)^{\frac{1}{2}} \sqrt{\log \frac{1}{K}}$$

as $K \to 0$, which in its turn is equivalent to

$$I_G(K) \sim \left(\frac{\psi(\tilde{p})}{T}\right)^{\frac{1}{2}} \sqrt{\log K}$$

as $K \to \infty$. Since G is a call pricing function, we can apply Theorem 10.24, and obtain one more equivalent condition:

$$f_1(K) \le G(K) \le f_2(K), \quad K > K_0, \tag{10.57}$$

for some functions $f_1 \in R_{-\tilde{p}}$ and $f_2 \in R_{-\tilde{p}}$. Now, it is not hard to see, using (9.71) and (10.57), that the conclusion in Theorem 10.28 holds. □

The asymptotic equivalence in Lee's moment formulas is also valid under certain restrictions on the complementary cumulative distribution function or the distribution density of the stock price. The next corollary can be derived from Theorems 10.24 and 10.28. We leave the proof of this corollary as an exercise for the interested reader.

Corollary 10.29 *The following statements hold*:

- *Let* $C \in PF_\infty$, *and suppose* $0 \le \tilde{p} < \infty$ *for the stock price* X_T. *Suppose also that the complementary cumulative distribution function* $\overline{F} = \overline{F}_T$ *of* X_T *is of weak*

Pareto type near infinity with index $\alpha \le -1$. *Then*

$$I(K) \sim \left(\frac{\psi(-\alpha-1)}{T}\right)^{\frac{1}{2}} \sqrt{\log K} \quad \text{as } K \to \infty.$$

- *Let* $C \in PF_\infty$, *and suppose* $0 \le \tilde{p} < \infty$ *for the stock price* X_T. *Suppose also that the distribution density* D_T *of* X_T *is of weak Pareto type near infinity with index* $\alpha \le -2$. *Then*

$$I(K) \sim \left(\frac{\psi(-\alpha-2)}{T}\right)^{\frac{1}{2}} \sqrt{\log K} \quad \text{as } K \to \infty.$$

- *Let* $C \in PF_0$, *and suppose* $0 \le \tilde{q} < \infty$ *for the stock price* X_T. *Suppose also that the complementary cumulative distribution function* $\overline{F} = \overline{F}_T$ *of* X_T *is of weak Pareto type near zero with index* $\alpha \le 0$. *Then*

$$I(K) \sim \left(\frac{\psi(-\alpha)}{T}\right)^{\frac{1}{2}} \sqrt{\log \frac{1}{K}} \quad \text{as } K \to \infty.$$

- *Let* $C \in PF_0$, *and suppose* $0 \le \tilde{q} < \infty$ *for the stock price* X_T. *Suppose also that the distribution density* D_T *of* X_T *is of weak Pareto type near zero with index* $\alpha \le 1$. *Then*

$$I(K) \sim \left(\frac{\psi(-\alpha+1)}{T}\right)^{\frac{1}{2}} \sqrt{\log \frac{1}{K}} \quad \text{as } K \to 0.$$

10.7 Implied Volatility in Mixed Models

Mixed stochastic asset price models are informally defined as follows. In such a model, the price process can be represented as the product of two independent positive stochastic processes, and, in addition, the marginal distributions of these processes satisfy weak Pareto type conditions.

We will next prove the following technical lemma.

Lemma 10.30 *Let* X *be a random variable on a probability space* $(\Omega, \mathcal{F}, \mathbb{P})$, *and suppose* $X = X^{(1)}X^{(2)}$, *where* $X^{(1)}$ *and* $X^{(2)}$ *are independent positive random variables. Suppose also that for every* $i = 1, 2$ *the distribution of the random variable* $X^{(i)}$ *possesses a density* $D^{(i)}$ *of weak Pareto type near infinity with index* $\alpha_i \le -1$. *Then the distribution density* D *of the random variable* X *is of weak Pareto type near infinity with index* $\alpha = \max\{\alpha_1, \alpha_2\}$.

Proof Set $L^{(i)} = \log X^{(i)}$, $i = 1, 2$, and $L = \log X$. It is clear that the random variables $L^{(i)}$, $i = 1, 2$, are independent, and the distribution density $\widetilde{D}^{(i)}$ of $L^{(i)}$ is given by

$$\widetilde{D}^{(i)}(y) = e^y D^{(i)}\left(e^y\right), \quad y > 0, \ i = 1, 2. \tag{10.58}$$

Since $L = L^{(1)} + L^{(2)}$, the distribution density $\widetilde{D}$ of the random variable L satisfies

$$\widetilde{D}(y) = \int_{-\infty}^{\infty} \widetilde{D}^{(1)}(y - z)\widetilde{D}^{(2)}(z)\,dz, \quad y > 0. \tag{10.59}$$

It follows from (10.58) and from the weak Pareto type conditions in the formulation of Lemma 10.30 that there exist positive slowly varying functions l_j, $1 \le j \le 4$, such that

$$e^{(\alpha_1+1)y} l_1\left(e^y\right) \le \widetilde{D}^{(1)}(y) \le e^{(\alpha_1+1)y} l_2\left(e^y\right) \tag{10.60}$$

and

$$e^{(\alpha_2+1)y} l_3\left(e^y\right) \le \widetilde{D}^{(2)}(y) \le e^{(\alpha_2+1)y} l_4\left(e^y\right) \tag{10.61}$$

for large enough values of y. Since a slowly varying function cannot grow faster than a positive power, (10.60) and (10.61) imply that for every $\delta > 0$,

$$e^{(\alpha_1+1-\delta)y} \le \widetilde{D}^{(1)}(y) \le e^{(\alpha_1+1+\delta)y} \tag{10.62}$$

and

$$e^{(\alpha_2+1-\delta)y} \le \widetilde{D}^{(2)}(y) \le e^{(\alpha_2+1+\delta)y} \tag{10.63}$$

for $y > y_\delta$.

It is not hard to see that with no loss of generality, we may assume that $\alpha_2 \le \alpha_1$ and $\alpha = \alpha_1$. Now suppose $\delta < \frac{1}{2}$. Then (10.59), (10.62), and (10.63) imply that for $y > \tilde{y}_\delta$,

$$\begin{aligned}
\widetilde{D}(y) &\ge \int_{\delta y}^{(1-\delta)y} e^{(\alpha_1+1-\delta)(y-z)} e^{(\alpha_2+1-\delta)z}\,dz \\
&\ge \frac{1}{\delta - \alpha_2 - 1} e^{(\alpha_1+1-\delta)(1-\delta)y}\left[e^{(\alpha_2+1-\delta)\delta y} - e^{(\alpha_2+1-\delta)(1-\delta)y}\right] \\
&\ge \frac{1}{2(\delta - \alpha_2 - 1)} e^{(\alpha_1+1-\delta)(1-\delta)y} e^{(\alpha_2+1-\delta)\delta y} \\
&= \frac{1}{2(\delta - \alpha_2 - 1)} e^{(\alpha_1+1)y} e^{(\alpha_2-\alpha_1-1)\delta y}.
\end{aligned}$$

It follows that for every sufficiently small $\varepsilon > 0$ there exists $\hat{y}_\varepsilon > 0$ such that $\widetilde{D}(y) \ge e^{(\alpha+1-\varepsilon)y}$ for all $y > \hat{y}_\varepsilon$. Since

$$D(x) = x^{-1} \log \widetilde{D}(\log x), \tag{10.64}$$

we have $D(x) \ge x^{\alpha-\varepsilon}$ for all $x > x_\varepsilon$. Next, applying Theorem 7.3 to the function $x \mapsto x^\alpha D(x)^{-1}$, we see that there exists a slowly varying function l_5 such that

$$D(x) \ge x^\alpha l_5(x), \quad x > x_0. \tag{10.65}$$

In order to obtain an estimate from above for the function D, we fix δ with $0 < \delta < \frac{1}{2}$, and rewrite (10.59) as follows:

$$\begin{aligned}\widetilde{D}(y) &= \int_{-\infty}^{\delta y} \widetilde{D}^{(1)}(y-z)\widetilde{D}^{(2)}(z)\,dz + \int_{\delta y}^{(1-\delta)y} \widetilde{D}^{(1)}(y-z)\widetilde{D}^{(2)}(z)\,dz \\ &\quad + \int_{(1-\delta)y}^{\infty} \widetilde{D}^{(1)}(y-z)\widetilde{D}^{(2)}(z)\,dz \\ &= I_{1,\delta}(y) + I_{2,\delta}(y) + I_{3,\delta}(y).\end{aligned} \tag{10.66}$$

It is not hard to see that (10.62) and (10.63) imply the following estimates:

$$\begin{aligned}I_{1,\delta}(y) + I_{3,\delta}(y) &\le e^{(\alpha_1+1+\delta)(1-\delta)y} \int_{-\infty}^{\delta y} \widetilde{D}^{(2)}(z)\,dz \\ &\quad + e^{(\alpha_2+1+\delta)(1-\delta)y} \int_{(1-\delta)y}^{\infty} \widetilde{D}^{(1)}(y-z)\,dz \\ &\le e^{(\alpha_1+1+\delta)(1-\delta)y} + e^{(\alpha_2+1+\delta)(1-\delta)y} \\ &\le 2e^{(\alpha-1)y} e^{\delta(1-\delta)y}, \quad y > \tilde{y}_\delta.\end{aligned} \tag{10.67}$$

In addition, (10.62) and (10.63) give

$$\begin{aligned}I_{2,\delta}(y) &\le \int_{\delta y}^{(1-\delta)y} e^{(\alpha_1+1+\delta)(y-z)} e^{(\alpha_2+1+\delta)z}\,dz \\ &\le y e^{(\alpha+1+\delta)y}, \quad y > \hat{y}_\delta.\end{aligned} \tag{10.68}$$

Next, using (10.66), (10.67), and (10.68), we see that for every sufficiently small $\varepsilon > 0$ there exists $s_\varepsilon > 0$ such that $\widetilde{D}(y) \le e^{(\alpha+1+\varepsilon)y}$ for all $y > s_\varepsilon$. Hence we have $D(x) \le x^{\alpha+\varepsilon}$ for all $x > x_\varepsilon$. Here we use (10.64). Next, applying Theorem 7.3 to the function $x \mapsto x^{-\alpha} D(x)$, we see that there exists a slowly varying function l_6 such that

$$D(x) \le x^\alpha l_6(x), \quad x > x_1. \tag{10.69}$$

Finally, combining (10.65) and (10.69), we see that the function D is of weak Pareto type at infinity with index α.

This completes the proof of Lemma 10.30. □

Let us consider a stochastic asset price model such that the price process in this model is given by $X = X^{(1)} X^{(2)}$, where $X^{(1)}$ and $X^{(2)}$ are independent positive adapted processes on $(\Omega, \mathcal{F}, \{\mathcal{F}_t\}, \mathbb{P}^*)$. Assume $\mathbb{P}^*$ is a risk-neutral measure.

The next corollary describes the asymptotics of the implied volatility in a mixed model.

Corollary 10.31 *Suppose for all $T > 0$ and every $i = 1, 2$ the distribution of $X_T^{(i)}$ admits a density $D_T^{(i)}$ of weak Pareto type near infinity with index $\alpha_{i,T} \leq -2$. Then for every $T > 0$ the following asymptotic formula holds for the implied volatility $K \mapsto I(K) = I(T, K)$:*

$$I(K) \sim \left(\frac{\psi(-\alpha_T - 2)}{T}\right)^{\frac{1}{2}} \sqrt{\log K}, \quad K \to \infty, \tag{10.70}$$

where $\alpha_T = \max\{\alpha_{1,T}, \alpha_{2,T}\}$.

It is not hard to see that Corollary 10.31 follows from part 2 of Corollary 10.29 and Lemma 10.30.

Remark 10.32 The restriction $\alpha_{i,T} \leq -2$ is imposed in Corollary 10.31 because we assume that $\mathbb{E}^*[X_T] < \infty$ for all $T \geq 0$.

Remark 10.33 Lemma 10.30 is also valid under milder restrictions on one of the random variables $X^{(i)}$, $i = 1, 2$. For instance, the lemma holds if there exist a number $c > 0$ and a nonnegative function $G^{(2)}$, satisfying the following conditions. The function $G^{(2)}$ is of weak Pareto type at infinity with index $\alpha_2 \leq -2$, and the distribution $\mu^{(2)}$ of the random variable $X^{(2)}$ can be represented in the following form:

$$\mu^{(2)}(A) = c\delta_1(A) + \int_A G^{(2)}(z)\,dz \tag{10.71}$$

for all Borel subsets A of $[0, \infty)$. In (10.71), δ_1 is the Dirac measure at $z = 1$. The restrictions on the random variable $X^{(1)}$ remain the same as in Lemma 10.30.

Remark 10.34 Put $\widehat{X}^{(2)} = \log X^{(2)}$, and denote the distribution of $\widehat{X}^{(2)}$ by $\hat{\mu}^{(2)}$. Then condition (10.71) is equivalent to the following:

$$\hat{\mu}^{(2)}(B) = c\delta_0(B) + \int_B \widehat{G}^{(2)}(y)\,dy \tag{10.72}$$

for all Borel subsets B of $\mathbb{R}^1$. In (10.72), $\widehat{G}^{(2)}(y) = e^y G^{(2)}(e^y)$. Taking this equivalence into account, we can prove the statement in Remark 10.33 using the same reasoning as in the proof of Lemma 10.30 with only minor differences in the details. Corollary 10.31 can be generalized similarly. Here it suffices to assume that the density $D_T^{(1)}$ of $X_T^{(1)}$ is of weak Pareto type near infinity with index $\alpha_{1,T} \leq -2$, and the distribution $\mu_T^{(2)}$ of $X_T^{(2)}$ satisfies

$$\mu_T^{(2)}(A) = c_T\delta_1(A) + \int_A G_T^{(2)}(z)\,dz \tag{10.73}$$

with a function G_T of weak Pareto type near infinity with index $\alpha_{2,T} \le -2$. The generalization of Corollary 10.31 mentioned above states that formula (10.70) holds under the previous assumptions.

For a mixed stochastic asset price model, the asymptotic behavior of the implied volatility $K \mapsto I(K)$ at small strikes can be described using the same methods as in the case of large strikes.

Corollary 10.35 *Suppose the asset price process X in a mixed model has the following representation: $X = X^{(1)} X^{(2)}$, where $X^{(1)}$ and $X^{(2)}$ are independent positive adapted processes on a filtered probability space $(\Omega, \mathcal{F}, \{\mathcal{F}_t\}, \mathbb{P}^*)$. Let us assume $\mathbb{P}^*$ is a risk-neutral measure, and suppose for every $T > 0$ the distribution of the random variable $X_T^{(1)}$ possesses a density $D_T^{(1)}$ of weak Pareto type near zero with index $\beta_{1,T} \le 1$, while the distribution $\mu_T^{(2)}$ of the random variable $X_T^{(2)}$ satisfies the condition in* (10.73) *with the function G_T of weak Pareto type near zero with index $\beta_{2,T} \le 1$. Then the distribution μ_T of X_T possesses a density D_T of weak Pareto type near zero with the index given by $\beta_T = \max\{\beta_{1,T}, \beta_{2,T}\}$. Moreover,*

$$I(K) \sim \left(\frac{\psi(-\beta_T + 1)}{T}\right)^{\frac{1}{2}} \sqrt{\log \frac{1}{K}}, \quad K \to 0. \tag{10.74}$$

The proof of the previous statement is left as an exercise for the reader.

10.8 Asset Price Models with Jumps

Asset price models, in which price processes are not necessarily continuous, are frequently encountered in financial mathematics. Many such models are mixtures of diffusion models, or stochastic volatility models, with exponential Lévy models. Note that the price process in an exponential Lévy model is given by e^J, where J is a Lévy process. Numerous asset price models with jumps have been developed in the last decades. For example, Merton's jump-diffusion model is the mixture of the Black–Scholes model with the exponential Lévy model, for which the process J is the compound Poisson process with the Gaussian law for jump sizes (see [Mer76]). In the Bates model, the same process J and the Heston model are used (see [Bat96]), while Kou's model is a mixture of the Black–Scholes model with the exponential Lévy model, in which the process J is the compound Poisson process with the double exponential law for jump sizes (see [Kou02], see also [KW03]). More information can be found in [CT04, Sch03].

In this book, asset price models with jumps appear only in the present section. We restrict ourselves to a special mixed model, in which the stock price process is a mixture of the jump process used in Kou's model and the stock price process in the correlated Heston model. However, many other stochastic volatility models with jumps can be studied using the techniques developed in this section.

Recall that a nonnegative random variable U on a probability space $(\Omega, \mathcal{F}, \mathbb{P})$ is called exponentially distributed with parameter $\lambda > 0$ if the distribution of U admits a density d_λ given by $d_\lambda(y) = \lambda e^{-\lambda y}\mathbb{1}_{\{y\geq 0\}}$. It is said that a nonnegative integer-valued random variable N is distributed according to the Poisson law with parameter λ if $\mathbb{P}(N = n) = e^{-\lambda}\lambda^n(n!)^{-1}$, $n \geq 1$. Let τ_k, $k \geq 1$, be a sequence of independent exponentially distributed with parameter λ random variables, and set

$$T_n = \sum_{k=1}^{n} \tau_k, \quad n \geq 1.$$

Then the stochastic process N given by

$$N_t = \sum_{n=1}^{\infty} \mathbb{1}_{\{t\geq T_n\}}, \quad t \geq 0,$$

is called the Poisson process with intensity λ. It is clear that for any $t \geq 0$, the random variable N_t is distributed according to the Poisson law with parameter λt. This random variable counts the number of random times T_n between 0 and t.

Definition 10.36 Let ρ be a distribution on $\mathbb{R}$. The compound Poisson process with intensity $\lambda > 0$ and jump size distribution ρ is the process J defined by

$$J_t = \sum_{k=1}^{N_t} Y_k, \quad t \geq 0. \tag{10.75}$$

In (10.75), Y is a sequence of independent identically distributed variables such that ρ is the law of Y_k for every $k \geq 1$, and N is a Poisson process with intensity λ independent of the process Y.

A compound Poisson process that will be used in a mixed model considered below is as follows:

$$J_t = \sum_{i=1}^{N_t} (V_i - 1), \quad t \geq 0, \tag{10.76}$$

where V_i are positive independent identically distributed random variables, which are independent of the process N. We assume that for every $i \geq 1$ the distribution density f of the random variable $U_i = \log V_i$ is double exponential, that is,

$$f(u) = p\eta_1 e^{-\eta_1 u}\mathbb{1}_{\{u\geq 0\}} + q\eta_2 e^{\eta_2 u}\mathbb{1}_{\{u<0\}}, \tag{10.77}$$

where $\eta_1 > 1$, $\eta_2 > 0$, and p and q are positive numbers such that $p + q = 1$.

We will next consider a model that is a mixture of the correlated Heston model and a special exponential Lévy model. This mixed model is as follows:

$$\begin{cases} d\widetilde{X}_t = \mu\widetilde{X}_{t-}\,dt + \sqrt{Y_t}\,\widetilde{X}_{t-}\,dW_t + \widetilde{X}_{t-}\,dJ_t \\ dY_t = (a - bY_t)\,dt + c\sqrt{Y_t}\,dZ_t. \end{cases} \tag{10.78}$$

It is assumed in (10.78) that $\mu \in \mathbb{R}$, $a \geq 0$, $b \geq 0$, and $c > 0$. The Brownian motions W and Z in (10.78) are correlated with the correlation coefficient $\rho \in (-1, 0]$, and the process J is given by (10.76). Recall that we set $U_i = \log V_i$. The stock price process $\widetilde{X}$ in the model described by (10.78) is given by the following formula:

$$\widetilde{X}_t = x_0 \exp\left\{\mu t - \frac{1}{2}\int_0^t Y_s\,ds + \int_0^t \sqrt{Y_s}\,dW_s + \widetilde{J}_t\right\} \tag{10.79}$$

where $\widetilde{J}_t = \sum_{i=1}^{N_t} U_i$. Formula (10.79) can be obtained from the Doléans–Dade formula (see, for example, [Pro04]). Now put

$$\widetilde{X}_t^{(1)} = x_0 \exp\left\{\mu t - \frac{1}{2}\int_0^t Y_s\,ds + \int_0^t \sqrt{Y_s}\,dW_s\right\}$$

and $\widetilde{X}^{(2)} = \exp\{\widetilde{J}_t\}$. The process $\widetilde{X}^{(1)}$ is the stock price process in the correlated Heston model. Hence the distribution density $\widetilde{D}_t^{(1)}$ of $\widetilde{X}_t^{(1)}$ is a function of Pareto type

$$\alpha_{1,t} = -A_3 < -2 \tag{10.80}$$

near infinity (see Sect. 6.1) and of Pareto type

$$\beta_{1,t} = -\widetilde{A}_3 < 1 \tag{10.81}$$

near zero (see Sect. 6.1.7). Note that the numbers A_3 and $\widetilde{A}_3$ depend on t.

Next, we turn our attention to the distribution $\hat{\mu}_t$ of the random variable $\widetilde{J}_t = \log \widetilde{X}^{(2)}$. It is not hard to see that the following formula holds:

$$\hat{\mu}_t(B) = \pi_{0,t}\delta_0(B) + \int_B \widehat{D}_t^{(2)}(y)\,dy \tag{10.82}$$

where

$$\widehat{D}_t^{(2)}(y) = \sum_{n=1}^{\infty} \pi_{n,t} f^{*(n)}(y). \tag{10.83}$$

In (10.82) and (10.83), $\pi_{0,t} = e^{-\lambda t}$, $\pi_{n,t} = e^{-\lambda t}(\lambda t)^n (n!)^{-1}$ for all $n \geq 1$, B is a Borel subset of $\mathbb{R}^1$, and f is given by (10.77). The star in (10.83) denotes the convolution.

Lemma 10.37 *Let f be the density of the double exponential law (see (10.77)). Then for every $n > 1$ the following formula holds*:

$$f^{*(n)}(y) = e^{-\eta_1 y} \sum_{k=1}^{n} P_{n,k}\eta_1^k \frac{1}{(k-1)!} y^{k-1} \mathbb{1}_{\{y \geq 0\}} + e^{\eta_2 y} \sum_{k=1}^{n} Q_{n,k}\eta_2^k \frac{1}{(k-1)!} (-y)^{k-1} \mathbb{1}_{\{y < 0\}},$$

where

$$P_{n,k} = \sum_{i=k}^{n-1} \binom{n-k-1}{i-k} \binom{n}{i} \left(\frac{\eta_1}{\eta_1+\eta_2}\right)^{i-k} \left(\frac{\eta_2}{\eta_1+\eta_2}\right)^{n-i} p^i q^{n-i}$$

for all $1 \le k \le n-1$, *and*

$$Q_{n,k} = \sum_{i=k}^{n-1} \binom{n-k-1}{i-k} \binom{n}{i} \left(\frac{\eta_1}{\eta_1+\eta_2}\right)^{n-i} \left(\frac{\eta_2}{\eta_1+\eta_2}\right)^{i-k} p^{n-i} q^i$$

for all $1 \le k \le n-1$. *In addition,* $P_{n,n} = p^n$ *and* $Q_{n,n} = q^n$.

Lemma 10.37 was obtained by S. Kou (see Proposition B.1 in [Kou02]).

Lemma 10.38 *For every Borel set* $B \subset \mathbb{R}$,

$$\hat{\mu}_t(B) = \pi_{0,t}\delta_0(B) + \int_{B\cap[0,\infty)} G_{1,t}(y)e^{-\eta_1 y}\, dy + \int_{B\cap(-\infty,0)} G_{2,t}(y)e^{\eta_2 y}\, dy,$$

where

$$G_{1,t}(y) = \sum_{k=0}^{\infty} \left[\frac{\eta_1^{k+1}}{k!} \sum_{n=k+1}^{\infty} \pi_{n,t} P_{n,k+1}\right] y^k \tag{10.84}$$

and

$$G_{2,t}(y) = \sum_{k=0}^{\infty} \left[\frac{\eta_2^{k+1}}{k!} \sum_{n=k+1}^{\infty} \pi_{n,t} Q_{n,k+1}\right] (-y)^k. \tag{10.85}$$

Lemma 10.38 follows from Lemma 10.37 and formula (10.82).

The next lemma estimates the rate of growth of the functions $G_{1,t}$ and $G_{2,t}$ defined by (10.84) and (10.85), respectively.

Lemma 10.39 *For every* $\varepsilon > 0$ *the function* $G_{1,t}$ *grows slower than the function* $y \mapsto e^{\varepsilon y}$ *as* $y \to \infty$. *Similarly, the function* $G_{2,t}$ *grows slower than the function* $y \mapsto e^{-\varepsilon y}$ *as* $y \to -\infty$.

Proof We will prove the lemma by comparing the Taylor coefficients

$$a_k = \frac{1}{k!}\eta_1^{k+1} \sum_{n=k+1}^{\infty} \pi_{n,t} P_{n,k+1}, \quad k \ge 0,$$

of the function $G_{1,t}$ and the Taylor coefficients $b_k = \frac{1}{k!}\varepsilon^k$, $k \geq 0$, of the function $e^{\varepsilon y}$. It is not hard to see that $a_k \leq b_k$ for $k > k_0$. Here we use the estimate

$$\eta_1^{k+1} \sum_{n=k+1}^{\infty} \pi_{n,t} P_{n,k+1} \leq \eta_1^{k+1} \sum_{n=k+1}^{\infty} \pi_{n,t}.$$

Finally, taking into account the fast decay of the complementary cumulative distribution function of the Poisson distribution, we complete the proof of Lemma 10.39 for the function $G_{1,t}$. The proof for the function $G_{2,t}$ is similar. □

It follows from (10.82) that the distribution $\tilde{\mu}_t$ of $\widetilde{X}_t^{(2)}$ is given by

$$\tilde{\mu}_t(A) = \pi_{0,t}\delta_1(A) + \int_A \widetilde{D}_t^{(2)}(z)\,dz$$

where A is a Borel subset of $[0, \infty)$ and

$$\widetilde{D}_t^{(2)}(z) = z^{-1}\widehat{D}_t^{(2)}(\log z) \tag{10.86}$$

for all $z \in (0, \infty)$. Now, using Lemmas 10.38 and 10.39, the estimate

$$\widehat{D}_t^{(2)}(y) \geq \pi_{1,t} f(y), \quad y \in \mathbb{R}^1,$$

and (10.86), we see that the function $\widetilde{D}_t^{(2)}$ is of weak Pareto type near infinity with index

$$\alpha_{2,t} = -\eta_1 - 1 < -2 \tag{10.87}$$

and near zero with index

$$\beta_{2,t} = 1 - \eta_2 < 1. \tag{10.88}$$

Our next goal is to study the asymptotic behavior of the implied volatility in the correlated Heston model with jumps, given by (10.78). We will first find risk-neutral measures for this model. The following well-known lemma (see part 2 of Proposition 3.17 in [CT04]) is often helpful in the search for risk-neutral measures in models with jumps.

Lemma 10.40 *Let H be a real-valued stochastic process with independent increments. Then the condition*

$$\mathbb{E}\left[\exp\{uH_t\}\right] < \infty$$

for some $u \in \mathbb{R}$ and all $t \geq 0$ implies that the process defined by

$$\exp\{uH_t\}\left\{\mathbb{E}\left[\exp\{uH_t\}\right]\right\}^{-1}, \quad t \geq 0,$$

is a martingale.

Proof The independence of increments of the process H and the properties of conditional expectations imply that for $s < t$,

$$\begin{aligned}\mathbb{E}\big[\exp\{uH_t\}\big|\mathcal{F}_s\big] &= \mathbb{E}\big[\exp\big\{u(H_t - H_s + H_s)\big\}\big|\mathcal{F}_s\big] \\ &= \exp\{uH_s\}\mathbb{E}\big[\exp\big\{u(H_t - H_s)\big\}\big].\end{aligned}$$

Moreover,

$$\mathbb{E}\big[\exp\{uH_t\}\big] = \mathbb{E}\big[\exp\big\{u(H_t - H_s)\big\}\big]\mathbb{E}\big[\exp\{uH_s\}\big].$$

This completes the proof of Lemma 10.40. □

The process $\widetilde{J}$ appearing in (10.79) is a Lévy process. Hence, it has independent increments. Recall that we denoted by f the double exponential density defined in (10.77). It is not hard to prove using (10.82) that

$$\mathbb{E}\big[\exp\{\widetilde{J}_t\}\big] = \exp\{\lambda\eta t\}, \quad t \geq 0, \tag{10.89}$$

where η is given by

$$\eta = \left[\int_{\mathbb{R}} e^u f(u)\, du - 1\right],$$

and λ is the intensity of the Poisson process N in (10.76). In addition, we have

$$\eta = \frac{p}{\eta_1 - 1} - \frac{q}{\eta_2 + 1}$$

(use (10.77)). Next, applying Lemma 10.40 with $u = 1$ and $H = \widetilde{J}$, and taking into account formula (10.89), we see that the process

$$t \mapsto \exp\{\widetilde{J}_t - \lambda\eta t\}, \quad t \geq 0, \tag{10.90}$$

is a martingale. We will next show that in order the physical measure $\mathbb{P}$ to be a risk-neutral measure, we have to choose the drift μ in (10.78) as follows:

$$\mu = r - \lambda\eta, \tag{10.91}$$

where r is the interest rate. The previous restriction on the drift in the stock price equation is imposed by the no-arbitrage condition. If μ satisfies (10.91), then the discounted stock price process is given by

$$e^{-rt}\widetilde{X}_t = x_0 \exp\left\{\int_0^t \sqrt{Y_s}\, dW_s - \frac{1}{2}\int_0^t Y_s\, ds + \widetilde{J}_t - \lambda\eta t\right\} \tag{10.92}$$

(formula (10.92) follows from (10.79)). Now it is not hard to prove that the process in (10.92) is a martingale with respect to the physical measure $\mathbb{P}$, using the following facts: the discounted stock price in the Heston model and the process in (10.90) are martingales, and the process $\widetilde{J}$ is independent from the other processes appearing on the right-hand side of formula (10.92).

Let us assume that condition (10.91) holds, and denote by $\mathbb{P}^*$ the corresponding risk-neutral measure. The next statement describes the asymptotic behavior of the implied volatility $K \mapsto I(K)$ in the Heston model with jumps, defined by (10.78).

Theorem 10.41 *The following formula holds as* $K \to \infty$:

$$I(K) \sim \left(\frac{\psi(\gamma_T)}{T}\right)^{\frac{1}{2}} \sqrt{\log K} \tag{10.93}$$

where $\gamma_T = \min\{A_3 - 2, \eta_1 - 1\}$.

The following formula holds as $K \to 0$:

$$I(K) \sim \left(\frac{\psi(\delta_T)}{T}\right)^{\frac{1}{2}} \sqrt{\log \frac{1}{K}} \tag{10.94}$$

where $\delta_T = \min\{\widetilde{A}_3 + 1, \eta_2\}$.

Proof It follows from (10.80), (10.81), (10.87), (10.88), Remarks 10.33 and 10.34, and Corollary 10.35 that the asset price density $\widetilde{D}_T$ is of weak Pareto type near infinity with index $\alpha_T = \max\{-A_3, -\eta_1 - 1\}$ and near zero with index $\beta_T = \max\{-\widetilde{A}_3, 1 - \eta_2\}$. Now using Remarks 10.33 and 10.34, and Corollary 10.35 again, we see that formulas (10.93) and (10.94) hold.

This completes the proof of Theorem 10.41. □

Remark 10.42 In uncorrelated stochastic asset price models, the implied volatility as a function of the log-strike $k = \log K$ is symmetric (even) on $\mathbb{R}^1$. In correlated models, the implied volatility may have a skewed shape. Similar skewness may appear when we pass from an uncorrelated continuous stochastic asset price model to a corresponding model with jumps.

10.9 Volatility Smile

The expression "volatility smile" was coined to describe the following pattern observed in option market data. Implied volatilities associated with in-the-money and out-of-the-money options are often greater than in the case of at-the money options. Recall that for in-the-money options we have $K \in (0, x_0 e^{rT})$, while out-of-the-money options are such that $K \in (x_0 e^{rT}, \infty)$. The option is at-the-money if $K = x_0 e^{rT}$.

It is important to determine which option pricing models exhibit a "smiling" implied volatility. In [RT96], E. Renault and N. Touzi proved that any uncorrelated stochastic volatility model has this property. The next assertion is the main result of [RT96].

Theorem 10.43 *Let $C \in PF_\infty \cap PF_0$ be a call pricing function in a stochastic volatility model defined by* (2.3). *If the model is uncorrelated, then the implied volatility $K \mapsto I(K)$ decreases on the interval $(0, x_0 e^{rT})$, increases on the interval $(x_0 e^{rT}, \infty)$, and attains its minimum at the point $K_{\min} = x_0 e^{rT}$.*

Proof The methods used in the proof of Theorem 10.43 are mostly borrowed from [RT96]. An alternative proof can be found in [FPS00].

For the sake of simplicity, we assume that $x_0 = 1$, $r = 0$, and $T = 1$. Recall that

$$C'(K) = -\int_K^\infty D_1(u)\,du \quad \text{and} \quad C''(K) = D_1(K).$$

Differentiating the equality $C_{\mathrm{BS}}(T, K, I(T, K)) = C(K)$ with respect to K, we obtain

$$I'(K) = \left[\frac{\partial C_{\mathrm{BS}}}{\partial \sigma}\big(K, I(K)\big)\right]^{-1}\left[C'(K) - \frac{\partial C_{\mathrm{BS}}}{\partial K}\big(K, I(K)\big)\right]. \tag{10.95}$$

By the symmetry condition in Lemma 9.25,

$$I(K) = I\big(K^{-1}\big), \quad K \neq 0.$$

Therefore

$$I'(K) = -K^{-2} I'\left(\frac{1}{K}\right), \quad K \neq 0, \tag{10.96}$$

and hence $K_{\min} = 1$ is a critical point of the function $K \mapsto I(K)$. Moreover, by (10.96), it suffices to prove that the implied volatility I increases on the interval $(1, \infty)$.

We will use the following equalities in the sequel:

$$\frac{\partial C_{\mathrm{BS}}}{\partial x}(x, K, \sigma) = \Phi\big(d_1(x, K, \sigma)\big)$$

where

$$\Phi(s) = \frac{1}{\sqrt{2\pi}} \int_{-\infty}^{s} \exp\left\{-\frac{y^2}{2}\right\} dy,$$

and

$$C_{\mathrm{BS}} = \frac{\partial C_{\mathrm{BS}}}{\partial x} + K \frac{\partial C_{\mathrm{BS}}}{\partial K}.$$

Since the denominator of the fraction on the right-hand side of (10.95) is positive, we have

$$\operatorname{sign}\big[I'(K)\big] = \operatorname{sign}\left[\Phi\big(d_1\big(K, I(K)\big)\big) - \int_K^\infty u D_1(u)\,du\right]. \tag{10.97}$$

For every $K > 1$ the function $\sigma \mapsto d_1(K,\sigma)$ is strictly increasing and maps $(0,\infty)$ onto $(-\infty,\infty)$. Therefore, the function $\sigma \mapsto \Phi(d_1(K,\sigma))$ is strictly increasing and maps $(0,\infty)$ onto $(0,1)$. It follows that for every $K > 1$ there exists a unique number $H(K)$ such that

$$\int_K^\infty uD_1(u)\,du = \Phi\bigl(d_1\bigl(K, H(K)\bigr)\bigr). \tag{10.98}$$

The function $K \mapsto H(K)$ determined from (10.98) is called the hedging volatility (see [RT96]).

Our next goal is to prove that

$$H(K) \le I(K) \quad \text{for all } K > 1. \tag{10.99}$$

It is not hard to see using (10.96), (10.97), (10.98), and (10.99) that the function I increases on $(1,\infty)$.

Set

$$h(K,\sigma) = C_{\mathrm{BS}}(K,\sigma) - C(K), \qquad C_1(K) = \int_K^\infty uD_1(u)\,du,$$

$$C_2(K) = \int_K^\infty D_1(u)\,du,$$

and recall that the mixing distribution density m_t is the density of the random variable

$$\alpha_1 = \left\{\int_0^1 \widetilde{Y}_s^2\,ds\right\}^{\frac{1}{2}}.$$

Here $\widetilde{Y}_s = f(Y_t)$ is the volatility process in the model described by (2.3). It follows from the relation between the stock price distribution density D_1 and the mixing distribution density m_1 that

$$C_1(K) = \mathbb{E}^*\left[\Phi\left(\frac{-\log K}{\alpha_1} + \frac{\alpha_1}{2}\right)\right] \tag{10.100}$$

and

$$C_2(K) = \mathbb{E}^*\left[\Phi\left(\frac{-\log K}{\alpha_1} - \frac{\alpha_1}{2}\right)\right]. \tag{10.101}$$

For every $K > 0$ the function $\sigma \mapsto h(\sigma)$ is strictly increasing, and since $h(K, I(K)) = 0$, the condition in (10.99) is equivalent to the following: $h(K, H(K)) \le 0$ for all $K > 1$. It is not hard to prove that every condition listed below is also equivalent to (10.99):

$$C_2(K) - \Phi\bigl(d_2\bigl(K, H(K)\bigr)\bigr) \le 0,$$

$$\Phi^{-1}\bigl(C_2(K)\bigr) - d_2\bigl(K, H(K)\bigr) \le 0,$$

$$\Phi^{-1}\big(C_2(K)\big) - d_1\big(K, H(K)\big) + H(K) \le 0,$$

and

$$\Phi^{-1}\big(C_2(K)\big) - \Phi^{-1}\big(C_1(K)\big) + H(K) \le 0,$$

for all $K > 1$. Since

$$H(K) = \Phi^{-1}\big(C_1(K)\big) + \sqrt{\big[\Phi^{-1}\big(C_1(K)\big)\big]^2 + 2\log K},$$

we see that in order to establish (10.99), it suffices to prove that

$$\Phi^{-1}\big(C_2(K)\big) + \sqrt{\big[\Phi^{-1}\big(C_1(K)\big)\big]^2 + 2\log K} \le 0 \tag{10.102}$$

for all $K > 1$.

The inequality in (10.102) is equivalent to the inequality

$$\Phi^{-1}\mathbb{E}^*\left[\Phi\left(\frac{-\log K}{\alpha_1} - \frac{\alpha_1}{2}\right)\right] + \sqrt{\left(\Phi^{-1}\mathbb{E}^*\left[\Phi\left(\frac{-\log K}{\alpha_1} + \frac{\alpha_1}{2}\right)\right]\right)^2 + 2\log K} \le 0 \tag{10.103}$$

(this follows from (10.100) and (10.101)). Since the first term on the left-hand side of (10.103) is negative, the estimate in (10.103) is equivalent to the following inequality:

$$\left(\Phi^{-1}\mathbb{E}^*\left[\Phi\left(\frac{-\log K}{\alpha_1} - \frac{\alpha_1}{2}\right)\right]\right)^2 - \left(\Phi^{-1}\mathbb{E}^*\left[\Phi\left(\frac{-\log K}{\alpha_1} + \frac{\alpha_1}{2}\right)\right]\right)^2 - 2\log K \ge 0. \tag{10.104}$$

It is clear that the random variable α_1 is square integrable with respect to the measure $\mathbb{P}^*$. We will next prove a lemma that contains (10.104) as a special case.

Lemma 10.44 *For every $k > 0$ and every nonnegative random variable U that is positive on a set of positive probability, the following inequality holds*:

$$\left(\Phi^{-1}\mathbb{E}^*\left[\Phi\left(\frac{-k}{U} - \frac{U}{2}\right)\right]\right)^2 - \left(\Phi^{-1}\mathbb{E}^*\left[\Phi\left(\frac{-k}{U} + \frac{U}{2}\right)\right]\right)^2 - 2k \ge 0. \tag{10.105}$$

Proof The proof of Lemma 10.44 consists of three parts. We start with a strictly positive Bernoulli random variable U and obtain estimate (10.105) with a strict inequality for U. The second step in the proof extends the previous estimate to all simple random variables by induction, and the final step establishes Lemma 10.44 in full generality by approximation.

Let $0 \le p \le 1$, and suppose that the random variable U takes the value $u_1 > 0$ with probability p and the value $u_2 > 0$, different from u_1, with probability $1 - p$. Put $k = \log K$, and define the function ψ by

$$\psi(k,p) = \left(\Phi^{-1}\left[p\Phi\left(\frac{-k}{u_1} - \frac{u_1}{2}\right) + (1-p)\Phi\left(\frac{-k}{u_2} - \frac{u_2}{2}\right)\right]\right)^2 - \left(\Phi^{-1}\left[p\Phi\left(\frac{-k}{u_1} + \frac{u_1}{2}\right) + (1-p)\Phi\left(\frac{-k}{u_2} + \frac{u_2}{2}\right)\right]\right)^2 - 2k,$$

where $k > 0$. Let us put

$$\varphi(s) = \frac{1}{\sqrt{2\pi}} \exp\left\{-\frac{s^2}{2}\right\}.$$

We will next differentiate the function ψ with respect to p. This ingenious method was suggested by Renault and Touzi in [RT96].

It is easy to see that

$$\frac{\partial \psi}{\partial p}(k,p) = \frac{2\Phi^{-1}}{\varphi \circ \Phi^{-1}} \mathbb{E}^*\left[\Phi\left(\frac{-k}{U} - \frac{U}{2}\right)\right]\Delta_1(k) - \frac{2\Phi^{-1}}{\varphi \circ \Phi^{-1}} \mathbb{E}^*\left[\Phi\left(\frac{-k}{U} + \frac{U}{2}\right)\right]\Delta_2(k), \tag{10.106}$$

where

$$\Delta_1(k) = \Phi\left(\frac{-k}{u_1} - \frac{u_1}{2}\right) - \Phi\left(\frac{-k}{u_2} - \frac{u_2}{2}\right)$$

and

$$\Delta_2(k) = \Phi\left(\frac{-k}{u_1} + \frac{u_1}{2}\right) - \Phi\left(\frac{-k}{u_2} + \frac{u_2}{2}\right).$$

Moreover,

$$\frac{\partial^2 \psi}{\partial p^2}(k,p) = \frac{2(1 + (\Phi^{-1})^2)}{(\varphi \circ \Phi^{-1})^2} \mathbb{E}^*\left[\Phi\left(\frac{-k}{U} - \frac{U}{2}\right)\right]\Delta_1(k)^2 - \frac{2(1 + (\Phi^{-1})^2)}{(\varphi \circ \Phi^{-1})^2} \mathbb{E}^*\left[\Phi\left(\frac{-k}{U} + \frac{U}{2}\right)\right]\Delta_2(k)^2. \tag{10.107}$$

It is clear that for any $k > 0$, we have $\psi(k,0) = \psi(k,1) = 0$. Our next goal is to prove the inequality $\psi(k,p) > 0$ for all $0 < p < 1$. It suffices to establish that

$$\frac{\partial^2 \psi}{\partial p^2}(k,p) < 0 \tag{10.108}$$

for all p satisfying the equation

$$\frac{\partial \psi}{\partial p}(k, p) = 0. \tag{10.109}$$

Suppose (10.109) holds for some $k > 0$ and all p with $0 < p < 1$. Then (10.106) and (10.107) imply the following equality:

$$\frac{\partial^2 \psi}{\partial p^2}(k, p) = 2\frac{\Delta_1(k)^2}{(\varphi \circ \Phi^{-1})^2 \mathbb{E}^*[\Phi(\frac{-k}{U} - \frac{U}{2})]} \times \left(1 - \left[\frac{\Phi^{-1}\mathbb{E}^*[\Phi(\frac{-k}{U} - \frac{U}{2})]}{\Phi^{-1}\mathbb{E}^*[\Phi(\frac{-k}{U} + \frac{U}{2})]}\right]^2\right). \tag{10.110}$$

Since

$$\mathbb{E}^*\left[\Phi\left(\frac{-k}{U} - \frac{U}{2}\right)\right] + \mathbb{E}^*\left[\Phi\left(\frac{-k}{U} + \frac{U}{2}\right)\right] < 1,$$

we have

$$\left|\Phi^{-1}\mathbb{E}^*\left[\Phi\left(\frac{-k}{U} - \frac{U}{2}\right)\right]\right| > \left|\Phi^{-1}\mathbb{E}^*\left[\Phi\left(\frac{-k}{U} + \frac{U}{2}\right)\right]\right|$$

$\mathbb{P}^*$-a.s. It follows from (10.110) that (10.108) holds for any Bernoulli random variable U.

Next, assume that the estimate in the lemma holds (with strict inequality) for all random variables, taking n positive values, where $n \geq 2$ is a given integer. Let U be a random variable taking the values $u_i > 0$ with probability p_i, where $1 \leq i \leq n+1$ and $\sum_{i=1}^{n+1} p_i = 1$. Fix the probabilities $p_1, \dots, p_{n-1}$ and define a function of the variable p_n by

$$\psi(k, p_n) = \left[\Phi^{-1}\left(\sum_{i=1}^{n+1} p_i \Phi\left(\frac{-k}{u_i} - \frac{u_i}{2}\right)\right)\right]^2 - \left[\Phi^{-1}\left(\sum_{i=1}^{n+1} p_i \Phi\left(\frac{-k}{u_i} + \frac{u_i}{2}\right)\right)\right]^2 - 2k,$$

where $k > 0$ and $0 \leq p_n \leq 1 - \sum_{i=1}^{n-1} p_i$. Then it is clear that for every p_n satisfying the previous condition we have

$$p_{n+1} = 1 - \sum_{i=1}^{n-1} p_i - p_n.$$

It is not hard to see that

$$\psi(k, 0) > 0 \quad \text{and} \quad \psi\left(k, 1 - \sum_{i=1}^{n-1} p_i\right) > 0. \tag{10.111}$$

We will next prove that the inequality

$$\psi(k, p_n) > 0 \tag{10.112}$$

holds for all $0 < p_n < 1 - \sum_{i=1}^{n-1} p_i$. We have

$$\psi(k, p_n) = \left[\Phi^{-1}\left(\Phi\left(\frac{-k}{u_{n+1}} - \frac{u_{n+1}}{2}\right) + \sum_{i=1}^{n} p_i \Delta_{i,n+1}(k)\right)\right]^2$$
$$- \left[\Phi^{-1}\left(\Phi\left(\frac{-k}{u_{n+1}} + \frac{u_{n+1}}{2}\right) + \sum_{i=1}^{n} p_i \widetilde{\Delta}_{i,n+1}(k)\right)\right]^2 - 2k,$$

where

$$\Delta_{i,n+1}(k) = \Phi\left(\frac{-k}{u_i} - \frac{u_i}{2}\right) - \Phi\left(\frac{-k}{u_{n+1}} - \frac{u_{n+1}}{2}\right)$$

and

$$\widetilde{\Delta}_{i,n+1}(k) = \Phi\left(\frac{-k}{u_i} + \frac{u_i}{2}\right) - \Phi\left(\frac{-k}{u_{n+1}} + \frac{u_{n+1}}{2}\right).$$

Now, reasoning as in the proof of Lemma 10.44 in the case where U is a Bernoulli random variable, we can show that

$$\frac{\partial \psi}{\partial p_n}(k, p) = 0 \Longrightarrow \frac{\partial^2 \psi}{\partial p_n^2}(k, p) < 0.$$

It is clear that the previous conclusion combined with (10.111) implies (10.112).

This completes the proof of Lemma 10.100 (with strict inequality) for all simple positive random variables U.

It remains to prove Lemma 10.44 in the general case. Let U be a random variable such as in the formulation of Lemma 10.44. Then there exists a sequence of strictly positive simple random variables U_j, $j \geq 1$, such that $U_j \to U$ $\mathbb{P}^*$-a.s. as $j \to \infty$. Fix $k > 0$, and recall that the following inequality has already been established:

$$\left(\Phi^{-1}\mathbb{E}^*\left[\Phi\left(\frac{-k}{U_j} - \frac{U_j}{2}\right)\right]\right)^2$$
$$- \left(\Phi^{-1}\mathbb{E}^*\left[\Phi\left(\frac{-k}{U_j} + \frac{U_j}{2}\right)\right]\right)^2 - 2k > 0 \tag{10.113}$$

for all $j \geq 1$. The function Φ is bounded and can be extended to a continuous function on $[-\infty, \infty]$. Taking the limit as $j \to \infty$ in (10.113) and using the bounded convergence theorem, we obtain

$$\lim_{j\to\infty} \mathbb{E}^*\left[\Phi\left(\frac{-k}{U_j} - \frac{U_j}{2}\right)\right] = \mathbb{E}^*\left[\Phi\left(\frac{-k}{U} - \frac{U}{2}\right)\right] \tag{10.114}$$

and

$$\lim_{j\to\infty} \mathbb{E}^*\left[\Phi\left(\frac{-k}{U_j}+\frac{U_j}{2}\right)\right]=\mathbb{E}^*\left[\Phi\left(\frac{-k}{U}+\frac{U}{2}\right)\right]. \tag{10.115}$$

Since the values of the limits in (10.114) and (10.115) are different from 0 and 1, and the function Φ^{-1} is continuous on (0, 1), we see that Lemma 10.44 holds. □

This completes the proof of Theorem 10.43. □

10.10 Gatheral's SVI Parameterization of Implied Variance

For a given call pricing function C, the implied variance V is defined by

$$V(T,K)=I(T,K)^2, \quad T>0,\ K>0.$$

For fixed maturity T, the implied variance can be considered either as a function $K \mapsto V(T,K)$ of the strike price K, or as a function $k \mapsto V(T,k)$ of the log-moneyness k given by $k=\log\frac{K}{x_0}$. For stochastic volatility models with moment explosions, the implied variance $k \mapsto V(T,k)$ is approximately linear in the wings, according to Lee's moment formulas, and curved in the middle as suggested by the observed volatility smile. A simple expression, which can be used to parametrize maturity slices of the implied variance surface was suggested by J. Gatheral (see [Gat04] and [Gat06]). This expression is as follows:

$$V_{\mathrm{SVI}}(k;a,b,\sigma,\rho,m)=a+b\left[\rho(k-m)+\sqrt{(k-m)^2+\sigma^2}\right], \tag{10.116}$$

where a, b, σ, ρ, and m are real parameters depending on T. In [Gat04] and [Gat06], the function V_{SVI} is called the SVI ("stochastic volatility inspired") parameterization of the implied variance.

It is clear that we have to impose certain restrictions on the parameters appearing in the expression on the right-hand side of (10.116). For example, in order the function V_{SVI} to be an approximation to the implied variance V as $k\to\infty$, one has to assume $b(1+\rho)>0$, since V is a positive function. A similar condition in the case where $k\to-\infty$ is the following: $b(1-\rho)>0$. The SVI parameterization is widely used in the industry. Note that in practical applications one has to calibrate the parameters in (10.116) to the implied total variance observed in equity markets.

The right asymptote of the function V_{SVI} is given by

$$V_{\mathrm{SVI}}^{+}(k)=a+b(1+\rho)(k-m),$$

while the left asymptote is

$$V_{\mathrm{SVI}}^{-}(k)=a-b(1-\rho)(k-m).$$

The constants in the SVI parameterization can be interpreted as follows: a is the overall level of the variance, b determines the angle between the left and right asymptotes, σ characterizes the smoothness of the vertex, ρ determines the orientation of the graph of the function V_{SVI}, and changing m translates the graph (see [Gat04]).

Expanding near infinity, we obtain a more precise relation between the function V_{SVI} and its right asymptote:

$$V_{\mathrm{SVI}}(k; a, b, \sigma, \rho, m) = kb(1+\rho) + \bigl(a - bm(1+\rho)\bigr) + O\bigl(k^{-1}\bigr)$$

as $k \to \infty$. We can also prove that if $b(1+\rho) > 0$, then

$$\sqrt{V_{\mathrm{SVI}}(k; a, b, \sigma, \rho, m)} = k^{\frac{1}{2}}\sqrt{b(1+\rho)} + k^{-\frac{1}{2}}\frac{(a - bm(1+\rho))}{2\sqrt{b(1+\rho)}} + O\bigl(k^{-\frac{3}{2}}\bigr) \tag{10.117}$$

as $k \to \infty$. In addition, if $b(1-\rho) > 0$ and $k \to -\infty$, then

$$V_{\mathrm{SVI}}(k; a, b, \sigma, \rho, m) = |k|b(1-\rho) + \bigl(a + bm(1-\rho)\bigr) + O\bigl(|k|^{-1}\bigr)$$

as $k \to -\infty$, and

$$\sqrt{V_{\mathrm{SVI}}(k; a, b, \sigma, \rho, m)} = |k|^{\frac{1}{2}}\sqrt{b(1-\rho)} + |k|^{-\frac{1}{2}}\frac{(a + bm(1-\rho))}{2\sqrt{b(1-\rho)}} + O\bigl(|k|^{-\frac{3}{2}}\bigr) \tag{10.118}$$

as $k \to -\infty$.

Consider a new family of parameters (w_1, w_2, ρ), where $w_1, w_2 \in \mathbb{R}$ and $\rho \in [-1, 1]$, and suppose the family of SVI parameters is given by the following formulas:

$$a = \frac{w_1}{2}\bigl(1 - \rho^2\bigr), \qquad b = \frac{w_1 w_2}{2}, \qquad \rho = \rho, \qquad m = -\frac{\rho}{w_2}, \qquad \sigma = \frac{\sqrt{1-\rho^2}}{w_2}. \tag{10.119}$$

It is simple to check that if the SVI parameters and the new parameters are related as in (10.119), then

$$V_{\mathrm{SVI}}(k; a, b, \sigma, \rho, m) = \frac{w_1}{2}\Bigl(1 + w_2\rho k + \sqrt{(w_2 k + \rho)^2 + 1 - \rho^2}\Bigr). \tag{10.120}$$

The SVI parameterization of the implied variance defined on the right-hand side of equality (10.120) was considered in [GJ11]. We will denote this parameterization by $V_{\mathrm{SVI}}(k; w_1, w_2, \rho)$.

J. Gatheral observed in [Gat04] that the SVI parameterization is consistent with the Heston implied variance in the large maturity regime. This observation was made rigorous in [GJ11]. We will next formulate and discuss the main result obtained in [GJ11]. Consider the Heston model given by

$$\begin{cases} dX_t = \sqrt{Y_t} X_t \left(\sqrt{1-\rho^2}\, d\widetilde{W}_t + \rho\, dZ_t\right), \\ dY_t = q(m - Y_t)\, dt + c\sqrt{Y_t}\, dZ_t \end{cases} \tag{10.121}$$

with $q > 0$, $m > 0$, $c > 0$, and $\rho \in (-1, 1)$. The initial conditions for the processes X and Y will be denoted by x_0 and y_0, respectively. It will also be assumed that the CIR-process Y does not reach zero almost surely. This is true if and only if $2qm \geq c^2$ (see Theorem 2.27). Another restriction imposed in [GJ11] is as follows: $q - \rho c > 0$. Note that the previous inequality holds when $\rho \leq 0$.

Let us assume that the new parameter ρ in (10.120) coincides with the correlation coefficient in (10.121), and put

$$w_1 = \frac{4qm}{c^2(1-\rho^2)}\left(\sqrt{(2q - \rho c)^2 + c^2(1-\rho^2)} - (2q - \rho c)\right) \tag{10.122}$$

and

$$w_2 = c(qm)^{-1}. \tag{10.123}$$

Denote by x the time-scaled log-moneyness defined by $x = T^{-1} \log \frac{K}{x_0}$. Then $K = x_0 e^{xT}$ is a maturity-dependent strike.

The next formula is the main result in [GJ11]:

$$\lim_{T\to\infty} V\left(T, x_0 e^{xT}\right) = V_{\mathrm{SVI}}(x; w_1, w_2, \rho) \tag{10.124}$$

for all $x \in \mathbb{R}$. In formula (10.124), the parameters w_1 and w_1 are related to the Heston model parameters by (10.122) and (10.123). Formula (10.124) shows that large maturity limits of the implied Heston variance along special paths coincide with the SVI parameterization of the implied variance. For every $x \in \mathbb{R}$ the special path is given by $T \mapsto (T, x_0 e^{xT})$, $T > 0$.

For the sake of shortness, put

$$\bar{q} = q - \rho c, \quad \bar{m} = qm(\bar{q})^{-1}, \quad \text{and} \quad \eta = \sqrt{4q^2 + c^2 - 4\rho q c}.$$

The parameters $\bar{q}$, $\bar{m}$, and η are defined in terms of the Heston model parameters appearing in (10.121). J. Gatheral and A. Jacquer derived formula (10.124) from the following formula obtained by M. Forde, A. Jacquier, and A. Mijatović in [FJM10]:

$$\begin{aligned} &\frac{1}{2}\lim_{T\to\infty} V\left(T, x_0 e^{xT}\right) \\ &\quad = 2V^*(x) - x \\ &\qquad + 2\left(\mathbb{1}_{\{(-\frac{m}{2}, \frac{\bar{m}}{2})\}}(x) - \mathbb{1}_{\{\mathbb{R}\setminus(-\frac{m}{2}, \frac{\bar{m}}{2})\}}(x)\right)\sqrt{V^*(x)^2 - V^*(x)x} \end{aligned} \tag{10.125}$$

for all $x \in \mathbb{R}$. The function V^* in (10.125) is defined as follows:

$$V^*(x) = p^*(x)x - V\big(p^*(x)\big),$$

where

$$p^*(x) = \frac{c - 2\rho q + \eta(\rho q m + cx)(c^2x^2 + 2\rho q m c x + q^2 m^2)^{-\frac{1}{2}}}{2c(1-\rho^2)},$$

while the function V is given by

$$V(p) = \frac{qm}{c^2}\Big(q - \rho c p - \sqrt{(q-\rho c p)^2 + c^2 p(1-p)}\Big).$$

It is also shown in [FJM10] that the function p^* maps $\mathbb{R}$ onto the interval (p_-, p_+) with the endpoints given by

$$p_- = \frac{-2\rho q + c - \eta}{2c\sqrt{1-\rho^2}} \quad \text{and} \quad p_+ = \frac{-2\rho q + c + \eta}{2c\sqrt{1-\rho^2}}.$$

We refer the reader to [FJM10] for more details and to [GJ11] for the proof of the fact that formula (10.124) follows from formula (10.125).

It is worth mentioning that Gatheral's parameterization of the Heston smile is not compatible with formula (10.44). Indeed, we can rewrite this formula using the log-moneyness k instead of the strike K as follows:

$$I(k) = \beta_1\sqrt{k} + \beta_2 + \beta_3\frac{\log k}{\sqrt{k}} + \beta_4\frac{1}{\sqrt{k}} + O\big(k^{-1}\big) \tag{10.126}$$

as $k \to \infty$. The constants β_1, β_2, β_3, and β_4 in (10.126) are given by (10.41), (10.42), (10.43), and (10.45), respectively. By comparing (10.117) and (10.126), it becomes clear that the SVI parameterization of the Heston smile is inconsistent with the correct expansion given in (10.126). This shows that the SVI type parameterizations could well benefit from additional terms appearing in sharp asymptotic formulas for the implied volatility in the Heston model. The previous observation was made in [FGGS11].

10.11 Notes and References

- The material contained in Sects. 10.1–10.4 is adapted from [Gul10].
- The book [H-L09] by P. Henry-Labordère contains several results concerning the wing behavior of the implied volatility. In Chap. 10 of this book, asymptotic formulas (without error estimates) are obtained for the implied volatility at extreme strikes in certain local and stochastic volatility models. Schrödinger semigroups with Kato class potentials, especially Gaussian estimates for the kernels of such

semigroups, play an important role in the study of the wing behavior of the implied volatility in [H-L09]. The reader, interested in Schrödinger semigroups and Kato class potentials, can consult the book [GvC06].

- Section 10.5 is devoted to sharp asymptotic formulas for the implied volatility in classical stochastic volatility models. For the uncorrelated Hull–White model such formulas were obtained in [GS09], while for the Hull–White model with nonzero correlation no such formulas are known. The asymptotic behavior of the implied volatility in the uncorrelated Heston model was characterized in [GS10b] for the uncorrelated model, and in [FGGS11] for the correlated one. Sharp asymptotic formulas for the implied volatility in the uncorrelated Stein–Stein model were obtained in [GS10b]. In the case of nonzero correlation, similar formulas were established in [DFJV11].
- The results discussed in Sects. 10.6 and 10.7 are taken from [Gul10], while the material included in Sect. 10.8 is contained in [GV11].
- The presentation in Sect. 10.9 mostly follows that in [RT96].
- A substantial part of this book is devoted to the asymptotic behavior of the implied volatility at extreme strikes. However, the book does not touch upon the asymptotics with respect to the other parameters, appearing in option pricing models. For example, asymptotic expansions at extreme maturities, which play an important role in calibration problems for stochastic asset price models, are not discussed in the present book. We only provide below a list of selected publications related to the implied volatility asymptotics. This list is incomplete, and we apologize for many papers which are omitted.

  Books: [FPS00, Lew00, Fen05, H-L09].

  Large-time asymptotics: [Teh09b, GL11, For11, FJ11a].

  Small-time and small-noise asymptotics: [HKLW02, BBF02, BBF04, RR09, Pau09, FJ09, FPS00, FFK10, FFF10, BAHLOW10, BAL10, GHLOW10, FJL11, MN11, GL11, FJ11b, DFJV11].

  Maturity-dependent strikes and mixed regimes: [FJM10, GJ11, GL11].

  It is worth mentioning here that asymptotic analysis of the implied volatility at extreme maturities draws on special tools from probability theory, mathematical analysis, and geometry. For example, Laplace principles, large deviation principles, and heat kernel estimates on Riemannian and sub-Riemannian manifolds are used to study small-time or large-time behavior of the implied volatility.
- Time-dependent SVI type parameterizations of the implied volatility in symmetric models were introduced and studied in [DM10, DMM10].

Chapter 11
Implied Volatility in Models Without Moment Explosions

This chapter studies the implied volatility $K \mapsto I(T,K)$ in stochastic asset price models under the assumption that the moments of all positive (all negative) orders of the asset price X_T are finite. Note that R. Lee's moment formula for large (small) strikes fails to describe the asymptotic behavior of the implied volatility in such models. Therefore, it is natural to ask the following question: How does the implied volatility behave at extreme strikes in models without moment explosions?

The present chapter addresses several problems suggested by the previous question. In Sect. 11.1, we obtain general asymptotic formulas for the implied volatility in stochastic asset price models without moment explosions (see Theorems 11.1 and 11.2). Section 11.5 is devoted to Piterbarg's conjecture. In an unpublished working paper [Pit04], V.V. Piterbarg formulated an asymptotic formula, which can replace R. Lee's moment formula at large strikes in asset price models without moment explosions. In Sect. 11.5, we show that Piterbarg's formula is valid in a slightly modified form, and also confirm Piterbarg's original conjecture under very mild restrictions.

Several sections of this chapter concern the asymptotic behavior of the implied volatility at extreme strikes in special asset price models without moment explosions. The list of these models includes Rubinstein's displaced diffusion model, the constant elasticity of variance model, the finite moment log-stable model of P. Carr and L. Wu, and SV1 and SV2 models of L.C.G. Rogers and L.A.M. Veraart.

11.1 General Asymptotic Formulas in Models Without Moment Explosions

Let C be a call pricing function. Recall that in Sect. 10.1 we defined the following numbers:

$$\tilde{p} = \sup\{p \ge 0 : \mathbb{E}^*[X_T^{1+p}] < \infty\}$$

A. Gulisashvili, *Analytically Tractable Stochastic Stock Price Models*, Springer Finance, DOI 10.1007/978-3-642-31214-4_11,
© Springer-Verlag Berlin Heidelberg 2012

and

$$\tilde{q} = \sup\{q \geq 0 : \mathbb{E}^*[X_T^{-q}] < \infty\}$$

(see Theorem 10.2). It is clear that if $\tilde{p} = \infty$, then the moments of all positive orders of the asset price X_T are finite. On the other hand, if $\tilde{q} = \infty$, then the moments of all negative orders of X_T are finite.

The next two theorems characterize the asymptotic behavior of the implied volatility in stochastic asset price models with no moment explosions.

Theorem 11.1 *Let $C \in PF_\infty$, and suppose $\widetilde{C}$ is a positive function such that $\widetilde{C}(K) \approx C(K)$ as $K \to \infty$. Suppose also that $\tilde{p} = \infty$. Then*

$$I(K) = \frac{1}{\sqrt{2T}} \log K \left(\log \frac{1}{\widetilde{C}(K)} \right)^{-\frac{1}{2}} + O\left((\log K)^2 \left(\log \frac{1}{\widetilde{C}(K)} \right)^{-\frac{3}{2}} \right)$$
$$+ O\left(\left(\log \frac{1}{\widetilde{C}(K)} \right)^{-\frac{1}{2}} \right) \tag{11.1}$$

as $K \to \infty$.

Proof It follows from formula (9.33) that

$$I(K) = \frac{\sqrt{2}}{\sqrt{T}} \left(\log \frac{1}{\widetilde{C}(K)} \right)^{\frac{1}{2}} \left[\sqrt{1 + u_1(K)} - \sqrt{1 + u_2(K)} \right]$$
$$+ O\left(\left(\log \frac{1}{\widetilde{C}(K)} \right)^{-\frac{1}{2}} \right) \tag{11.2}$$

as $K \to \infty$, where

$$u_1(K) = \log K \left(\log \frac{1}{\widetilde{C}(K)} \right)^{-1} - \log\log \frac{1}{\widetilde{C}(K)} \left(2 \log \frac{1}{\widetilde{C}(K)} \right)^{-1}$$

and

$$u_2(K) = -\log\log \frac{1}{\widetilde{C}(K)} \left(2 \log \frac{1}{\widetilde{C}(K)} \right)^{-1}.$$

By Lemma 10.5,

$$\tilde{p} = l = \liminf_{K \to \infty} (\log K)^{-1} \log \frac{1}{C(K)}.$$

Therefore

$$\log K \left(\log \frac{1}{\widetilde{C}(K)} \right)^{-1} \to 0$$

as $K \to \infty$. Next, using (11.2) and the formula

$$\sqrt{1+u} = 1 + \frac{1}{2}u + O(u^2) \tag{11.3}$$

as $u \to 0$, we get

$$I(K) = \frac{1}{\sqrt{2T}} \left(\log \frac{1}{\widetilde{C}(K)} \right)^{\frac{1}{2}} A(K) + O\left(\left(\log \frac{1}{\widetilde{C}(K)} \right)^{-\frac{1}{2}} \right)$$

where

$$A(K) = (\log K) \left(\log \frac{1}{\widetilde{C}(K)} \right)^{-1} + O\left((\log K)^2 \left(\log \frac{1}{\widetilde{C}(K)} \right)^{-2} \right)$$
$$+ O\left(\left(\log \log \frac{1}{\widetilde{C}(K)} \right)^2 \left(\log \frac{1}{\widetilde{C}(K)} \right)^{-2} \right).$$

Now it is easy to see that formula (11.1) holds.

This completes the proof of Theorem 11.1. □

Theorem 11.2 *Let $C \in PF_0$, and let P be the corresponding put pricing function. Suppose $P(K) \approx \widetilde{P}(K)$ as $K \to 0$, where $\widetilde{P}$ is a positive function. Suppose also that $\tilde{q} = \infty$. Then*

$$I(K) = \frac{1}{\sqrt{2T}} \log \frac{1}{K} \left(\log \frac{K}{\widetilde{P}(K)} \right)^{-\frac{1}{2}} + O\left(\left(\log \frac{1}{K} \right)^2 \left(\log \frac{K}{\widetilde{P}(K)} \right)^{-\frac{3}{2}} \right)$$
$$+ O\left(\left(\log \frac{K}{\widetilde{P}(K)} \right)^{-\frac{1}{2}} \right) \tag{11.4}$$

as $K \to 0$.

Proof By Lemma 10.7,

$$\tilde{q} = m - 1 = \liminf_{K \to 0} \left(\log \frac{1}{K} \right)^{-1} \log \frac{1}{P(K)} - 1.$$

Therefore

$$\lim_{K \to \infty} \log \frac{1}{K} \left(\log \frac{K}{P(K)} \right)^{-1} = 0.$$

Next, using formula (9.93) and reasoning as in the proof of Theorem 11.1, we obtain formula (11.4).

The proof of Theorem 11.2 is thus completed. □

It follows from Theorem 11.1 that for $C \in PF_\infty$ with $\tilde{p} = \infty$,

$$I(K) \sim \frac{1}{\sqrt{2T}}(\log K)\left(\log \frac{1}{C(K)}\right)^{-\frac{1}{2}} \tag{11.5}$$

as $K \to \infty$. Similarly, if $C \in PF_0$ and $\tilde{q} = \infty$, then Theorem 11.2 implies that

$$I(K) \sim \frac{1}{\sqrt{2T}}\left(\log \frac{1}{K}\right)\left(\log \frac{K}{P(K)}\right)^{-\frac{1}{2}} \tag{11.6}$$

as $K \to 0$.

Corollary 11.3 *Suppose the conditions in Theorem* 11.1 *hold. Then the following are true:*

- *If there exist $\alpha > 0$ and $K_1 > 0$ such that*

$$\widetilde{C}(K) \leq \exp\{-\alpha(\log K)^2\} \tag{11.7}$$

for all $K > K_1$, then

$$I(K) = \frac{1}{\sqrt{2T}}(\log K)\left(\log \frac{1}{\widetilde{C}(K)}\right)^{-\frac{1}{2}} + O\left(\left(\log \frac{1}{\widetilde{C}(K)}\right)^{-\frac{1}{2}}\right)$$

as $K \to \infty$.

- *If there exist $\beta_1 > 0$, $\beta_2 > 0$, and $K_2 > 0$ such that*

$$\exp\{-\beta_1(\log K)^2\} \leq \widetilde{C}(K) \leq \exp\{-\beta_2(\log K)^{\frac{3}{2}}\}$$

for all $K > K_2$, then

$$I(K) = \frac{1}{\sqrt{2T}} \log K \left(\log \frac{1}{\widetilde{C}(K)}\right)^{-\frac{1}{2}} - \frac{1}{2\sqrt{2T}}(\log K)^2\left(\log \frac{1}{\widetilde{C}(K)}\right)^{-\frac{3}{2}} + O\left(\left(\log \frac{1}{\widetilde{C}(K)}\right)^{-\frac{1}{2}}\right)$$

as $K \to \infty$.

- *If there exist $\gamma > 0$ and $K_3 > 0$ such that*

$$\widetilde{C}(K) \geq \exp\{-\gamma(\log K)^{\frac{3}{2}}\}$$

for all $K > K_3$, then

$$I(K) = \frac{1}{\sqrt{2T}} \log K \left(\log \frac{1}{\widetilde{C}(K)}\right)^{-\frac{1}{2}} - \frac{1}{2\sqrt{2T}}(\log K)^2\left(\log \frac{1}{\widetilde{C}(K)}\right)^{-\frac{3}{2}}$$

$$+ O\left((\log K)^3 \left(\log \frac{1}{\widetilde{C}(K)}\right)^{-\frac{5}{2}}\right)$$

as $K \to \infty$.

Corollary 11.4 *Suppose the conditions in Theorem* 11.2 *hold. Then the following are true*:

- *If there exist* $\delta > 0$ *and* $K_4 > 0$ *such that*

$$\widetilde{P}(K) \le K \exp\left\{-\delta \left(\log \frac{1}{K}\right)^2\right\}$$

for all $K < K_4$, *then*

$$I(K) = \frac{1}{\sqrt{2T}} \left(\log \frac{1}{K}\right) \left(\log \frac{K}{\widetilde{P}(K)}\right)^{-\frac{1}{2}} + O\left(\left(\log \frac{K}{\widetilde{P}(K)}\right)^{-\frac{1}{2}}\right)$$

as $K \to 0$.

- *If there exist* $\eta_1 > 0$, $\eta_2 > 0$, *and* $K_5 > 0$ *such that*

$$K \exp\left\{-\eta_1 \left(\log \frac{1}{K}\right)^2\right\} \le \widetilde{P}(K) \le K \exp\left\{-\eta_2 \left(\log \frac{1}{K}\right)^{\frac{3}{2}}\right\}$$

for all $K < K_5$, *then*

$$I(K) = \frac{1}{\sqrt{2T}} \log \frac{1}{K} \left(\log \frac{K}{\widetilde{P}(K)}\right)^{-\frac{1}{2}} - \frac{1}{2\sqrt{2T}} \left(\log \frac{1}{K}\right)^2 \left(\log \frac{K}{\widetilde{P}(K)}\right)^{-\frac{3}{2}}$$
$$+ O\left(\left(\log \frac{K}{\widetilde{P}(K)}\right)^{-\frac{1}{2}}\right)$$

as $K \to 0$.

- *If there exist* $\rho > 0$ *and* $K_6 > 0$ *such that*

$$\widetilde{P}(K) \ge K \exp\left\{-\gamma \left(\log \frac{1}{K}\right)^{\frac{3}{2}}\right\}$$

for all $K < K_6$, *then*

$$I(K) = \frac{1}{\sqrt{2T}} \log \frac{1}{K} \left(\log \frac{K}{\widetilde{P}(K)}\right)^{-\frac{1}{2}} - \frac{1}{2\sqrt{2T}} \left(\log \frac{1}{K}\right)^2 \left(\log \frac{K}{\widetilde{P}(K)}\right)^{-\frac{3}{2}}$$
$$+ O\left(\left(\log \frac{1}{K}\right)^3 \left(\log \frac{K}{\widetilde{P}(K)}\right)^{-\frac{5}{2}}\right)$$

as $K \to 0$.

The first statements in Corollaries 11.3 and 11.4 follow from Theorems 11.1 and 11.2, respectively. The second and the third statements can established by reasoning as in the proofs of Theorems 11.1 and 11.2, and using the formula $\sqrt{1+u} = 1 + \frac{1}{2}u - \frac{1}{4}u^2 + O(u^3)$ as $u \to 0$ instead of formula (11.3).

11.2 Constant Elasticity of Variance Model

The constant elasticity of variance model (the CEV model) was developed by J.C. Cox and S.A. Ross (see [CR76]). In the first part of the present section, we gather various definitions and facts related to the CEV model, while the second part concerns the implied volatility in this model. We omit the proofs of the known facts, and refer the reader to [JYC09, BL10] where all the necessary details can be found.

The asset price process in the CEV model satisfies the following stochastic differential equation:

$$dS_t = \sigma S_t^{\rho} \, dW_t. \tag{11.8}$$

We assume that $r = 0$, $0 < \rho < 1$, $\sigma > 0$, and $s_0 > 0$, where s_0 is the initial price of the asset. For $\frac{1}{2} \le \rho < 1$, the boundary at $x = 0$ is naturally absorbing for the CEV process. If $0 < \rho < \frac{1}{2}$, then we can either impose an absorbing or a reflecting boundary condition. In this section, we will only consider the absorbing boundary conditions. The CEV model is a local volatility model. Indeed, the volatility of the asset in this model is described by the expression $\sigma S_t^{\rho-1}$ up to the stopping time $\tau = \inf_{t>0}\{S_t = 0\}$. Note that the CEV model takes into account the leverage effect. It is clear that the volatility decreases if the asset price increases, and the volatility increases if the asset price decreases.

The transformation

$$X = \frac{S^{2(1-\rho)}}{\sigma^2(1-\rho)^2} \tag{11.9}$$

reduces the stochastic differential equation in (11.8) to the following equation:

$$dX_t = \delta \, dt + 2\sqrt{X_t} \, dW_t \tag{11.10}$$

with $\delta = \frac{1-2\rho}{1-\rho}$. The initial condition for the process X in (11.10) is given by

$$x_0 = \frac{s_0^{2(1-\rho)}}{\sigma^2(1-\rho)^2}. \tag{11.11}$$

Therefore, the process X is nothing else but the squared Bessel process $BESQ_{x_0}^{\delta}$.

The index of the process X is defined by

$$\nu = \frac{\delta}{2} - 1 = -\frac{1}{2(1-\rho)},$$

and the distribution of the random variable X_T is given by the following formula:

$$\mu_T(A) = \left[1 - \Gamma\left(-\nu; \frac{x_0}{2T}\right)\right]\delta_0(A) + \frac{1}{2T}\int_A \left(\frac{x}{x_0}\right)^{\frac{\nu}{2}} \exp\left\{-\frac{x+x_0}{2T}\right\} I_{-\nu}\left(\frac{\sqrt{x_0 x}}{T}\right) dx \qquad (11.12)$$

for every Borel subset A of $[0, \infty)$ (see Sects. 4.1 and 4.2 in [BL10]). The function I in (11.12) is the I-Bessel function, δ_0 is the delta-function at $x = 0$, and Γ is the normalized incomplete gamma function given by

$$\Gamma(a, y) = \frac{1}{\Gamma(a)} \int_0^y t^{a-1} e^{-t}\, dt, \quad a > 0,\ y \geq 0.$$

Let us denote by $p_T(x)$ the density of the absolutely continuous component of μ_T. This density is given by

$$p_T(x) = \frac{1}{2T}\left(\frac{x}{x_0}\right)^{\frac{\nu}{2}} \exp\left\{-\frac{x+x_0}{2T}\right\} I_{-\nu}\left(\frac{\sqrt{x_0 x}}{T}\right). \qquad (11.13)$$

Recall that the following formulas hold:

$$I_\alpha(x) \sim \frac{1}{\Gamma(\alpha+1)}\left(\frac{x}{2}\right)^\alpha \quad \text{as } x \to 0 \qquad (11.14)$$

for all $\alpha \neq -1, -2, \ldots$ (see [AS72], 9.6.7). Moreover,

$$I_\alpha(x) \sim \frac{e^x}{\sqrt{2\pi x}} \quad \text{as } x \to \infty \qquad (11.15)$$

(see [AS72], 9.7.1).

Using formulas (11.9) and (11.11), we see that the density of the absolutely continuous component $d_T(x)$ of the distribution of the asset price S_T satisfies the equality

$$d_T(x) = c x^{(\nu+2)(1-\rho)-1} \exp\left\{-\frac{x^{2(1-\rho)}}{2T\sigma^2(1-\rho)^2}\right\} I_{-\nu}\left(\frac{s_0^{1-\rho} x^{1-\rho}}{T\sigma^2(1-\rho)^2}\right), \qquad (11.16)$$

where $c > 0$ is a constant depending on the model parameters. Therefore, (11.14) and (11.16) give

$$d_T(x) \sim c_1 x^{1-2\rho} \quad \text{as } x \to 0, \qquad (11.17)$$

where $c_1 > 0$ depends on the model parameters.

It is not hard to see that the singular component of μ_T does not affect the behavior of the put pricing function P near zero (use the definition of P). Integrating the function d_T near zero twice and using (11.14), we obtain

$$P(K) \approx K^{3-2\rho} \quad \text{as } K \to 0. \qquad (11.18)$$

Next, we turn our attention to the call pricing function C. It is clear that the singular component of μ_T does not influence the behavior of $C(K)$ as $K \to \infty$. Using (11.9), (11.11), (11.13), and (11.15), we see that

$$d_T(x) \sim c_2 x^{-\frac{3}{2}\rho} \exp\left\{\frac{s_0^{1-\rho} x^{1-\rho}}{T\sigma^2(1-\rho)^2}\right\} \exp\left\{-\frac{x^{2(1-\rho)}}{2T\sigma^2(1-\rho)^2}\right\} \tag{11.19}$$

where $c_2 > 0$ is a constant depending on the model parameters. Integrating (11.19) over a neighborhood of infinity twice, we obtain

$$C(K) \approx K^{\frac{5\rho-4}{2}} \exp\left\{\frac{s_0^{1-\rho} K^{1-\rho}}{T\sigma^2(1-\rho)^2}\right\} \exp\left\{-\frac{K^{2(1-\rho)}}{2T\sigma^2(1-\rho)^2}\right\} \tag{11.20}$$

as $K \to \infty$.

It follows from (11.17) and (11.19) that for the asset price S_T in the CEV model we have $\tilde{p} = \infty$ and $\tilde{q} = 2(1-\rho)$. This means that some moments of negative order of the asset price in the CEV model explode, while the moments of positive order do not.

Theorem 11.5 *The following formulas hold for the implied volatility in the CEV model*:

$$I(K) = \sigma(1-\rho)\frac{\log K}{K^{1-\rho}} + O\left(\frac{1}{K^{1-\rho}}\right) \tag{11.21}$$

as $K \to \infty$, *and*

$$\begin{aligned} I(K) = \frac{\sqrt{2}}{\sqrt{T}}\Bigg[&\sqrt{(3-2\rho)\log\frac{1}{K} - \frac{1}{2}\log\log\frac{1}{K}} \\ &- \sqrt{(2-2\rho)\log\frac{1}{K} - \frac{1}{2}\log\log\frac{1}{K}}\Bigg] \\ &+ O\left(\left(\log\frac{1}{K}\right)^{-\frac{1}{2}}\right) \end{aligned} \tag{11.22}$$

as $K \to 0$.

Proof The asymptotic formula in (11.22) follows from formula (11.18), Theorem 9.29 with $\widetilde{P}(K) = K^{3-2\rho}$, and the mean value theorem. As for the asymptotic formula in (11.21), it can be derived from (11.20) and Corollary 11.3 as follows. Set

$$\widetilde{C}(K) = K^{\frac{5\rho-4}{2}} \exp\left\{\frac{s_0^{1-\rho} K^{1-\rho}}{T\sigma(1-\rho)}\right\} \exp\left\{-\frac{K^{2(1-\rho)}}{2T\sigma^2(1-\rho)^2}\right\}.$$

Then

$$\log\frac{1}{\widetilde{C}(K)} = \frac{K^{2(1-\rho)}}{2T\sigma^2(1-\rho)^2} - \frac{s_0^{1-\rho} K^{1-\rho}}{T\sigma(1-\rho)} - \frac{5\rho-4}{2}\log K \tag{11.23}$$

and

$$\log \frac{1}{\widetilde{C}(K)} \approx K^{2(1-\rho)} \quad \text{as } K \to \infty. \tag{11.24}$$

Next, using (11.20), (11.23), (11.24), Corollary 11.3 and the mean value theorem, we obtain (11.21). □

11.3 Displaced Diffusion Model

The displaced diffusion model was introduced and studied by Rubinstein in [Rub83]. The stock price process in Rubinstein's model is a convex combination of a risky asset, following a driftless geometric Brownian motion, and a riskless asset. Let us suppose $r = 0$. Then the stock price process S in Rubinstein's model is given by

$$S_t = s_0 \left[\eta \exp\left\{ -\frac{1}{2}\sigma^2 t + \sigma W_t \right\} + (1-\eta) \right] \tag{11.25}$$

where $0 \le \eta \le 1$, $s_0 > 0$ is the initial price, and $\sigma > 0$ is the volatility parameter. More general displaced diffusion models are described by the following stochastic differential equation:

$$dS_t = \sigma(S_t + a)\, dW_t, \quad S_0 = s_0 \text{ a.s.} \tag{11.26}$$

where $s_0 > 0$, $\sigma > 0$, and $a \neq 0$. It is clear that if the process S satisfies (11.26), then the process $X_t = S_t + a$ is a driftless geometric Brownian motion with the volatility equal to σ and the initial condition given by $x_0 = s_0 + a$. It follows that

$$S_t = (s_0 + a) \exp\left\{ -\frac{1}{2}\sigma^2 t + \sigma W_t \right\} - a. \tag{11.27}$$

Since we are studying only positive stock price processes, it is natural to suppose that $a < 0$ and $s_0 > |a|$. Then the process S defined by (11.27) coincides with the stock price process in Rubinstein's model (11.25) with $S_0 = s_0$ and $\eta = \frac{s_0 + a}{s_0}$.

It is not hard to see that any moment of positive order of the asset price in the displaced diffusion model is finite. Let C be the call pricing function associated with this model. Then $C \in PF_\infty$, but $C \notin PF_0$. The next assertion characterizes the asymptotic behavior of the implied volatility at large strikes in the displaced diffusion model.

Theorem 11.6 *Let I be the implied volatility in model* (11.26) *with $a < 0$ and $s_0 > |a|$. Then*

$$I(K) = \sigma + O\left(\frac{\log \log K}{\log K} \right) \tag{11.28}$$

as $K \to \infty$.

Proof The distribution density D_T of the random variable X_T is log-normal. More precisely,

$$D_T(x) \approx x^{-\frac{3}{2}} \exp\left\{-\frac{1}{2T\sigma^2}\left(\log \frac{x}{s_0+a}\right)^2\right\}$$

as $x \to \infty$.

Denote by $\widetilde{D}_T$ the distribution density of the random variable S_T. It is not hard to see that $\widetilde{D}_T(x) = D_T(x-a)$ as $x \to \infty$. Moreover, $\widetilde{D}_T(x) \approx D_T(x)$ as $x \to \infty$. Next, using the last formula in Sect. 5 of [Gul10], we obtain the following asymptotic formula for the implied volatility in Rubinstein's model:

$$\begin{aligned}\sqrt{T} I(K) = {} & \sqrt{\frac{1}{T\sigma^2}\left(\log \frac{K}{s_0+a}\right)^2 + \log K} - \sqrt{\frac{1}{T\sigma^2}\left(\log \frac{K}{s_0+a}\right)^2 - \log K} \\ & + O\left(\frac{\log \log K}{\log K}\right)\end{aligned}$$

as $K \to \infty$. It follows that

$$\begin{aligned}I(K) = {} & \frac{1}{T\sigma} \log \frac{K}{s_0+a}\left[\sqrt{1+T\sigma^2(\log K)\left(\log \frac{K}{s_0+a}\right)^{-2}}\right. \\ & \left. - \sqrt{1-T\sigma^2(\log K)\left(\log \frac{K}{s_0+a}\right)^{-2}}\right] \\ & + O\left(\frac{\log \log K}{\log K}\right)\end{aligned}$$

as $K \to \infty$. Next, taking into account formula (11.3) and making simplifications, we see that Theorem 11.6 holds. □

Remark 11.7 Formula $I(K) \to \sigma$ as $K \to \infty$ was obtained in [LW12]. The formula in (11.28) includes the error term.

The case where $a > 0$ and $s_0 + a > 0$ in (11.26) is also interesting. In this case, the random variable S_t may take negative values with positive probability. However, we can still define (informally) the call pricing function C by

$$C(K) = \mathbb{E}\left[(S_T - K)^+\right] = \mathbb{E}\left[\left(X_T - (K+a)\right)^+\right].$$

The implied volatility $I(T, K)$ in this case is determined from the following equality:

$$C_{\mathrm{BS}}\big(T, K, s_0, I(T, K)\big) = C_{\mathrm{BS}}(T, K+a, s_0+a, \sigma).$$

In addition, it can be shown that formula (11.28) remains true for $a > 0$ and $s_0 + a > 0$.

11.4 Finite Moment Log-Stable Model

The finite moment log-stable model was developed by P. Carr and L. Wu in [CW03]. They used α-stable Lévy processes with skew parameter $\beta = -1$ to model the log-returns associated with the spot levels of S&P 500 index. We will next briefly discuss α-stable laws. More information and proofs can be found in [Zol86, ST94, JW94].

The family of α-stable distributions depends on four parameters:

$$0 < \alpha \leq 2, \quad -1 \leq \beta \leq 1, \quad \lambda > 0, \quad \text{and} \quad -\infty < \gamma < \infty.$$

For fixed values of the parameters, the α-stable law will be denoted by $L_\alpha(\gamma, \lambda, \beta)$. We will restrict ourselves to the case where $1 < \alpha < 2$. Then the law $L_\alpha(\gamma, \lambda, \beta)$ is defined through its characteristic function as follows:

$$\int_{-\infty}^{\infty} e^{itx} \, dL_\alpha(\gamma, \lambda, \beta) = \exp\{it\gamma - \lambda|t|^\alpha \Lambda(t, \alpha, \beta)\}$$

where

$$\Lambda(t, \alpha, \beta) = \exp\left\{-i\frac{\pi}{2}\beta\left[\alpha - 1 + \operatorname{sign}(1-\alpha)\right] \operatorname{sign} t\right\}.$$

Let $1 < \alpha < 2$, and let $L^{\alpha,-1}$ be a standardized Lévy α-stable motion. This is a Lévy process such that for every $t > 0$ the random variable $L_t^{\alpha,-1}$ is distributed according to the law $L_\alpha(0, t^{\frac{1}{\alpha}}, -1)$. It follows that the random variables $t^{-\frac{1}{\alpha}} L_t^{\alpha,-1}$ and $L_1^{\alpha,-1}$ are identically distributed. We will denote by g_α the distribution density of $L_\alpha(0, 1, -1)$ and by f_α the distribution density of $L_\alpha(0, 1, 1)$. It is known that

$$g_\alpha(y) \approx y^{\frac{2-\alpha}{2(\alpha-1)}} \exp\left\{-(\alpha - 1)\left(\frac{y}{\alpha}\right)^{\frac{\alpha}{\alpha-1}}\right\} \tag{11.29}$$

as $y \to \infty$. Formula (11.29) follows from a sharper formula due to V. Zolotarev (see [Zol86], Theorem 2.5.2).

The behavior of g_α near $-\infty$ is characterized by the formula

$$g_\alpha(-y) \approx y^{-\alpha-1} \tag{11.30}$$

as $y \to \infty$. Formula (11.30) follows from the equality $g_\alpha(-x) = f_\alpha(x)$ (see Sect. 2.2 in [Zol86]) and from the asymptotic formula for $f_\alpha(x)$ as $x \to \infty$ (see Corollary 2 to Theorem 2.5.1 in [Zol86]).

Carr and Wu suggested to model the spot index level by the process S satisfying the following stochastic differential equation under the risk-neutral measure $\mathbb{P}^*$:

$$dS_t = S_t\left(r\,dt + \sigma\,dL_t^{\alpha,-1}\right), \tag{11.31}$$

where $1 < \alpha < 2$, $\sigma > 0$, and $r > 0$ is the interest rate. The model described by (11.31) is called the finite moment log-stable model. The solution S to (11.31) can be represented as follows:

$$S_t = S_0 \exp\{(r+\mu)t + \sigma L_t^{\alpha,-1}\} \tag{11.32}$$

where

$$\mu = \sigma^\alpha \sec\frac{\pi\alpha}{2} \tag{11.33}$$

(see [CW03], p. 763). The parameter μ in (11.33) is chosen so that the martingality condition

$$\mathbb{E}^*\left[\exp\{\mu t + \sigma L_t^{\alpha,-1}\}\right] = 1$$

holds. It follows from (11.32) that the random variable defined by $s = \phi(S_T)$ with

$$\phi(u) = \sigma^{-1} T^{-\frac{1}{\alpha}} \left[\log\frac{u}{S_0} - (r+\mu)T\right], \quad u > 0,$$

has the function g_α as its distribution density. Denote the distribution density of S_T by D_T. Then the previous statement implies that

$$D_T(x) = g_\alpha\big(\phi(x)\big)\sigma^{-1} T^{-\frac{1}{\alpha}} x^{-1}. \tag{11.34}$$

Indeed, since $S_T = \phi^{-1}(s_T)$, we have $D_T(x) = \phi'(x) g_\alpha(\phi(x))$, and formula (11.34) follows. Therefore, (11.29) and (11.30) give

$$D_T(x) \approx x^{-1}(\log x)^{\frac{2-\alpha}{2(\alpha-1)}} \times \exp\left\{-\frac{\alpha-1}{(\sigma\alpha)^{\frac{\alpha}{\alpha-1}} T^{\frac{1}{\alpha-1}}}\left[\log\frac{x}{S_0} - (r+\mu)T\right]^{\frac{\alpha}{\alpha-1}}\right\} \tag{11.35}$$

as $x \to \infty$, and also

$$D_T(x) \approx x^{-1}\left(\log\frac{1}{x}\right)^{-\alpha-1} \tag{11.36}$$

as $x \downarrow 0$. Integrating the functions in formula (11.35) twice and taking into account the equalities

$$C(K) = e^{-rT}\int_K^\infty \overline{F}_T(y)\,dy \quad \text{and} \quad \overline{F}(y) = \int_y^\infty D_T(x)\,dx, \tag{11.37}$$

we obtain

$$C(K) \approx K(\log K)^{-\frac{\alpha+2}{2(\alpha-1)}} \times \exp\left\{-\frac{\alpha-1}{(\sigma\alpha)^{\frac{\alpha}{\alpha-1}} T^{\frac{1}{\alpha-1}}}\left[\log\frac{K}{S_0} - (r+\mu)T\right]^{\frac{\alpha}{\alpha-1}}\right\} \tag{11.38}$$

as $K \to \infty$.

For small strikes the definition of the put pricing function and (11.36) imply the asymptotic formula

$$
\begin{aligned}
P(K) &= e^{-rT} K \int_0^K D_T(x)\,dx - e^{-rT} \int_0^K x D_T(x)\,dx \\
&\approx K\left(\log\frac{1}{K}\right)^{-\alpha}
\end{aligned}
\tag{11.39}
$$

as $K \to 0$.

Theorem 11.8 *The following formulas hold for the implied volatility in the finite moment log-stable model*:

$$
I(K) = \frac{(\sigma\alpha)^{\frac{\alpha}{2(\alpha-1)}} T^{\frac{1-\alpha}{2\alpha}}}{\sqrt{2(\alpha-1)}} (\log K)^{-\frac{2-\alpha}{2(\alpha-1)}} + O\big((\log K)^{-\frac{\alpha}{2(\alpha-1)}}\big)
\tag{11.40}
$$

as $K \to \infty$, *and*

$$
\begin{aligned}
I(K) = \frac{\sqrt{2}}{\sqrt{T}}&\left[\sqrt{\log\frac{1}{K} + \alpha \log\log\frac{1}{K} - \frac{1}{2}\log\log\log\frac{1}{K}}\right. \\
&\left. - \sqrt{\alpha \log\log\frac{1}{K} - \frac{1}{2}\log\log\log\frac{1}{K}}\right] \\
&+ O\left(\left(\log\log\frac{1}{K}\right)^{-\frac{1}{2}}\right)
\end{aligned}
\tag{11.41}
$$

as $K \to 0$.

Proof Put

$$
\widetilde{C}(K) = K(\log K)^{-\frac{\alpha+2}{2(\alpha-1)}} \exp\left\{-\frac{\alpha-1}{(\sigma\alpha)^{\frac{\alpha}{\alpha-1}} T^{\frac{1}{\alpha-1}}}\left[\log\frac{K}{S_0} - (r+\mu)T\right]^{\frac{\alpha}{\alpha-1}}\right\}.
$$

Then we have

$$
\begin{aligned}
\log\frac{1}{\widetilde{C}(K)} &= -\log K + \frac{\alpha+2}{2(\alpha-1)}\log\log K \\
&\quad + \frac{\alpha-1}{(\sigma\alpha)^{\frac{\alpha}{\alpha-1}} T^{\frac{1}{\alpha-1}}}\left[\log\frac{K}{S_0} - (r+\mu)T\right]^{\frac{\alpha}{\alpha-1}}.
\end{aligned}
$$

Now it is clear that condition (11.7) in Corollary 11.3 holds (use (11.38)). Next, applying this corollary and making simplifications in the resulting formula, we get (11.40). Finally, it is not hard to see that (11.41) follows from Theorem 9.29 and (11.39). □

11.5 Piterbarg's Conjecture

The asymptotic formulas for the implied volatility discussed in Chaps. 9 and 10 show that for models with moment explosions a typical behavior of the implied volatility near infinity is described by the function $c_1\sqrt{\log K}$ and near zero by the function $c_2\sqrt{\log \frac{1}{K}}$. However, if the moments of the asset price do not explode, then the class of approximating functions is wider.

Let X be the asset price process, and assume that the corresponding call pricing function C satisfies the condition $C \in PF_\infty$. Suppose that for fixed maturity $T > 0$ the moments of positive orders of X_T do not explode. Let w be a positive increasing function on $(0, \infty)$ satisfying the condition $w(y) \to \infty$ as $y \to \infty$. In this section, we discuss the asymptotic behavior of the function

$$\Lambda(K) = \frac{I(K)\sqrt{w(K)}}{\log K} \tag{11.42}$$

as $K \to \infty$. In (11.42), I is the implied volatility corresponding to the call pricing function C. Set

$$\gamma_w = \limsup_{K\to\infty} \Lambda(K) = \limsup_{K\to\infty} \frac{I(K)\sqrt{w(K)}}{\log K}.$$

Our goal is to compute the number γ_w.

In [Pit04], V.V. Piterbarg suggested an explicit formula for the number γ_w. Piterbarg's conjecture will be discussed below. For the sake of simplicity, we exclude functions w with irregular behavior. Let us assume that the limit

$$M = \lim_{y\to\infty} \frac{w(y)}{\log y}$$

exists (finite or infinite). If $M < \infty$, then we have

$$\gamma_w = \sqrt{M} \limsup_{K\to\infty} \frac{I(K)}{\sqrt{\log K}} = \sqrt{\frac{M\psi(\tilde{p})}{T}},$$

by Lee's moment formula (10.2). However, in the case where $M = \infty$, Lee's formula does not provide enough information about the asymptotic behavior of the implied volatility near infinity.

In the remaining part of the present section, we assume that $\tilde{p} = \infty$ for the asset price X_T. We also assume w is a positive increasing function on $(0, \infty)$ satisfying the condition

$$\lim_{y\to\infty} \frac{w(y)}{\log y} = \infty. \tag{11.43}$$

Recall that the symbol $\overline{F}_T$ stands for the complementary cumulative distribution function of X_T.

Consider a call pricing functions $C \in PF_\infty$ and set

$$r_w^* = \sup\{r \geq 0 : C(K) = O(e^{-rw(K)}) \text{ as } K \to \infty\},$$

$$\hat{p}_w = \sup\left\{p \geq 0 : \mathbb{E}^*\left[\int_0^{X_T} e^{pw(y)}\, dy\right] < \infty\right\}, \tag{11.44}$$

$$\tilde{p}_w = \sup\{p \geq 0 : \mathbb{E}^*[\exp\{pw(X_T)\}] < \infty\}, \tag{11.45}$$

and

$$l_w = \liminf_{K \to \infty} \frac{\log \frac{1}{C(K)}}{w(K)}. \tag{11.46}$$

It is not hard to see that

$$\hat{p}_w = \sup\left\{p \geq 0 : \int_0^\infty \overline{F}_T(u) e^{pw(u)}\, du < \infty\right\}$$

and

$$\tilde{p}_w = \sup\left\{p \geq 0 : \int_0^\infty e^{pw(u)}\, d[-\overline{F}_T(u)] < \infty\right\}.$$

Piterbarg's conjecture is the following equality:

$$\limsup_{K \to \infty} \frac{I(K)\sqrt{w(K)}}{\log K} = \frac{1}{\sqrt{2T\tilde{p}_w}}. \tag{11.47}$$

It will be shown below that formula (11.47) holds if we replace the number $\tilde{p}_w$ by the number $\hat{p}_w$ (see Theorem 11.10 below). Moreover, under a very mild additional restriction on the function w, formula (11.47) is valid without any modifications (see Remark 11.13).

Our first goal is to discuss various relations between the constants introduced above.

Lemma 11.9 *Suppose w is a positive increasing function on $(0, \infty)$ satisfying* (11.43). *Then $l_w = r_w^* = \hat{p}_w$ and $\tilde{p}_w \leq \hat{p}_w$.*

Proof Let $0 < l_w < \infty$. Then for every small $\epsilon > 0$ there exists $K_\varepsilon > 0$ such that for all $K > K_\varepsilon$,

$$\left(w(K)\right)^{-1} \log \frac{1}{C(K)} > l - \varepsilon.$$

It follows that $C(K) \leq \exp\{(-l + \varepsilon)w(K)\}$ for all $K > K_\varepsilon$, which implies the inequality $l_w \leq r_w^*$. For $l_w = \infty$ the proof is similar, while the case $l_w = 0$ is trivial.

Next, let $r_w^* > 0$ and let r with $0 < r < r_w^*$ be such that

$$C(K) = O(\exp\{-rw(K)\})$$

as $K \to \infty$. Then we have

$$\frac{\log \frac{1}{C(K)}}{w(K)} \geq r + \frac{\log c}{w(K)}$$

where $c > 0$ does not depend on K. Now it is clear that $r_w^* \leq l_w$. The case where $r_w^* = 0$ is trivial. This establishes the equality $l_w = r_w^*$.

We will next prove the equality $\hat{p}_w = r_w^*$. Suppose $r_w^* > 0$ and let $r > 0$ be such that $r < r_w^*$. Then we have $C(K) = O(\exp\{-rw(K)\})$ as $K \to \infty$. Let $\varepsilon < r$. Using the integration by parts formula for Stieltjes integrals and the formula

$$C(K) = e^{-rT} \int_K^\infty \overline{F}_T(u)\, du, \tag{11.48}$$

we obtain

$$\begin{aligned}\int_0^\infty \overline{F}_T(u) e^{(r-\varepsilon)w(u)}\, du &= c + e^{rT} \int_0^\infty C(y)\, de^{(r-\varepsilon)w(y)} \\ &\leq c_1 + c_2 \int_a^\infty e^{-rw(y)}\, de^{(r-\varepsilon)w(y)} < \infty,\end{aligned}$$

which implies the estimate $r_w^* \leq \hat{p}_w$.

Next, suppose $\hat{p}_w > 0$ and let $p > 0$ be such that $p < \hat{p}_w$. Then using (11.48), we see that, for every $K > 0$,

$$\infty > \int_0^\infty \overline{F}_T(u) e^{pw(u)}\, du \geq e^{pw(K)} \int_K^\infty \overline{F}_T(u)\, du = e^{rT} e^{pw(K)} C(K).$$

It follows that $C(K) = O(\exp\{-pw(K)\})$ as $K \to \infty$ and hence $\hat{p}_w \leq r_w^*$. This establishes the equality $\hat{p}_w = r_w^*$.

It remains to prove the inequality $\tilde{p}_w \leq \hat{p}_w$. For all $x > 0$ and $p \geq 0$, we have

$$\int_0^x e^{pw(y)}\, dy \leq x e^{pw(x)}.$$

Therefore, (11.43) shows that for every $\varepsilon > 0$ there exists $x_\varepsilon > 0$ such that

$$\int_0^x e^{pw(y)}\, dy \leq e^{(p+\varepsilon)w(x)}, \ x > x_\varepsilon.$$

Now, it is not hard to see that (11.44) and (11.45) imply the inequality $\tilde{p}_w \leq \hat{p}_w$.

This completes the proof of Lemma 11.9. □

The next assertion provides an explicit formula for the number γ_w. This formula shows that Piterbarg's conjecture is valid in a slightly modified form.

Theorem 11.10 *Let $C \in PF_\infty$ be a call pricing function, and suppose $\tilde{p} = \infty$. Then for every positive increasing function w on $(0, \infty)$ satisfying condition* (11.43),

$$\limsup_{K\to\infty} \frac{I(K)\sqrt{w(K)}}{\log K} = \frac{1}{\sqrt{2T\hat{p}_w}}. \tag{11.49}$$

Proof Using (11.5) and (11.46), we see that

$$\limsup_{K\to\infty} \frac{I(K)\sqrt{w(K)}}{\log K} = \left(2T \liminf_{K\to\infty} \frac{\log \frac{1}{C(K)}}{w(K)}\right)^{-\frac{1}{2}} = \frac{1}{\sqrt{2Tl_w}}.$$

Therefore, formula (11.49) follows from the equality $l_w = \hat{p}_w$ in Lemma 11.9. □

It is not hard to see that Piterbarg's conjecture (formula (11.47)) is equivalent to the validity of the inequality $\hat{p}_w \le \tilde{p}_w$ (use Theorem 11.10 and the inequality $\tilde{p}_w \le \hat{p}_w$ in Lemma 11.9). Our next goal is to prove the equality $\hat{p}_w = \tilde{p}_w$ under certain additional restrictions on the function w.

Lemma 11.11 *Let w be an increasing positive function on $(0, \infty)$ satisfying condition* (11.43). *Suppose for any $0 < \varepsilon < 1$ there exists a number $x_\varepsilon > 0$ such that*

$$\int_0^x e^{w(u)}\, du \ge e^{(1-\varepsilon)w(x)} \tag{11.50}$$

for all $x > x_\varepsilon$. Then $\hat{p}_w = \tilde{p}_w$.

Proof It suffices to prove the estimate $\hat{p}_w \le \tilde{p}_w$. Let us assume that the conditions in the formulation of Lemma 11.11 hold. We will prove that actually the following stronger condition is valid. For all $0 < p < \infty$ and $0 < \varepsilon < p$ there exists a number $x_{p,\varepsilon} > 0$ such that

$$\int_0^x e^{pw(u)}\, du \ge e^{(p-\varepsilon)w(x)} \tag{11.51}$$

for all $x > x_{p,\varepsilon}$.

If $p = 1$, then inequality (11.51) is simply inequality (11.50). Now let $p > 1$. Then Hölder's inequality and (11.50) imply that

$$\int_0^x e^{pw(u)}\, du \ge x^{-\frac{p}{q}} e^{(p-p\varepsilon)w(x)}$$

for all $x > x_\varepsilon$ where $\frac{1}{p} + \frac{1}{q} = 1$. It follows from condition (11.43) that for every $\delta > 0$ and $r > 0$ the estimate $x^r \le e^{\delta w(x)}$ holds when x is large enough. Therefore, the estimate

$$\int_0^x e^{pw(u)}\, du \ge e^{(p-p\varepsilon-\delta)w(x)}$$

also holds for sufficiently large values of x. It is clear that the previous statement implies (11.51) for $p > 1$.

Next, let $0 < p < 1$. Then using (11.50) we see that

$$\int_0^x e^{pw(u)}\,du = \int_0^x e^{(p-1)w(u)} e^{w(u)}\,du \ge e^{(p-1)w(u)} \int_0^x e^{w(u)}\,du \ge e^{(p-\varepsilon)w(u)}\,du$$

for $x \ge x_\varepsilon$. This establishes (11.51) for all $0 < p < 1$. It follows that (11.51) holds for all $p > 0$. Now, it is not hard to see that the inequality $\hat{p}_w \le \tilde{p}_w$ can be obtained from (11.44), (11.45), and (11.51). □

Corollary 11.12 *Let w be an increasing positive function on $(0,\infty)$ satisfying condition (11.43). Suppose there exists a number $c > 0$ such that w is absolutely continuous on every compact subinterval of (c,∞), and for every $0 < \varepsilon < 1$ there exists $y_\varepsilon > c$ such that*

$$w'(y) \le e^{\varepsilon w(y)} \tag{11.52}$$

almost everywhere on (y_ε,∞) with respect to the Lebesgue measure. Then $\hat{p}_w = \tilde{p}_w$.

Proof We will show that the conditions in the formulation of Corollary 11.12 imply estimate (11.50). Indeed, it follows from (11.52) that for all $0 < \varepsilon < 1$ and $x > y_\varepsilon$,

$$\int_0^x e^{w(y)}\,dy \ge \int_{y_\varepsilon}^x e^{w(y)} w'(y) \frac{1}{w'(y)}\,dy \ge \int_{y_\varepsilon}^x e^{(1-\varepsilon)w(y)} w'(y)\,dy.$$

Therefore, there exist $c_\varepsilon > 0$ and $x_\varepsilon > 0$ such that

$$\int_0^x e^{w(y)}\,dy \ge c_\varepsilon e^{(1-\varepsilon)w(x)}$$

for all $x > x_\varepsilon$. It is not hard to see that the previous inequality implies (11.50). Now Corollary 11.12 follows from Lemma 11.11. □

Remark 11.13 It is clear that under the conditions in Lemma 11.11 or Corollary 11.12, Piterbarg's formula (11.47) holds.

Let w be an increasing positive function on $(0,\infty)$, and suppose there exists a number $c > 0$ such that w is absolutely continuous on every compact subinterval of (c,∞). Put

$$\hat{s}_w = \sup\bigl\{s \ge 0 : \overline{F}_T(y) = O\bigl(e^{-sw(y)} w'(y)\bigr) \text{ a.e., as } y \to \infty\bigr\},$$

where $\overline{F}_T$ is the complementary cumulative distribution function of the stock price X_T.

Lemma 11.14 *Let w be an increasing positive function on $(0,\infty)$ and assume w is absolutely continuous on every compact subinterval of (c,∞) for some $c\geq 0$. If for every $0<\varepsilon<1$ there exists $y_\varepsilon>c$ such that*

$$e^{-\varepsilon w(y)}\leq w'(y)\leq e^{\varepsilon w(y)} \tag{11.53}$$

almost everywhere on (y_ε,∞) with respect to the Lebesgue measure, then $r_w^*=\hat{s}_w$.

Proof Suppose $r_w^*>0$ and let $r>0$ be such that $r<r_w^*$. Then

$$C(K)\leq c_r e^{-rw(K)}$$

for all $K>K_r$. Set $\lambda_\varepsilon(y)=e^{-\varepsilon w(y)}$ where $\varepsilon>0$. It follows from (10.12) that

$$c_r e^{-rw(y-\lambda_\varepsilon(y))}\geq \int_{y-\lambda_\varepsilon(y)}^{y}\overline{F}_T(u)\,du\geq \overline{F}_T(y)e^{-\varepsilon w(y)},\quad y>y_{\varepsilon,r}.$$

Therefore, condition (11.52) implies that

$$\begin{aligned}\overline{F}_T(y)&\leq c_r e^{(-r+\varepsilon)w(y)}\exp\bigl\{r\bigl[w(y)-w\bigl(y-\lambda_\varepsilon(y)\bigr)\bigr]\bigr\}\\&=c_r e^{(-r+\varepsilon)w(y)}\exp\left\{r\int_{y-\lambda_\varepsilon(y)}^{y}w'(u)\,du\right\}\\&\leq c_r e^{(-r+\varepsilon)w(y)}\exp\left\{r\int_{y-\lambda_\varepsilon(y)}^{y}e^{\varepsilon w(u)}\,du\right\}\\&\leq c_r e^{r}e^{(-r+\varepsilon)w(y)}\end{aligned}\tag{11.54}$$

for almost all $y>\tilde{y}_{\varepsilon,r}$. Using (11.53) and (11.54), we see that for every $\varepsilon>0$,

$$\overline{F}_T(y)=O\bigl(w'(y)e^{(-r+2\varepsilon)w(y)}\bigr)$$

as $y\to\infty$. Now it is clear that $r_w^*\leq\hat{s}_w$.

Next, suppose $\hat{s}_w>0$ and let $s>0$ be such that $s<\hat{s}_w$. Then

$$\overline{F}_T(y)=O\bigl(e^{-sw(y)}w'(y)\bigr)$$

a.e., as $y\to\infty$. Therefore

$$C(K)\leq c\int_K^\infty e^{-sw(y)}w'(y)\,dy=O\bigl(e^{-sw(K)}\bigr)$$

as $K\to\infty$. Now it is not hard to see that the previous reasoning implies the estimate $\hat{s}_w\leq r_w^*$.

This completes the proof of Lemma 11.14. □

Lemma 11.15 *Let w be an increasing positive function on $(0,\infty)$ such that the following conditions hold:*

1. *The function w is absolutely continuous on every compact subinterval of (c,∞) for some $c \geq 0$.*
2. *The function w is such that $w(y)(\log y)^{-1} \uparrow \infty$ as $y \to \infty$.*
3. *For any $0 < \varepsilon < 1$ there exists $y_\varepsilon > c$ such that $w'(y) \leq e^{\varepsilon w(y)}$ a.e. on (y_ε, ∞).*

Then for any $0 < \varepsilon < 1$ there exists $\tilde{y}_\varepsilon > c$ such that $e^{-\varepsilon w(y)} \leq w'(y)$ a.e. on $(\tilde{y}_\varepsilon, \infty)$.

Proof There exists $y_0 > c$ such that

$$0 \leq \left(\frac{w(y)}{\log y}\right)' = \frac{w'(y)\log y - y^{-1}w(y)}{\log^2 y}$$

a.e. on (y_0,∞). Therefore, $w'(y) \geq (y\log y)^{-1}w(y)$ a.e. on (y_0,∞). It is clear that for every $\varepsilon > 0$ there exists $\tilde{y}_\varepsilon > c$ such that

$$w(y) \geq \exp\left\{-\frac{\varepsilon}{2}w(y)\right\}$$

and

$$y\log y \leq \exp\left\{\frac{\varepsilon}{2}w(y)\right\}$$

for almost all $y > \tilde{y}_\varepsilon$. Hence $w'(y) \geq \exp\{-\varepsilon w(y)\}$ for almost all $y > \tilde{y}_\varepsilon$.

This completes the proof of Lemma 11.15. □

It follows from Theorem 11.10 and Lemma 11.11 that the estimate in (11.50) is a sufficient condition for the validity of Piterbarg's conjecture. It is tempting to try to prove that the estimate in (11.50) always holds. Unfortunately, this is not the case. We will next provide a counterexample.

Let $\{a_n\}_{n\geq 0}$ and $\{\delta_n\}_{n\geq 0}$ be sequences of positive numbers such that $a_n \uparrow \infty$, $\delta_n \downarrow 0$ as $n \to \infty$, and $\delta_n < 1$ for all $n \geq 0$ (these sequences will be chosen later). Define a function on $[0,\infty)$ by $w(u) = a_n$ if $u \in [n, n+1-\delta_n]$ and

$$w(u) = a_n + \frac{a_{n+1}-a_n}{\delta_n}\big(u-(n+1-\delta_n)\big)$$

if $u \in [n+1-\delta_n, n+1]$.

Let $n > 0$ and $n+1-\delta_n \leq x < n+1$. Then

$$\begin{aligned}\int_0^x e^{w(u)}\,du &\leq \sum_{k=0}^n e^{a_k} + \sum_{k=0}^n e^{a_k}\int_0^{\delta_k}\exp\left\{\frac{a_{k+1}-a_k}{\delta_k}y\right\}dy \\ &\leq ne^{a_n} + \sum_{k=0}^n e^{a_k}\frac{\delta_k}{a_{k+1}-a_k}\big(e^{a_{k+1}-a_k}-1\big)\end{aligned}$$

$$= ne^{a_n} + \sum_{k=0}^{n} \frac{\delta_k}{a_{k+1} - a_k} \left(e^{a_{k+1}} - e^{a_k}\right)$$

$$\leq ne^{a_n} + \sum_{k=0}^{n} \delta_k e^{a_{k+1}}. \tag{11.55}$$

We will next choose the sequences $\{a_n\}_{n \geq 0}$ and $\{\delta_n\}_{n \geq 0}$. Set $a_0 = 1$, and let a_n with $n \geq 1$ be defined by the formula

$$a_{n+1} = 3a_n + 4\log(2n).$$

Then we have

$$2ne^{a_n} = \exp\left\{\frac{a_n + a_{n+1}}{4}\right\}, \quad n \geq 1.$$

Put $\delta_n = e^{-a_{n+1}}$, $n \geq 0$. It follows from (11.55) that for all $n > 0$ and $n + 1 - \delta_n \leq x < n + 1$

$$\int_0^x e^{w(u)}\, du \leq 2ne^{a_n} = \exp\left\{\frac{a_n + a_{n+1}}{4}\right\}. \tag{11.56}$$

Now suppose that $n > 0$ and $n + 1 - \frac{\delta_n}{2} < x < n + 1$. Then

$$e^{\frac{1}{2}w(x)} = \exp\left\{\frac{1}{2}a_n\right\} \exp\left\{\frac{a_{n+1} - a_n}{2\delta_n}\left(x - (n + 1 - \delta_n)\right)\right\}$$

$$> \exp\left\{\frac{a_n + a_{n+1}}{4}\right\}. \tag{11.57}$$

It follows from (11.56) and (11.57) that

$$\int_0^x e^{w(u)}\, du < e^{\frac{1}{2}w(x)}$$

for all $x \in A$ where

$$A = \bigcup_{n=1}^{\infty} \left(n + 1 - \frac{\delta_n}{2}, n + 1\right).$$

Therefore, the estimate in (11.50) does not hold for the function w defined above.

11.6 Asymptotic Equivalence and Piterbarg's Conjecture

In this section, we take up the issue of determining when the upper limit in formula (11.49) can be replaced by the true limit. Let w be a positive increasing function on $(0, \infty)$, for which condition (11.43) holds, and assume the number $\hat{p}_w$ defined by (11.44) satisfies $0 < \hat{p}_w < \infty$. It follows from (11.5) that the existence of the limit

$$M_1 = \lim_{K \to \infty} \frac{I(K)\sqrt{w(K)}}{\log K}$$

is equivalent to the existence of

$$M_2 = \lim_{K \to \infty} \frac{\log \frac{1}{C(K)}}{w(K)}.$$

Moreover,

$$M_1 = (2TM_2)^{-\frac{1}{2}} \quad \text{and} \quad M_2 = \hat{p}_w.$$

Applying Theorem 11.10, we see that the existence of the limits M_1 and M_2 is equivalent to the validity of the following condition:

$$I(K) \sim \frac{\log K}{\sqrt{2T\hat{p}_w}\sqrt{w(K)}} \tag{11.58}$$

as $K \to \infty$. If the function w satisfies the conditions in Corollary 11.12, then $\hat{p}_w$ in (11.58) can be replaced by $\tilde{p}_w$ defined in (11.45). Thus we have

$$I(K) \sim \frac{\log K}{\sqrt{2T\tilde{p}_w}\sqrt{w(K)}}$$

as $K \to \infty$. Similarly, we can show that if there exists $0 < A < \infty$ such that

$$I(K) \sim \frac{\log K}{\sqrt{2TA}\sqrt{w(K)}} \tag{11.59}$$

as $K \to \infty$, then $A = \tilde{p}_w$. Therefore, the following lemma holds.

Lemma 11.16 *Suppose $C \in PF_\infty$ and $0 < A < \infty$. Suppose also w is a positive increasing function on* $(0, \infty)$ *satisfying condition* (11.43). *Then the asymptotic formula in* (11.59) *holds if and only if for every $\varepsilon > 0$ there exists $K_\varepsilon > 0$ such that*

$$\exp\{(-A-\varepsilon)w(K)\} \le C(K) \le \exp\{(-A+\varepsilon)w(K)\} \tag{11.60}$$

for all $K > K_\varepsilon$.

We will next obtain sufficient conditions for the validity of formula (11.59). These conditions are expressed in terms of the complementary cumulative distribution function $\overline{F}_T$ of the asset price X_T and the distribution density D_T of X_T.

Lemma 11.17 *Let $C \in PF_\infty$, and let w be a positive increasing function on* $(0, \infty)$ *satisfying condition* (11.43) *and the conditions in Lemma* 11.14. *Then the following statements are true:*

- *Let $0 < A < \infty$, and suppose for any $\varepsilon > 0$ there exists $y_\varepsilon > 0$ such that*

$$\exp\{(-A-\varepsilon)w(y)\} \le \overline{F}_T(y) \le \exp\{(-A+\varepsilon)w(y)\} \tag{11.61}$$

for all $y > y_\varepsilon$. Then the asymptotic formula in (11.59) *holds.*

- *Let $0 < A < \infty$, and suppose the distribution of the asset price X_T admits a density D_T. Suppose also that for every $\varepsilon > 0$ there exists $x_\varepsilon > 0$ such that*

$$\exp\{(-A-\varepsilon)w(x)\} \le D_T(x) \le \exp\{(-A+\varepsilon)w(x)\} \tag{11.62}$$

for all $x > x_\varepsilon$. Then the asymptotic formula in (11.59) *holds.*

Proof Suppose the assumptions in the first part of Lemma 11.17 hold. Then, integrating the functions in (11.61), and taking into account the first equality in (11.37) and the restrictions on the function w stated in Lemma 11.14, we see that the estimates in (11.60) hold. Therefore, Lemma 11.16 can be applied, and we obtain the estimates in (11.58). The proof of the estimates in (11.62) is similar. Here we use the second equality in (11.37). □

The following statements concern the case where $K \to 0$.

Lemma 11.18 *Let $C \in PF_0$, and let w be a positive increasing function on $(0, \infty)$ satisfying condition* (11.43) *and the conditions in Lemma* 11.14. *Then the formula*

$$I(K) \sim \frac{\log \frac{1}{K}}{\sqrt{2TA}\sqrt{w(\frac{1}{K})}}, \quad K \to 0, \tag{11.63}$$

holds with some constant $0 < A < \infty$ if and only if for every $\varepsilon > 0$ there exists $K_\varepsilon > 0$ such that

$$\exp\left\{(-A-\varepsilon)w\left(\frac{1}{K}\right)\right\} \le P(K) \le \exp\left\{(-A+\varepsilon)w\left(\frac{1}{K}\right)\right\}$$

for all $K < K_\varepsilon$.

Lemma 11.19 *Let $0 < A < \infty$, and suppose the distribution of the asset price X_T admits a density D_T. Suppose also that for every $\varepsilon > 0$ there exists $x_\varepsilon > 0$ such that*

$$\exp\left\{(-A-\varepsilon)w\left(\frac{1}{x}\right)\right\} \le D_T(x) \le \exp\left\{(-A+\varepsilon)w\left(\frac{1}{x}\right)\right\}$$

for all $0 < x < x_0$. Then formula (11.63) *holds.*

It is not hard to see that Lemma 11.18 can be obtained by applying Lemma 11.16 to the call pricing function G defined by (10.52) and the function δ given by $\delta(K) = w((x_0e^{rT})^2K^{-1})$. Similarly, Lemma 11.19 can be derived from the second part of Lemma 11.17.

11.7 SV1 and SV2 Models of Rogers and Veraart

In [RV08], L.C.G. Rogers and L.A.M. Veraart introduced new stochastic asset price models (SV1 and SV2 models), and suggested to use them as simple alternatives for the popular SABR model defined in [HKLW02]. Our goal in this section is to study the asymptotic behavior of the implied volatility at extreme strikes in the SV1 and SV2 models.

We will first briefly describe the SV1 model. Let us fix $\eta > 0$ and $1 < \gamma < 2$, and set

$$a_1 = 2(\gamma - 1)\gamma^{-1} \quad \text{and} \quad a_2 = (2 - \gamma)\eta^2\gamma^{-1}.$$

For the sake of simplicity, it will be assumed that the interest rate r is equal to zero. In the SV1 model, the asset price process X is defined by $X = X^{(1)}X^{(2)}$ where

$$X_t^{(1)} = \sigma_t^{\frac{2}{\gamma}}, \qquad X_t^{(2)} = z_t^{\frac{1}{\gamma}}, \quad t > 0,$$

and the processes σ and z are the solutions to the following stochastic differential equations:

$$d\sigma_t = \eta\sigma_t\, dB_t \tag{11.64}$$

and

$$dz_t = (a_1 - a_2 z_t)\, dt + 2\sqrt{z_t}\, dW_t. \tag{11.65}$$

In the previous equations, the driving standard Brownian motions B and W are independent. The initial conditions for the processes σ and z will be denoted by σ_0 and z_0, respectively. We assume that $\sigma_0 > 0$ and $z_0 > 0$. Thus, the initial asset price is given by $x_0 = (\sigma_0)^{\frac{2}{\gamma}}(z_0)^{\frac{1}{\gamma}}$. The process σ in (11.64) is a driftless geometric Brownian motion, while the process z in (11.65) is a CIR process. It is known that if the parameters are chosen as above, then the process X is a martingale (see Lemma 1 in [RV08]).

Remark 11.20 The parameter γ in [RV08] satisfies the condition $0 < \gamma < 2$. This implies that $a_2 > 0$, but a_1 may be negative. Since in this book only CIR processes with positive parameters a_1 and a_2 are considered, we restrict ourselves to the case where $1 < \gamma < 2$. In such a case the process z solving (11.65) reaches zero almost surely, but is instantaneously reflected. Hence it is not difficult to justify that we can use the log-process $\log z$ in the proofs below.

Theorem 11.21 *Let $K \mapsto I(K)$ be the implied volatility in the SV1 model. Then the following asymptotic formulas hold:*

$$I(K) \sim 2\eta\gamma^{-1} \tag{11.66}$$

as $K \to \infty$, *and*

$$I(K) \sim \left(\frac{\psi(\gamma-1)}{T}\right)^{\frac{1}{2}} \sqrt{\log\frac{1}{K}} \tag{11.67}$$

as $K \to 0$.

Proof We have

$$X_t^{(1)} = \sigma_0^{\frac{2}{\gamma}} \exp\left\{-\frac{\eta^2}{\gamma}t + \frac{2\eta}{\gamma}B_t\right\}.$$

Fix $T > 0$, and denote the distribution density of the random variable $X_T^{(1)}$ by $D_T^{(1)}$. Then

$$D_T^{(1)}(x) = \frac{\gamma}{2\eta\sqrt{2\pi T}} x^{-1} \exp\left\{-\frac{1}{2T}\left(\frac{\gamma}{2\eta}\log x - \frac{1}{\eta}\log\sigma_0 + \frac{\eta}{2}T\right)^2\right\} \tag{11.68}$$

for all $x > 0$. Hence, for every $\varepsilon > 0$ there exists $x_{1,\varepsilon} > 0$ such that

$$\exp\left\{\left(-\frac{\gamma^2}{8\eta^2 T} - \varepsilon\right)\log^2 x\right\} \le D_T^{(1)}(x) \le \exp\left\{\left(-\frac{\gamma^2}{8\eta^2 T} + \varepsilon\right)\log^2 x\right\} \tag{11.69}$$

for all $x > x_{1,\varepsilon}$.

The marginal distribution densities of the CIR process z are given by

$$\begin{aligned} \rho_t(y) &= \exp\left\{a_2\left(\frac{a_1}{4} + \frac{1}{2}\right)t\right\} \frac{a_2}{2(e^{a_2 t} - 1)} \left(\frac{y}{z_0}\right)^{\frac{a_1}{4} - \frac{1}{2}} \\ &\quad \times \exp\left\{-\frac{a_2(z_0 + e^{a_2 t} y)}{2(e^{a_2 t} - 1)}\right\} I_{\frac{a_1}{2} - 1}\left(\frac{a_2\sqrt{e^{a_2 t} z_0 y}}{e^{a_2 t} - 1}\right) \end{aligned} \tag{11.70}$$

for all $t > 0$ and $y > 0$, where I is the modified Bessel function of the first kind (see formula (1.76) in Theorem 1.45). Denote by $D_T^{(2)}$ the distribution density of the random variable $X_T^{(2)}$, and recall that $X_T^{(2)} = z_T^{\frac{1}{\gamma}}$. It is not hard to see, using (11.15) and (11.70), that for every $\varepsilon > 0$ there exists $x_{2,\varepsilon} > 0$ such that

$$\exp\left\{\left[-\frac{a_2 e^{a_2 T}}{2(e^{a_2 T} - 1)} - \varepsilon\right]x^\gamma\right\} \le D_T^{(2)}(x) \le \exp\left\{\left[-\frac{a_2 e^{a_2 T}}{2(e^{a_2 T} - 1)} + \varepsilon\right]x^\gamma\right\} \tag{11.71}$$

for all $x > x_{2,\varepsilon}$.

Our next goal is to show how to estimate the convolution in formula (10.59). We keep the same notation as in the proof of Lemma 10.30, but replace estimates (10.60) and (10.61) in this proof by the estimates

$$\exp\left\{\left(-\frac{\gamma^2}{8\eta^2 T} - \varepsilon\right)x^2\right\} \le \widetilde{D}_T^{(1)}(x) \le \exp\left\{\left(-\frac{\gamma^2}{8\eta^2 T} + \varepsilon\right)x^2\right\} \tag{11.72}$$

and

$$\exp\left\{\left[-\frac{a_2e^{a_2T}}{2(e^{a_2T}-1)}-\varepsilon\right]e^{\gamma x}\right\}\le\widetilde{D}_T^{(2)}(x)\le\exp\left\{\left[-\frac{a_2e^{a_2T}}{2(e^{a_2T}-1)}+\varepsilon\right]e^{\gamma x}\right\}. \tag{11.73}$$

The previous inequalities follow from (11.69) and (11.71). Since the density $\widetilde{D}_T^{(2)}$ decays much faster than the density $\widetilde{D}_T^{(1)}$ (see (11.72) and (11.73)), we can estimate the convolution in formula (10.59) exactly as in the proof of Lemma 10.30. The resulting inequalities are as follows:

$$\exp\left\{\left(-\frac{\gamma^2}{8\eta^2T}-\varepsilon\right)x^2\right\}\le\widetilde{D}_T(x)\le\exp\left\{\left(-\frac{\gamma^2}{8\eta^2T}+\varepsilon\right)x^2\right\}$$

for all $\varepsilon>0$ and all $x>x_{3,\varepsilon}$. Therefore

$$\exp\left\{\left(-\frac{\gamma^2}{8\eta^2T}-\varepsilon\right)\log^2 x\right\}\le D_T(x)\le\exp\left\{\left(-\frac{\gamma^2}{8\eta^2T}+\varepsilon\right)\log^2 x\right\}$$

for all $\varepsilon>0$ and all $x>x_{4,\varepsilon}$. Finally, applying the second part of Lemma 11.17 with $A=\frac{\gamma^2}{8\eta^2T}$ and $w(y)=\log^2 y$, we complete the proof of formula (11.66).

Our next goal is to estimate the asset price distribution density D_T near zero. It can be shown using (11.68) that for every $\varepsilon>0$ there exists $x_{5,\varepsilon}>0$ such that

$$\exp\left\{\left(-\frac{\gamma^2}{8\eta^2T}-\varepsilon\right)\log^2\frac{1}{x}\right\}\le D_T^{(1)}(x)\le\exp\left\{\left(-\frac{\gamma^2}{8\eta^2T}+\varepsilon\right)\log^2\frac{1}{x}\right\}$$

for all $0<x<x_{5,\varepsilon}$. Moreover, (11.14) and (11.70) imply that for every $\varepsilon>0$ there exists $x_{6,\varepsilon}>0$ such that

$$x^{\gamma-2+\varepsilon}\le D_T^{(2)}(x)\le x^{\gamma-2-\varepsilon}$$

for all $0<x<x_{6,\varepsilon}$. In addition, there exist $x_{7,\varepsilon}>0$ and $x_{8,\varepsilon}>0$ such that

$$\exp\left\{\left(-\frac{\gamma^2}{8\eta^2T}-\varepsilon\right)x^2\right\}\le\widetilde{D}_T^{(1)}(x)\le\exp\left\{\left(-\frac{\gamma^2}{8\eta^2T}+\varepsilon\right)x^2\right\}$$

for all $x<-x_{7,\varepsilon}$, and

$$e^{(\gamma-1+\varepsilon)x}\le\widetilde{D}_T^{(2)}(x)\le e^{(\gamma-1-\varepsilon)x}$$

for all $x<-x_{8,\varepsilon}$.

Estimating the convolution in (10.59), we see that for every $\varepsilon>0$ there exists $\tilde{x}_{1,\varepsilon}>0$ such that

$$e^{(\gamma-1+\varepsilon)x}\le\widetilde{D}_T(x)\le e^{(\gamma-1-\varepsilon)x}$$

for all $x < -\tilde{x}_{1,\varepsilon}$. Therefore there exists $\tilde{x}_{2,\varepsilon} > 0$ such that

$$x^{\gamma-2+\varepsilon} \le D_T(x) \le x^{\gamma-2-\varepsilon}$$

for all $0 < x < \tilde{x}_{2,\varepsilon}$. It follows from Theorem 7.3 that the density D_T is of weak Pareto type near zero with index $2-\gamma$. Next, using part 4 of Corollary (10.29) we establish (11.67).

This completes the proof of Theorem 11.21. □

Remark 11.22 It follows from the proof of Theorem 11.21 that for the asset price process in the SV1 model, all the moments of positive order are finite, while some moments of negative order explode. The previous observation explains why the asymptotic behavior of the implied volatility in the SV1 model is qualitatively different in the cases of large and small strikes.

The SV2 model developed in [RV08] is a generalization of the SV1 model. Here the asset price process X is given by

$$X = X^{(1)} X^{(2)}, \quad \text{where } X^{(1)} = \sigma,\ X^{(2)} = g(z),$$

and the processes σ and z solve the following stochastic differential equations:

$$d\sigma_t = \sigma_t(\mu\, dt + \eta\, dB_t)$$

and

$$dz_t = (a_1 - a_2 z_t)\, dt + 2\sqrt{z_t}\, dW_t.$$

In addition, the function g is a solution to the ordinary differential equation

$$2ug''(u) + (a_1 - a_2 z)g'(u) + \mu g(u) = 0. \tag{11.74}$$

The previous equation can be solved in terms of the Kummer functions M and U. The function M is given by

$$M(a,b,u) = \sum_{n=0}^{\infty} \frac{(a)_n u^n}{(b)_n n!}$$

where

$$(a)_n = a(a+1)\cdots(a+n-1)$$

is the Pochhammer symbol (see [AS72], 13.1.2). It is clear that if $a > 0$, $b > 0$ and $u > 0$, then $M(a,b,u) > 0$. The function U is defined by

$$U(a,b,u) = \frac{\pi}{\sin(\pi b)} \left[\frac{M(a,b,u)}{\Gamma(1+a-b)\Gamma(b)} - u^{1-b} \frac{M(1+a-b, 2-b, u)}{\Gamma(a)\Gamma(2-b)} \right]$$

(see [AS72], 13.1.3). For $a > 0$ and $u > 0$, the function U has the following Laplace integral representation:

$$U(a,b,u) = \frac{1}{\Gamma(a)} \int_0^\infty e^{-ut} t^{a-1} (1+t)^{b-a-1} \, dt$$

(see [AS72], 13.2.5). Hence, we have $U(a,b,u) > 0$.

The general solution to (11.74) is given by

$$g(u, a_1, a_2, \mu) = C_1 M\left(-\frac{\mu}{a_2}, \frac{a_1}{2}, \frac{a_2 u}{2}\right) + C_2 U\left(-\frac{\mu}{a_2}, \frac{a_1}{2}, \frac{a_2 u}{2}\right), \tag{11.75}$$

where C_1 and C_2 are real constants.

Let us suppose that $\mu < 0$, $a_1 > 2$, $a_2 > 0$, $C_1 \geq 0$, and $C_2 \geq 0$. The trivial case where $C_1 = C_2 = 0$ will be excluded. It is clear that the function g is positive and real-analytic on $(0, \infty)$. It is also true that under the conditions mentioned above, the CIR process z does not reach zero, and hence $t \mapsto g(z_t)$ is a positive stochastic process. Therefore, the SV2 model is a mixed model with $X^{(1)} = \sigma$ and $X^{(2)} = g(z)$.

The following formulas hold for the Kummer functions:

$$M(a,b,u) = \frac{\Gamma(b)}{\Gamma(a)} e^u u^{a-b} \left(1 + O\left(u^{-1}\right)\right) \tag{11.76}$$

and

$$U(a,b,u) = u^{-a} \left(1 + O\left(u^{-1}\right)\right) \tag{11.77}$$

as $u \to \infty$ (see [AS72], 13.1.4 and 13.1.5). Moreover

$$\frac{\partial M}{\partial u}(a,b,u) = \frac{a}{b} M(a+1, b+1, u) \tag{11.78}$$

and

$$\frac{\partial U}{\partial u}(a,b,u) = -aU(a+1, b+1, u) \tag{11.79}$$

for all $u > 0$ (see [AS72], 13.4.8 and 13.4.21). The asymptotic behavior of the function M near zero is as follows:

$$M(a,b,u) \to 1 \quad \text{as} \quad u \downarrow 0, \tag{11.80}$$

for all $a > 0$ and $b > 0$ (see [AS72], 13.5.5). In addition, for $a > 0$,

$$U(a,b,u) = \frac{\Gamma(b-1)}{\Gamma(a)} u^{1-b} + \begin{cases} O(u^{b-2}), & \text{if } b > 2, \\ O(|\log u|), & \text{if } b = 2, \\ O(1), & \text{if } 1 < b < 2, \end{cases} \tag{11.81}$$

if $u \downarrow 0$ (see [AS72], 13.5.6–13.5.8). It follows from (11.75)–(11.81) that the function $u \mapsto g(a,b,u)$ is increasing near infinity and decreasing near zero. Since g is strictly positive, we have $g(u, a_1, a_2, \mu) > \delta_0$ for some $\delta_0 > 0$ and all $u > 0$. Therefore,

$$D_T^{(2)}(x) = 0 \quad \text{for all } x < \delta_1 = \log \delta_0. \tag{11.82}$$

The next statement characterizes the asymptotic behavior of the implied volatility $K \mapsto I(K)$ in the SV2 model.

Theorem 11.23 *The following asymptotic formula holds as $K \to \infty$:*

$$I(K) \sim \left(\frac{\psi(\rho)}{T}\right)^{\frac{1}{2}} \sqrt{\log K} \tag{11.83}$$

where

$$\rho = \begin{cases} \min\{(\exp\{a_2 T\} - 1)^{-1}, 2(a_1 - 2)^{-1}\}, & \text{if } C_1 > 0 \text{ and } C_2 > 0, \\ (\exp\{a_2 T\} - 1)^{-1}, & \text{if } C_1 > 0 \text{ and } C_2 = 0, \\ 2(a_1 - 2)^{-1}, & \text{if } C_1 = 0 \text{ and } C_2 > 0. \end{cases}$$

On the other hand, if $K \downarrow 0$, then $I(K) \sim \eta$.

Proof We will prove Theorem 11.23 only in the case where $C_1 > 0$ and $C_2 > 0$. The remaining cases can be established similarly. Our first goal is to obtain two-sided estimates for the distribution densities of random variables $X_T^{(1)}$ and $X_T^{(2)}$. Since the process $X^{(1)}$ is a geometric Brownian motion, we have

$$\exp\left\{\left(-\frac{1}{2\eta^2 T} - \varepsilon\right)\log^2 x\right\} \le D_T^{(1)}(x) \le \exp\left\{\left(-\frac{1}{2\eta^2 T} + \varepsilon\right)\log^2 x\right\} \tag{11.84}$$

for all $\varepsilon > 0$ and $x > x_{1,\varepsilon}$.

It is more difficult to estimate the density $D_T^{(2)}$. We have $X_T^{(2)} = g(z_T)$. Therefore there exists $x_0 > 0$ such that for $x > x_0$,

$$D_T^{(2)}(x) = \frac{1}{g_1'(g_1^{-1}(x))}\rho_T\left(g_1^{-1}(x)\right) - \frac{1}{g_2'(g_2^{-1}(x))}\rho_T\left(g_2^{-1}(x)\right) \tag{11.85}$$

where g_1 coincides with g in a neighborhood of infinity, g_2 coincides with g in a right neighborhood of zero, and ρ_T is defined by (11.70). It follows from (11.75), (11.77), (11.80), and (11.81) that the input of the second term on the right-hand side of (11.85) into the estimate for $D_T^{(2)}(x)$ near infinity is approximately $x^{-\frac{\exp\{a_2 T\}}{\exp\{a_2 T\}-1}-1}$. To prove the previous statement, we need (11.70) and (11.15). On the other hand, it can be seen from (11.70) and (11.14) that the input of the first term on the right-hand side of (11.85) is approximately $x^{-\frac{a_1}{a_1-2}-1}$. It is not hard to make the informal reasoning above rigorous. The resulting estimates are as follows:

$$x^{-\rho-2-\varepsilon} \le D_T^{(2)}(x) \le x^{-\rho-2+\varepsilon} \tag{11.86}$$

for all $\varepsilon > 0$ and $x > x_{0,\varepsilon}$, where ρ is defined in the formulation of Theorem 11.23. Next using the same method as in the proof of Lemma 10.30, and taking into account (10.59), (11.84), and (11.86), we obtain similar estimates for the density D_T of the asset price. These estimates have the following form:

$$x^{-\rho-2-\varepsilon} \le D_T(x) \le x^{-\rho-2+\varepsilon}$$

for every $\varepsilon > 0$ and $x > x_{1,\varepsilon}$. The previous inequalities and Theorem 7.3 show that the function D_T is of weak Pareto type near infinity with index

$$\alpha = -\rho - 2 < -2.$$

Therefore, we can apply part 2 of Corollary 10.29 to finish the proof of (11.83) in the case where $C_1 > 0$ and $C_2 > 0$.

Our next goal is to prove that $I(K) \sim \eta$ as $K \to 0$. Since the process $X^{(1)}$ is a geometric Brownian motion, we have

$$\exp\left\{(-A-\varepsilon)\log^2\frac{1}{x}\right\} \le D_T^{(1)}(x) \le \exp\left\{(-A+\varepsilon)\log^2\frac{1}{x}\right\}$$

for all $\varepsilon > 0$ and $x < x_{4,\varepsilon}$, where $A = \frac{1}{2\eta^2 T}$. Hence

$$\exp\{(-A-\varepsilon)y^2\} \le \widetilde{D}_T^{(1)}(y) \le \exp\{(-A+\varepsilon)y^2\} \tag{11.87}$$

for all $\varepsilon > 0$ and $-\infty < y < y_{1,\varepsilon} < 0$.

It follows from (11.82) and (10.59) that

$$\widetilde{D}_T(y) = \int_{\delta_1}^{\infty} \widetilde{D}_T^{(1)}(y-z)\widetilde{D}_t^{(2)}(z)\,dz, \quad y \in \mathbb{R}. \tag{11.88}$$

Next, using (11.87) and (11.88), we get

$$\int_{\delta_1}^{\infty} \exp\{(-A-\varepsilon)(y-z)^2\}\widetilde{D}_t^{(2)}(z)\,dz$$

$$\le \widetilde{D}_T(y) \le \int_{\delta_1}^{\infty} \exp\{(-A+\varepsilon)(y-z)^2\}\widetilde{D}_t^{(2)}(z)\,dz$$

for all $\varepsilon > 0$ and $-\infty < y < y_{2,\varepsilon} < 0$. Now, it is not hard to prove that for every $\varepsilon > 0$ there exists $y_{3,\varepsilon} < 0$ such that

$$\exp\left\{\left(-\frac{1}{2\eta^2 T}-\varepsilon\right)y^2\right\} \le \widetilde{D}_T(y) \le \exp\left\{\left(-\frac{1}{2\eta^2 T}+\varepsilon\right)y^2\right\}$$

for all $-\infty < y < y_{3,\varepsilon}$. Hence, for every $\varepsilon > 0$ there exists $x_{5,\varepsilon} > 0$ such that

$$\exp\left\{\left(-\frac{1}{2\eta^2 T}-\varepsilon\right)\log^2\frac{1}{x}\right\} \le D_T(x) \le \exp\left\{\left(-\frac{1}{2\eta^2 T}+\varepsilon\right)\log^2\frac{1}{x}\right\}$$

for all $0 < x < x_{5,\varepsilon}$. It follows from Lemma 11.19 with $A = \frac{1}{2\eta^2 T}$ and $w(y) = \log^2 \frac{1}{y}$ that $I(K) \sim \eta$ as $K \to 0$.

This completes the proof of Theorem 11.23. □

11.8 Notes and References

- The material included in this chapter is adapted from [Gul12].
- In [BF09], S. Benaim and P. Friz obtained formulas (11.5) and (11.6) under certain restrictions on call pricing functions. The results presented in Sect. 10.2 show that no such restrictions are needed.
- W. Feller found in [Fel51] an explicit expression for the fundamental solution of the diffusion equation related to the CEV process (see [BL10] for more information and details). Additional facts concerning the CEV model can be found in [EM82, Cox96, DS02, JYC09].
- Formula (11.21) without an error estimate was reported in [For06]. The proof of this formula in [For06] uses the right-tail-wing formula from [BF09] and the asset price distribution estimates. See also [BFL09] where an alternative proof is given. One more proof can be found in [H-L09], Example 10.3. Note that formula (11.21) in the present book contains an error estimate.
- Displaced diffusion models with $a > 0$ provide first order approximations to more complicated stochastic volatility models (see [Mar99, Muc04, S-G09]). These approximations can be used to study call option prices for small time in the region of the at-the-money strike. In [S-G09], general local volatility models are approximated by displaced diffusion models. Note that the asymptotic behavior of the implied volatility at large strikes in the original and the approximating model may be qualitatively different. For the CEV model, this can be seen by comparing formula (11.21) with formula (11.28). Another application of displaced diffusion models was given in [LW12], where the authors use these models to reduce variance in Monte Carlo simulations of other models.
- For more information on stable distributions and stochastic models based on stable random processes see [Zol86, ST94, McC96, CH09].

References

[AS72] Abramowitz, M., Stegun, I. A. (Eds.), *Handbook of Mathematical Functions*, Applied Mathematics Series 55, National Bureau of Standards, Washington, 1972.

[AG97] Alili, L., Gruet, J. C., An explanation of a generalized Bboujerol's identity in terms of hyperbolic geometry, in: Yor, M. (Ed.), *Exponential Functionals and Principal Values Related to Brownian Motion*, pp. 15–33, Biblioteca de la Revista Matemàtica Ibero-Americana, Madrid, 2007.

[All07] Allen, E., *Modeling with Itô Stochastic Differential Equations*, Springer, Dordrecht, 2007.

[AP07] Andersen, L. B. G., Piterbarg, V. V., Moment explosions in stochastic volatility models, *Finance and Stochastics* 11 (2007), pp. 29–50.

[AAR99] Andrews, G. E., Askey, R., Roy, B., *Special Functions*, Cambridge University Press, Cambridge, 1999.

[B1900] Bachelier, L., Théorie de la spéculation, *Annales Scientifiques de l'École Normale Supérieure* 17 (1900), pp. 21–86. Les grands classiques Gautier-Villars, Éditions Jacques Gabay (1995), pp. 21–86.

[BR94] Ball, C. A., Roma, A., Stochastic volatility option pricing, *Journal of Financial and Quantitative Analysis* 29 (1994), pp. 589–607.

[B-NNS02] Barndorff-Nielsen, O. E., Nicolato, E., Shephard, N., Some recent developments in stochastic volatility modelling, *Quantitative Finance* 2 (2002), pp. 11–23.

[BRY04] Barrieu, P., Rouault, A., Yor, M., A study of the Hartman–Watson distribution motivated by numerical problems related to the pricing of Asian options, *Journal of Applied Probability* 41 (2004), pp. 1049–1058.

[Bat96] Bates, D., Jumps and stochastic volatility: the exchange rate processes implicit in Deutschemark options, *Revue of Financial Studies* 9 (1996), pp. 69–109.

[BA88a] Ben Arous, G., Dévepoppement asymptotique du noyau de la chaler hypoelliptique hors du cut-locus, *Annales Scientifiques de l'École Normale Supérieure* 4 (1988), pp. 307–331.

[BA88b] Ben Arous, G., Methods de Laplace et de la phase stationnaire sur l'espace de Wiener, *Stochastics* 25 (1988), pp. 125–153.

[BAL10] Ben Arous, G., Laurence, P., Second order expansion for implied volatility models and applications to the dynamic SABR model, preprint, 2010.

[BAHLOW10] Ben Arous, G., Hsu, E. P., Laurence, P., Ouyang, C., Wang, T.-H., Asymptotics of implied volatility in stochastic volatility models, preprint, 2010.

[BF09] Benaim, S., Friz, P. K., Regular variation and smile asymptotics, *Mathematical Finance* 19 (2009), pp. 1–12.

[BF08] Benaim, S., Friz, P. K., Smile asymptotics, II: models with known moment generating function, *Journal of Applied Probability* 45 (2008), pp. 16–32.

A. Gulisashvili, *Analytically Tractable Stochastic Stock Price Models*,
Springer Finance, DOI 10.1007/978-3-642-31214-4,
© Springer-Verlag Berlin Heidelberg 2012

[BFL09] Benaim, S., Friz, P. K., Lee, R., On Black–Scholes implied volatility at extreme strikes, in: Cont, R. (Ed.), *Frontiers in Quantitative Finance: Volatility and Credit Risk Modeling*, pp. 19–45, Wiley, Hoboken, 2009.

[BGM10] Benhamou, E., Gobet, E., Miri, M., Time dependent Heston model, *SIAM Journal on Financial Mathematics* 1 (2010), pp. 289–325.

[BBF02] Berestycki, H., Busca, J., Florent, I., Asymptotics and calibration of local volatility models, *Quantitative Finance* 2 (2002), pp. 61–69.

[BBF04] Berestycki, H., Busca, J., Florent, I., Computing the implied volatility in stochastic volatility models, *Communications on Pure and Applied Mathematics* LVII (2004), pp. 1–22.

[BGT87] Bingham, N. H., Goldie, C. M., Teugels, J. L., *Regular Variation*, Cambridge University Press, Cambridge, 1987.

[Bjö04] Björk, T., *Arbitrage Theory in Continuous Time*, Oxford University Press, Oxford, 2004.

[BS73] Black, F., Scholes, M., The pricing of options and corporate liabilities, *Journal of Political Economy* 81 (1973), pp. 635–654.

[BH95] Bleistein, N., Handelsman, R. A., *Asymptotic Expansions of Integrals*, Holt, Rinehart and Winston, New York, 1995.

[BLT06] Bollerslev, T., Litvinova, J., Tauchen, G., Leverage and volatility feedback effects in high-frequency data, *Journal of Financial Econometrics* 4 (2006), pp. 353–384.

[Bou83] Bougerol, Ph., Exemples des théorèmes locaux sur les groupes résolubles, *Annales de l'Institut Henry Poincaré* 19 (1983), pp. 369–391.

[BL10] Brecher, D. R., Lindsay, A. E., Results on the CEV process, past and present, preprint, 2010.

[Bue06] Buehler, H., Expensive martingales, *Quantitative Finance* 6 (2006), pp. 207–218.

[CN09] Carmona, R., Nadtochiy, S., Local volatility dynamic models, *Finance and Stochastics* 13 (2009), pp. 1–48.

[CL09] Carr, P., Lee, R., Put-call symmetry: extensions and applications, *Mathematical Finance* 19 (2009), pp. 523–560.

[CL10] Carr, P., Lee, R., Implied volatility in stochastic volatility models, in: *Encyclopedia of Quantitative Finance*, 2010.

[CM05] Carr, P., Madan, D. B., A note on sufficient conditions for no arbitrage, *Finance Research Letters* 2 (2005), pp. 125–130.

[CS04] Carr, P., Schröder, M., Bessel processes, the integral of geometric Brownian motion, and Asian options, *Theory of Probability and Its Applications* 48 (2004), pp. 400–425.

[CW03] Carr, P., Wu, L., The finite moment log-stable process and option pricing, *Journal of Finance* LVIII (2003), pp. 753–777.

[CH09] Cartea, Á., Howison, S., Option pricing with Lévy-stable processes generated by Lévy-stable integrated variance, *Quantitative Finance* 9 (2009), pp. 397–409.

[CKSI00] Chandrasekhar, S., Kac, M., Smoluchowski, R., Ingarden, R. S., *Marian Smoluchowski: His Life and Scientific Work*, Polish Scientific Publishers PWN, Warsaw, 2000.

[CE05] Cherny, A. S., Engelbert, H.-J., *Singular Stochastic Differential Equations*, Lecture Notes in Mathematics 1858, Springer, Berlin, 2005.

[CD65] Chung, K. L., Doob, J. L., Fields, optionality and measurability, *American Journal of Mathematics* 87 (1965), pp. 397–424.

[Con01] Cont, R., Empirical properties of asset returns: stylized facts and statistical issues, *Quantitative Finance* 1 (2001), pp. 223–236.

[CdF02] Cont, R., da Fonseca, J., Dynamics of implied volatility surfaces, *Quantitative Finance* 2 (2002), pp. 45–60.

[CT04] Cont, R., Tankov, P., *Financial Modeling with Jump Processes*, Chapman and Hall/CRC, Boca Raton, 2004.

[Con78] Conway, J. B., *Functions of One Complex Variable*, 2nd ed., Springer, New York, 1978.

[CGLS09] Corquera, J. M., Guillaume, F., Leoni, P., Schoutens, W., Implied Lévy volatility, *Quantitative Finance* 9 (2009), pp. 383–393.

[CKBC00] Courtault, J.-M., Kabanov, Yu., Bru, B., Crépel, P., Louis Bachelier on the centenary of théorie de la spéculation, *Mathematical Finance* 10 (2000), pp. 341–353.

[CK02] Courtault, J.-M., Kabanov, Yu., *Louis Bachelier: aux Origines de la Finance Mathématique*, Presses Universitaires Franc-Comtoises, Paris, 2002.

[Cou07] Cousot, L., Conditions on option prices for absence of arbitrage and exact calibration, *Journal of Banking and Finance* 31 (2007), pp. 3377–3397.

[Cox96] Cox, J. C., Notes on option pricing, I: constant elasticity of variance diffusions, *Journal of Portfolio Management* 22 (1996), pp. 15–17.

[CIR85] Cox, J. C., Ingersoll, J. E., Ross, S. A., A theory of the term structure of interest rates, *Econometrica* 53 (1985), pp. 385–407.

[CR76] Cox, J. C., Ross, S. A., The valuation of options for alternative stochastic processes, *Journal of Financial Economics* 3 (1976), pp. 145–166.

[CFT10] Cuciero, C., Filipović, D., Teichmann, J., Affine models, in: *Encyclopedia of Mathematical Finance*, pp. 16–20, Wiley, New York, 2010.

[DH07] Davis, M. H. A., Hobson, D. G., The range of traded option prices, *Mathematical Finance* 17 (2007), pp. 1–14.

[dBru81] de Bruijn, N. G., *Asymptotic Methods in Analysis*, Dover, New York, 1981.

[dBRF-CU10] del Baño Rollin, S., Ferreiro-Castilla, A., Utzet, F., On the density of log-spot in Heston volatility model, *Stochastic Processes and Their Applications* 120 (2010), pp. 2037–2062.

[DS06] Delbaen, F., Schachermayer, W., *The Mathematics of Arbitrage*, Springer, Berlin, 2006.

[DS02] Delbaen, F., Shirakawa, H., A note on option pricing for constant elasticity of variance model, *Asia-Pacific Financial Markets* 9 (2002), pp. 85–99.

[DM10] De Marco, S., On probability distributions of diffusions and financial models with non-globally smooth coefficients, Ph.D. Dissertation, Université Paris-Est Marne-la-Vallée and Scuola Normale Superiore di Pisa, 2010.

[DMM10] De Marco, S., Martini, C., The term structure of implied volatility in symmetric models with applications to Heston, preprint, 2010; available at papers.ssrn.com/sol3/papers.cfm?abstract_id=1622828.

[Den06] Denisov, D. E., On the existence of a regularly varying majorant of an integrable monotone function, *Mathematical Notes* 79 (2006), pp. 129–133.

[DS89] Deuschel, J.-D., Stroock, D. W., *Large Deviations*, Academic Press, Boston, 1989.

[DFJV11] Deuschel, J.-D., Friz, P. K., Jacquier, A., Violante, S., Marginal density expansions for diffusions and stochastic volatility, preprint, 2011; available at arXiv:1111.2462.

[DY02] Drăgulescu, A. A., Yakovenko, V. M., Probability distribution of returns in the Heston model with stochastic volatility, *Quantitative Finance* 2 (2002), pp. 443–453.

[DPS00] Duffie, D., Pan, J., Singleton, K., Transform analysis and asset pricing for affine jump-diffusions, *Econometrica* 68 (2000), pp. 1343–1376.

[DFS03] Duffie, D., Filipović, D., Schachermayer, W., Affine processes and applications in finance, *The Annals of Applied Probability* 13 (2003), pp. 984–1053.

[Duf01] Dufresne, D., The integral of geometric Brownian motion, *Advances in Applied Probability* 33 (2001), pp. 223–241.

[Duf05] Dufresne, D., Bessel processes and Asian options, in: Breton, M., Ben-Ameur, H. (Eds.), *Numerical Methods in Finance*, pp. 35–57, Springer, Berlin, 2005.

[Dup06] Duplantier, B., Brownian motion, "Diverse and undulating", in: Damour, T., Darrigol, O., Duplantier, B., Rivasseau, V. (Eds.), *Einstein, 1905–2005, Poincaré Seminar 1, 2005*, Progress in Mathematical Physics 47, pp. 201–293, Birkhäuser, Basel, 2006.

[Dur04] Durrleman, V., From implied to spot volatilities, Ph.D. Dissertation, Princeton University, 2004.

[Ein56] Einstein, A., *Investigations on the Theory of Brownian Movement*, Dover, New York, 1956. (1926 English translation of Einstein's five papers on Brownian motion; edited with notes by R. Fürth.)

[Ein08] Galison, P. L., Holton, G., Schweber, S. S. (Eds.), *Einstein for the 21st Century: His Legacy in Science, Art, and Modern Culture*, Princeton University Press, Princeton, 2008.

[EKM97] Embrechts, P., Klüppelberg, C., Mikosch, T., *Modelling Extremal Events for Insurance and Finance*, Springer, Berlin, 1997.

[EM82] Emmanuel, D., MacBeth, J., Further results on the constant elasticity of variance call option pricing model, *Journal of Financial and Quantitative Analysis* 17 (1982), pp. 533–554.

[Fel51] Feller, W., Two singular diffusion problems, *Annals of Mathematics* 54 (1951), pp. 173–182.

[Fel66] Feller, W., *An Introduction to Probability Theory and Its Applications, Vol. II*, Wiley, New York, 1966.

[FFF10] Feng, J., Forde, M., Fouque, J.-P., Short maturity asymptotics for a fast mean-reverting Heston stochastic volatility model, *SIAM Journal on Financial Mathematics* 1 (2010), pp. 126–141.

[FFK10] Feng, J., Fouque, J.-P., Kumar, R., Small-time asymptotics for fast mean-reverting stochastic volatility models, preprint; to appear in *Annals of Applied Probability*; available at arXiv:1009.2782.

[Fen05] Fengler, M. R., *Semiparametric Modeling of Implied Volatility*, Springer, Berlin, 2005.

[FW01] Figlewski, S., Wang, X., Is the "Leverage effect"' a Leverage effect? Working paper, 2001.

[FM09] Filipović, D., Mayerhofer, E., Affine diffusion processes: theory and applications, *Radon Series on Computational and Applied Mathematics* 8 (2009), pp. 1–40.

[FS09] Flajolet, P., Sedgewick, R., *Analytic Combinatorics*, Cambridge University Press, Cambridge, 2009.

[For06] Forde, M., Tail asymptotics for diffusion processes, with applications to local volatility and CEV-Heston models, preprint, 2006; available at arXiv:math/0608634.

[For11] Forde, M., Large-time asymptotics for a general stochastic volatility model, preprint, 2011.

[FJ09] Forde, M., Jacquer, A., Small-time asymptotics for implied volatility under the Heston model, *International Journal of Theoretical and Applied Finance* 12 (2009), pp. 861–876.

[FJ11a] Forde, M., Jacquier, A., The large-maturity smile for the Heston model, *Finance and Stochastics* 15 (2011), pp. 755–780.

[FJ11b] Forde, M., Jacquier, A., Small-time asymptotics for an uncorrelated local-stochastic volatility model, *Applied Mathematical Finance* 18 (2011), pp. 517–535.

[FJL11] Forde, M., Jacquer, A., Lee, R., The small-time smile and term structure of implied volatility under the Heston model, preprint, 2011.

[FJM10] Forde, M., Jacquier, A., Mijatovich, A., Asymptotic formulae for implied volatility in the Heston model, *Proceedings of the Royal Society A* 466 (2010), pp. 3593–3620.

[FPS00] Fouque, J.-P., Papanicolaou, G., Sircar, R., *Derivatives in Financial Markets with Stochastic Volatility*, Cambridge University Press, Cambridge, 2000.

[FPSS11] Fouque, J.-P., Papanicolaou, G., Sircar, R., Sølna, K., *Multiscale Stochastic Volatility for Equity, Interest Rate, and Credit Derivatives*, Cambridge University Press, Cambridge, 2011.

[Fri10] Friz, P., Implied volatility: large strike asymptotics, in: Cont, R. (Ed.), *Encyclopedia of Quantitative Finance*, pp. 909–913, Wiley, Chichester, 2010.

[FGGS11] Friz, P., Gerhold, S., Gulisashvili, A., Sturm, S., On refined volatility smile expansion in the Heston model, *Quantitative Finance* 11 (2011), pp. 1151–1164.

[FK-R10] Friz, P., Keller-Ressel, M., Moment explosions, in: Cont, R. (Ed.), *Encyclopedia of Quantitative Finance*, pp. 1247–1253, Wiley, Chichester, 2010.

[GL11] Gao, K., Lee, R., Asymptotics of implied volatility to arbitrary order, preprint, 2011; available at http://ssrn.com/abstract=1768383.

[Gat04] Gatheral, J., A parsimonious arbitrage-free implied volatility parametrization with application to the valuation of volatility derivatives, in: *Global Derivatives and Risk Management, Madrid, May 26*, 2004.

[Gat06] Gatheral, J., *The Volatility Surface: A Practitioner's Guide*, Wiley, Hoboken, 2006.

[GJ11] Gatheral, J., Jacquier, A., Convergence of Heston to SVI, *Quantitative Finance* 11 (2011), pp. 1129–1132.

[GHLOW10] Gatheral, J., Hsu, E., Laurence, P., Ouyang, C., Wang, T.-H., Asymptotics of implied volatility in local volatility models; to appear in *Mathematical Finance*; available at http://ssrn.com/abstract=1542077.

[GY93] Geman, H., Yor, M., Bessel processes, Asian options, and perpetuities, *Mathematical Finance* 4 (1993), pp. 349–375.

[Ger11] Gerhold, S., The Hartman–Watson distribution revisited: asymptotics for pricing Asian options, *Journal of Applied Probability* 48 (2011), pp. 892–899.

[GHR96] Ghysels, E., Harvey, A., Renault, E., Stochastic volatility, in: Maddala, S. G. (Ed.), *Handbook of Statistics*, Statistical Methods in Finance 14, North Holland, Amsterdam, 1996.

[Gir02] Girlich, H.-J., Bachelier's predecessors, preprint, Mathematisches Institut (Leipzig), Univ. Leipzig, Fak. für Mathematik u Informatik, 2002.

[GK10] Glasserman, P., Kim, K.-K., Moment explosions and stationary distributions in affine diffusion modles, *Mathematical Finance* 20 (2010), pp. 1–33.

[Gob11] Gobet, E., Asymptotic methods in option pricing, Journées George Papanicolaou, Université Paris Diderot, 1er décembre 2011 (based on a joint work with M. Miri).

[G-JY03] Göing-Jaeschke, A., Yor, M., A survey and some generalizations of Bessel processes, *Bernoulli* 9 (2003), pp. 313–349.

[GS87] Goldie, C. M., Smith, R. L., Slow variation with remainder: theory and applications, *Quarterly Journal of Mathematics* 38 (1987), pp. 45–71.

[Gul10] Gulisashvili, A., Asymptotic formulas with error estimates for call pricing functions and the implied volatility at extreme strikes, *SIAM Journal on Financial Mathematics* 1 (2010), pp. 609–641.

[Gul12] Gulisashvili, A., Asymptotic equivalence in Lee's moment formulas for the implied volatility, asset price models without moment explosions, and Piterbarg's conjecture, *International Journal of Theoretical and Applied Finance* 15 (2012), 1250020.

[GS06] Gulisashvili, A., Stein, E. M., Asymptotic behavior of the distribution of the stock price in models with stochastic volatility: the Hull–White model, *Comptes Rendus de l'Académie des Sciences de Paris, Série I* 343 (2006), pp. 519–523.

[GS09] Gulisashvili, A., Stein, E. M., Implied volatility in the Hull–White model, *Mathematical Finance* 19 (2009), pp. 303–327.

[GS10a] Gulisashvili, A., Stein, E. M., Asymptotic behavior of distribution densities in models with stochastic volatility, I, *Mathematical Finance* 20 (2010), pp. 447–477.

[GS10b] Gulisashvili, A., Stein, E. M., Asymptotic behavior of the stock price distribution density and implied volatility in stochastic volatility models, *Applied Mathematics and Optimization* 61 (2010), pp. 287–315.

[GvC06] Gulisashvili, A., van Casteren, J. A., *Non-Autonomous Kato Classes and Feynman–Kac Propagators*, World Scientific, Singapore, 2006.

[GV11] Gulisashvili, A., Vives, J., Two-sided estimates for distribution densities in models with jumps, in: Zili, M., Filatova, D. V. (Eds.), *Stochastic Differential Equations and Processes*, Springer Proceedings in Mathematics 7, pp. 237–252, Springer, Berlin, 2011.

[Haf04] Hafner, R., *Stochastic Implied Volatility: A Factor-Based Model*, Springer, Berlin, 2004.

[HKLW02] Hagan, P. S., Kumar, D., Lesniewski, A. S., Woodward, D. E., Managing smile risk, *Wilmott Magazine* November (2002), pp. 84–108.

[HW74] Hartman, P., Watson, G. S., "Normal" distribution functions on spheres and the modified Bessel functions, *Annals of Probability* 2 (1974), pp. 593–607.

[H-L09] Henry-Labordère, P., *Analysis, Geometry, and Modeling in Finance: Advanced Methods in Option Pricing*, Chapman & Hall/CRC, Boca Raton, 2009.

[Hes93] Heston, S. L., A closed-form solution for options with stochastic volatility, with applications to bond and currency options, *Review of Financial Studies* 6 (1993), pp. 327–343.

[HLW07] Heston, S. L., Loewenstein, M., Willard, G. A., Options and bubbles, *Review of Financial Studies* 20 (2007), pp. 359–390.

[HPRY11] Hirsh, F., Profeta, C., Roynette, B., Yor, M., *Peacocks and Associated Martingales, with Explicit Constructions*, Springer, Italia, 2011.

[Hob98] Hobson, D. G., Stochastic volatility, in: Hand, D., Jacka, S. (Eds.), *Statistics in Finance*, Applications of Statistical Series, Arnold, London, 1998.

[HR83] Huberman, G., Ross, S., Portfolio turnpike theorems, risk aversion, and regularly varying utility functions, *Econometrica* (1983), pp. 1345–1361.

[HW87] Hull, J., White, A., The pricing of options on assets with stochastic volatilities, *Journal of Finance* 42 (1987), pp. 281–300.

[IW77] Ikeda, N., Watanabe, S., A comparison theorem for solutions of stochastic differential equations and its applications, *Osaka Journal of Mathematics* 14 (1977), pp. 619–633.

[IW81] Ikeda, N., Watanabe, S., *Stochastic Differential Equations and Diffusion Processes*, North Holland, Amsterdam, 1981.

[IM65] Itô, K., McKean, H. P. Jr., *Diffusion Processes and Their Sample Paths*, Springer, Berlin, 1965.

[Jäc03] Jäckel, P., Stochastic volatility models—past, present, and future, presentation at the *Quantitative Finance Review* conference in November 2003 in London.

[Jac96] Jacobsen, M., Laplace and the origin of the Ornstein–Uhlenbeck process, *Bernoulli* 2 (1996), pp. 271–286.

[JW94] Janicki, A., Weron, A., *Simulation and Chaotic Behavior of α-Stable Stochastic Processes*, Marcel Dekker, New York, 1994.

[JYC09] Jeanblanc, M., Yor, M., Chesney, M., *Mathematical Methods for Financial Markets*, Springer, London, 2009.

[Jef96] Jefferies, B., *Evolution Processes and the Feynman–Kac Formula*, Kluwer Academic, Dordrecht, 1996.

[JM06] Jessen, A. H., Mikosch, T., Regularly varying functions, *Publications de i'Institut Mathématique* 80 (2006), pp. 171–192.

[JL00] Johnson, G. W., Lapidus, M. L., *The Feynman Integral and Feynman's Operational Calculus*, Oxford University Press, New York, 2000.

[Kah97] Kahane, J.-P., A century of interplay between Taylor series, Fourier series and Brownian motion, *Bulletin of the London Mathematical Society* 29 (1997), pp. 257–279.

[Kah98] Kahane, J.-P., Le mouvement brownien - un essai sur les origines de la théorie mathematique, *Séminaires et Congrès de la SMF* 3 (1998), pp. 123–155.

[Kah06] Kahane, J.-P., Le mouvement brownien et son histoire, réponses à quelques questions, *Images des Mathématiques, CNRS* (2006).

[KS91] Karatzas, I., Shreve, S. E., *Brownian Motion and Stochastic Calculus*, 2nd ed., Springer, New York, 1991.

[Kel72] Kellerer, H. G., Markov-Komposition und eine Anwendung auf Martingale, *Mathematische Annalen* 198 (1972), pp. 99–122.

[K-RST11] Keller-Ressel, M., Schachermayer, W., Teichmann, J., Affine processes are regular, *Probability Theory and Related Fields* 151 (2011), pp. 591–611.

[K-R11] Keller-Ressel, M., Moment explosions and long-term behavior of affine stochastic volatility models, *Mathematical Finance* 21 (2011), pp. 73–98.

[Ken53] Kendall, M. S., The analysis of economic time series, part I: prices, *Journal of the Royal Statistical Society* 116 (1953), pp. 11–25.

[Kor04] Korevaar, J., *Tauberian Theory, a Century of Developments*, Springer, Berlin, 2004.

[Kou02] Kou, S., A jump-diffusion model for option pricing, *Management Science* 48 (2002), pp. 1086–1101.

[KW03] Kou, S., Wang, H., Option pricing under a double exponential jump diffusion model, *Management Science* 50 (2003), pp. 1178–1192.

[LL08] Lamberton, D., Lapeyre, B., *Introduction to Stochastic Calculus Applied to Finance*, Chapman & Hall/CRC, Boca Raton, 2008.

[LR76] Latané, H. A., Rendleman, R. J. Jr., Standard deviations of stock price ratios implied in option prices, *Journal of Finance* 31 (1976), pp. 369–381.

[Lau02] Lauritzen, S. L., *Thiele*: *Pioneer in Statistics*, Clarendon, Oxford, 2002.

[LeG83] Le Gall, G. F., Applications du temps local aux equations differentielles stochastiques unidimensionnelles, in: *Lecture Notes in Mathematics* 983, pp. 15–31, Springer, Berlin, 1983.

[Lee01] Lee, R., Implied and local volatilities under stochastic volatility, *International Journal of Theoretical and Applied Finance* 4 (2001), pp. 45–89.

[Lee04a] Lee, R., Implied volatility: statics, dynamics, and probabilistic interpretation, in: Baeza-Yates, R., Glaz, J., Ghyz, H., et al. (Eds.), *Recent Advances in Applied Probability*, Springer, Berlin, 2004.

[Lee04b] Lee, R., The moment formula for implied volatility at extreme strikes, *Mathematical Finance* 14 (2004), pp. 469–480.

[LW12] Lee, R., Wang, D., Displaced lognormal volatility skews: analysis and applications to stochastic volatility simulations, *Annals of Finance* 8 (2012), pp. 159–181.

[Lev64] Levin, B. Ya., *Distribution of Zeros of Entire Functions*, Translations of Mathematical Monographs 5, AMS, Providence, 1964.

[Lew00] Lewis, A. L., *Option Valuation Under Stochastic Volatility: with Mathematica Code*, Finance Press, Newport Beach, 2000.

[LM07] Lions, P.-L., Musiela, M., Correlations and bounds for stochastic volatility models, *Annales de l'Institut Henry Poincaré* 24 (2007), pp. 1–16.

[Lip01] Lipton, A., *Mathematical Methods for Foreign Exchange: a Financial Engineer's Approach*, World Scientific, River Edge, 2001.

[LS08] Lipton, A., Sepp, A., Stochastic volatility models and Kelvin waves, *Journal of Physics A: Mathematical and Theoretical* 41, (2008), 344012.

[Luc07] Lucic, V., On singularities in the Heston models, working paper, 2007.

[Mag07] Maghsoodi, Y., Exact solution of a martingale stochastic volatility option problem and its empirical evaluation, *Mathematical Finance* 17 (2007), pp. 249–265.

[Man63] Mandelbrot, B. B., The variation of certain speculative prices, *Journal of Business* 36 (1963), pp. 394–419.

[MR06] Marcus, M. B., Rosen, J., *Markov Processes, Gaussian Processes, and Local Times*, Cambridge University Press, Cambridge, 2006.

[Mar99] Marris, D., Financial option pricing and skewed volatility, MPhil Thesis, Statistical Laboratory, University of Cambridge, 1999.

[MY05a] Matsumoto, H., Yor, M., Exponential functionals of Brownian motion, I: probability laws at fixed time, *Probability Surveys* 2 (2005), pp. 312–347.

[MY05b] Matsumoto, H., Yor, M., Exponential functionals of Brownian motion, II: some related diffusion processes, *Probability Surveys* 2 (2005), pp. 348–384.

[McC96] McCulloch, J. H., Financial applications of stable distributions, in: Maddala, G. S., Rao, C. R. (Eds.), *Handbook of Statistics* 14, pp. 393–425, Elsevier, Amsterdam, 1996.

[Mer76] Merton, R., Option pricing when underlying stock returns are discontinuous, *Journal of Financial Economics* 3 (1976), pp. 125–144.

[Mil06] Miller, P. D., *Applied Asymptotic Analysis*, American Mathematical Society, Providence, 2006.

[Muc04] Muck, M., On the similarity between constant elasticity of variance and displaced diffusion market models of the term structure, Working paper, University of Bamberg, 2004.

[MN11] Muhle-Karbe, J., Nutz, M., Small-time asymptotics of option prices and first absolute moments, *Journal of Applied Probability* 48 (2011), pp. 1003–1020.

[Mur84] Murray, J. D., *Asymptotic Analysis*, Springer, New York, 1984.

[MR05] Musiela, M., Rutkowski, M., *Martingale Methods in Financial Modelling*, Springer, Berlin, 2005.

[Nel67] Nelson, E., *Dynamical Theories of Brownian Motion*, Princeton University Press, Princeton, 1967.

[Nik02] Nikolić, A., Jovan Karamata (1902–1967), *Novi Sad Journal of Mathematics* 32 (2002), pp. 1–5.

[OU30] Ornstein, L. S., Uhlenbeck, G. E., On the theory of Brownian motion, *Physics Reviews* 36 (1930), pp. 823–841.

[Osb59] Osborne, M. F. M., Brownian motion in the stock market, *Operations Research* 7 (1959), pp. 145–173.

[Øks03] Øksendal, B., *Stochastic Differential Equations. An Introduction with Applications*, 6th ed., Springer, Berlin, 2003.

[Pai05] Pais, A., *Subtle is the Lord: The Science and the Life of Albert Einstein*, Oxford University Press, Oxford, 2005.

[PK01] Paris, R. B., Kaminski, D., *Asymptotics and Mellin–Barnes Integrals*, Cambridge University Press, Cambridge, 2001.

[Pau09] Paulot, L., Asymptotic implied volatility at the second order with application to the SABR model, preprint, 2009; available at ssrn.com/abstract=1313649.

[Pit04] Piterbarg, V. V., Implied volatility smile asymptotics when all moments are finite, Working paper, 2004.

[PY82] Pitman, J., Yor, M., A decomposition of Bessel bridges, *Zeitschrift für Wahrscheinlichkeitstheorie und Verwandte Gebiete* 59 (1982), pp. 425–457.

[Pro04] Protter, P. E., *Stochastic Integration and Differential Equations*, 2nd ed., Springer, Berlin, 2004.

[RW02] Rasmussen, H., Wilmott, P., Asymptotic analysis of stochastic volatility models, in: Wilmott, P., Rasmussen, H. (Eds.), *New Directions in Mathematical Finance*, Wiley, Chichester, 2002.

[Reb04] Rebonato, R., *Volatility and Correlation: The Perfect Hedger and the Fox*, 2nd ed., Wiley Finance, New York, 2004.

[RT96] Renault, E., Touzi, N., Option hedging and implied volatilities in a stochastic volatility model, *Mathematical Finance* 6 (1996), pp. 279–302.

[Res87] Resnick, S. I., *Extreme Values, Regular Variation, and Point Processes*, Springer, New York, 1987.

[Res07] Resnick, S. I., *Heavy-Tail Phenomena: Probabilistic and Statistical Modeling*, Springer, New York, 2007.

[RY04] Revuz, D., Yor, M., *Continuous Martingales and Brownian Motion*, Springer, Berlin, 2004.

[RS87] Riccardi, L. M., Sacerdote, L., On the probability densities of the Ornstein–Uhlenbeck process with a reflecting boundary, *Journal of Applied Probability* 24 (1987), pp. 355–369.

[Rob59] Roberts, H. W., Stock market patterns and financial analysis: methodological suggestions, *Journal of Finance* 14 (1959), pp. 1–10.

[RV08] Rogers, L. C. G., Veraart, L. A. M., A stochastic volatility alternative to SABR, *Journal of Applied Probability* 45 (2008), pp. 1071–1085.

[Rop09] Roper, M., Implied volatility: general properties and asymptotics, Ph.D. thesis, The University of New South Wales, 2009.

[Rop10] Roper, M., Arbitrage free implied volatility surfaces, preprint, 2010.

[RR09] Roper, M., Rutkowski, M., On the relationship between the call price surface and the implied volatility surface close to expiry, *International Journal of Theoretical and Applied Finance* 12 (2009), pp. 427–441.

[Rub83] Rubinstein, M., Displaced diffusion option pricing, *Journal of Finance* 38 (1983), pp. 213–217.

[ST94] Samorodnitsky, G., Taqqu, M. S., *Stable Non-Gaussian Random Processes: Stochastic Models with Infinite Variance*, Chapman & Hall, New York, 1994.

[Sam65] Samuelson, P. A., Rational theory of warrant pricing, *Industrial Management Review* 6 (1965), pp. 13–32.

[Sam02] Samuelson, P. A., Modern finance theory within one lifetime, in: Geman, H., Madan, D., Pliska, S. R., Vorst, T. (Eds.), *Mathematical Finance—Bachelier Congress 2000, Selected Papers from the First World Congress of the Bachelier Finance Society*, pp. 41–45, Springer, Berlin, 2002.

[Sch03] Schoutens, W., *Lévy Processes in Finance, Pricing Financial Derivatives*, Wiley, Chichester, 2003.

[SZ99] Schöbel, R., Zhu, J., Stochastic volatility with an Ornstein–Uhlenbeck process: an extension, *European Finance Review* 3 (1999), pp. 23–46.
[Sco87] Scott, L. O., Option pricing when the variance changes randomly: theory, estimation and an application, *Journal of Financial and Quantitative Analysis* 22 (1987), pp. 419–438.
[Sen76] Seneta, E., *Functions of Regular Variation*, Springer, New York, 1976.
[She05] Shephard, N. (Ed.), *Stochastic Volatility: Selected Readings*, Oxford University Press, Oxford, 2005.
[She06] Shephard, N., Stochastic volatility, in: Durlauf, S., Blume, L. (Eds.), *New Palgrave Dictionary of Economics*, 2nd ed., 2006.
[SA09] Shephard, N., Andersen, T. G., Stochastic volatility: origins and overview, in: Andersen, T. G., Davis, R. A., Kreiss, J.-P., Mikosch, T. (Eds.), *Handbook of Financial Time Series, Part II*, pp. 233–254, Springer, Berlin, 2009.
[SW73] Shiga, T., Watanabe, S., Bessel diffusions as a one-parameter family of diffusion processes, *Zeitschrift für Wahrscheinlichkeitstheorie und Verwandte Gebiete* 27 (1973), pp. 37–46.
[Sin98] Sin, C. A., Complications with stochastic volatility models, *Advances in Applied Probability* 30 (1998), pp. 256–268.
[SP99] Sircar, K. R., Papanicolaou, G., Stochastic volatility, smile & asymptotics, *Applied Mathematical Finance* 6 (1999), pp. 107–145.
[Ski01] Skiadopoulos, G., Volatility smile consistent option models: a survey, *International Journal of Theoretical and Applied Finance* 4 (2001), pp. 403–438.
[SHK99] Skiadopoulos, G., Hodges, S., Klewlow, L., The dynamics of the SP500 implied volatility surface, *Review of Derivatives Research* 3 (1999), pp. 263–282.
[S1906] Smoluchowski, M., Zur kinetischen Theorie der Brownschen Molekularbewegung und der Suspensionen, *Annalen der Physik* 21 (1906), pp. 756–780.
[SS91] Stein, E. M., Stein, J., Stock price distributions with stochastic volatility: an analytic approach, *Review of Financial Studies* 4 (1991), pp. 727–752.
[SSh03] Stein, E. M., Shakarchi, R., *Complex Analysis*, Princeton University Press, Princeton, 2003.
[Str65] Strassen, V., The existence of probability measures with given marginals, *The Annals of Mathematical Statistics* 36 (1965), pp. 423–439.
[S-G09] Svoboda-Greenwood, S., Displaced diffusion as an approximation of the constant elasticity of variance, *Applied Mathematical Finance* 16 (2009), pp. 269–286.
[Tar04] Tarov, V. A., Smoothly varying functions and perfect proximate orders, *Mathematical Notes* 76 (2004), pp. 238–243.
[Taq02] Taqqu, M. S., Bachelier and his times: a conversation with Bernard Bru, in: *Mathematical Finance—Bachelier Congress 2000, Selected Papers from the First World Congress of the Bachelier Finance Society*, pp. 1–39, Springer, Berlin, 2002.
[Tau04] Tauchen, G., Recent developments in stochastic volatility: statistical modelling and general equilibrium analysis, Working paper, 2004.
[Teh09a] Tehranchi, M. R., Symmetric martingales and symmetric smiles, *Stochastic Processes and Their Applications* 119 (2009), pp. 3785–3797.
[Teh09b] Tehranchi, M. R., Asymptotics of implied volatility far from maturity, *Journal of Applied Probability* 46 (2009), pp. 629–650.
[Tom01] Tomić, M., Jovan Karamata: 1902–1967, *Bulletin T. CXXII de l'Académie Serbe des Sciences et des Arts, Sciences Mathématiques et Naturelles*, No. 26 (2001).
[Vui63] Vuilleumier, M., On asymptotic behaviour of linear transformations of slowly varying sequences and of sequences of regular asymptotic behaviour, Mathematical research center technical reports 435, Madison, WI, 1963.
[Wat95] Watson, G. N., *A Treatise on the Theory of Bessel Functions*, Cambridge University Press, Cambridge, 1995.
[Wen90] Wenocur, M. L., Ornstein-Uhlenbeck process with quadratic killing, *Journal of Applied Probability* 27 (1990), pp. 707–712.

[Wie23] Wiener, N., Differential space, *Journal of Mathematics and Physics* 2 (1923), pp. 455–498.

[Wie24] Wiener, N., The average value of a functional, *Proceedings of the London Mathematical Society* 22 (1924), pp. 454–467.

[Wig87] Wiggins, J., Option values under stochastic volatility, *Journal of Financial Economics* 19 (1987), pp. 351–372.

[WH04] Wong, B., Heyde, C. C., On the martingale property of stochastic exponentials, *Journal of Applied Probability* 41 (2004), pp. 654–664.

[WH06] Wong, B., Heyde, C. C., On changes of measure in stochastic volatility models, *Journal of Applied Mathematics and Stochastic Analysis* 2006, pp. 1–13.

[Yor80] Yor, M., Loi de l'indice du lacet Brownien, et distribution de Hartman–Watson, *Zeitschrift für Wahrscheinlichkeitstheorie und Verwandte Gebiete* 53 (1980), pp. 71–95.

[Yor92a] Yor, M., On some exponential functionals of Brownian motion, *Advances in Applied Probability* 24 (1992), pp. 509–531.

[Yor92b] Yor, M., Sur les lois des fonctionells exponentielles du mouvement brownien, considérées en certain instants aléatoires, *Comptes Rendus de l'Académie des Sciences de Paris* 314 (1992), pp. 951–956.

[Yor01] Yor, M., *Exponential Functionals of Brownian Motion and Related Processes*, Springer, Berlin, 2001.

[ZA98] Zhu, Y., Avellaneda, M., A risk-neutral stochastic volatility model, *International Journal of Theoretical and Applied Finance* 1 (1998), pp. 289–310.

[Zol86] Zolotarev, V., *One-Dimensional Stable Distributions*, AMS, Providence, 1986.

Index

A. Gulisashvili, *Analytically Tractable Stochastic Stock Price Models*, Springer Finance, DOI 10.1007/978-3-642-31214-4,

© Springer-Verlag Berlin Heidelberg 2012